Corporate Responses to Climate Change

To my wife Melinda and
our daughters Claire and Laura.

Corporate Responses to Climate Change

ACHIEVING EMISSIONS REDUCTIONS THROUGH REGULATION, SELF-REGULATION AND ECONOMIC INCENTIVES

EDITED BY RORY SULLIVAN

Greenleaf
PUBLISHING

This book has been compiled from the contributions of the named authors. The views expressed herein do not necessarily reflect the views of their respective companies. While all reasonable care has been taken in the preparation of this book, Greenleaf Publishing and the authors do not accept responsibility for any errors it may contain or for any loss sustained by any person placing reliance on its contents.

© 2008 Greenleaf Publishing Ltd

Published by Greenleaf Publishing Limited
Aizlewood's Mill
Nursery Street
Sheffield S3 8GG
UK
www.greenleaf-publishing.com

Printed in Great Britain on acid-free paper by CPI Antony Rowe, Chippenham, Wiltshire.

FSC
Mixed Sources
Product group from well-managed
forests and other controlled sources

Cert no. SGS-COC-2953
www.fsc.org
© 1996 Forest Stewardship Council

Cover by LaliAbril.com.

British Library Cataloguing in Publication Data:

Corporate responses to climate change : achieving emissions
reductions through regulation, self-regulation and economic
incentives
1. Greenhouse gas mitigation 2. Waste minimization
3. Climatic changes - Effect of human beings on 4. Social
responsibility of business
I. Sullivan, Rory, 1968-
363.7'38746

ISBN-13: 9781906093082

Contents

Part III: Non-state actors and their influence on corporate climate change performance

Part IV: Corporate responses and case studies

Part I
Introduction

1
Introduction

Rory Sullivan
Insight Investment, UK

Setting the scene

Climate change is the most serious environmental threat faced by our planet. The scientific consensus is that the increasing concentrations of greenhouse gases in the atmosphere — primarily due to the burning of fossil fuels and land-use change — are raising average global temperatures and changing the Earth's climate and prevailing weather patterns. The consequences (some of which are already starting to be observed) are expected to include rising sea levels, the melting of glaciers and the thawing of permafrost, changes in patterns of precipitation with increased risks of flooding and drought, and changes in the frequency and intensity of extreme weather events (e.g. heatwaves are becoming more common, storm intensities are likely to increase).[1]

It is widely agreed that dramatic reductions — probably 80% or more by 2050, against a 1990 baseline — will be required if we are to avert the worst consequences of climate change. It is also recognised that the process of reducing greenhouse gas (GHG) emissions cannot wait; significant emission reductions will be required over the next ten or 15 years. An example of what this might look like in practice has been provided by the European Union which, after its 2007 Spring Council, stated that EU-level reductions of greenhouse gas emissions would be at least 20% lower by 2020, against a 1990 baseline, and that it would increase this reduction target to 30% if other countries followed suit.

---

1 The most authoritative consolidated research on the scientific and economic impacts of climate change is contained in the reports of the Intergovernmental Panel on Climate Change (IPCC). See, in particular, the most recent IPCC reports (IPCC 2007a, 2007b, 2007c) and the summary of the IPCC's 2007 reports (IPCC 2007d).

There are strong economic arguments in support of prompt action, specifically the need to avoid or reduce the consequences of the physical impacts of climate change[2] and the wider benefits (in terms of reduced energy demand growth and improved energy security) that may accrue through using and producing energy more efficiently (see, for example, IEA 2006).

The implications for companies

From a business perspective, companies will be exposed to weather-related impacts such as droughts, floods, storms and rising sea levels. These impacts will particularly affect companies in sectors such as agriculture, forestry, fisheries, healthcare, insurance, tourism, water and property. However, the reality is that all sectors of the economy will be affected in some way.

Companies will also be affected by policy measures directed at limiting greenhouse gas emissions. Given the scale of the emissions reductions seen as necessary to avert the worst effects of climate change, policy action directed at reducing greenhouse gas emissions is likely to result in a complete reshaping of the world economy. Economic development over the past two hundred years has been predicated on access to relatively cheap and widely available sources of energy (in particular fossil fuels). Efforts to reduce greenhouse gas emissions therefore go to the heart of the modern industrial economy. The consequences are not confined to 'obvious' sectors such as power generation, transport and heavy industry; virtually every company's activities, business models and strategies will need to be completely rethought. There is another dimension to this, which is that companies have the potential to make important contributions to reducing greenhouse gas emissions through:

● The allocation of capital (e.g. in clean energy sources)

● Innovation and the development of new technologies

● Their influence on the actions taken by governments on climate change

This book has been complied at an important point (October 2007 to April 2008) in the climate change debate. The last five years in particular have seen a dramatic shift in public and policy attitudes to climate change, with the issue moving from something of a niche environmental concern to a central question for economic and energy policy. The reasons are various.

From the perspective of the public discourse around climate change, long-running climate change campaigns by non-governmental organisations (NGOs) such as Greenpeace and WWF, the personal leadership shown by Al Gore (in particular, through his film *An Inconvenient Truth*), the linking of extreme weather events such as the 2003 summer heatwave in Europe to the effects of climate change, the *Stern Review on the Economics*

---

2 For example, the influential *Stern Review on the Economics of Climate Change* suggested that the overall costs and risks of climate change could be equivalent to losing at least 5% of global gross domestic product (GDP) each year, now and forever (Stern 2006).

of Climate Change (Stern 2006) and the 2007 reports from the Intergovernmental Panel on Climate Change (IPCC 2007a, 2007b, 2007c) have all helped elevate climate change from the environmental pages to the front pages.

From the business perspective, the single most important event has been the leadership shown by the European Union and its Member States in introducing the EU Emission Trading Scheme (EU ETS), which came into force on 1 January 2005. The EU ETS sets limits on emissions of carbon dioxide (CO_2) – the most important of the greenhouse gases – from more than 12,000 installations, representing approximately half of the EU's total greenhouse gas emissions. While, perhaps inevitably, the operation of the EU ETS has been criticised (in particular the over-allocation of permits for the first phase, 2005–2007), the effect of the EU ETS must not be underestimated. Not only were emissions of carbon dioxide given a real (market) price, but the EU ETS signalled that governments were serious about responding to climate change and that they were willing to countenance public policy measures that imposed costs on companies as a necessary part of delivering on this objective. This message has been reinforced by the impressive rate of climate change-related policy development and implementation in Organisation for Economic Cooperation and Development (OECD) countries in particular, with many governments increasing their support for renewable energy, adopting new product standards, and providing incentives and other support for energy-saving measures.

Despite this progress, it is by no means a 'done deal' that climate change will remain high on the political agenda. There are many uncertainties in future climate change policy such as:

● The specific targets that will be adopted at the international and national levels

● The specific policy instruments that will be used and how these will vary between countries

● The future price of carbon dioxide and other GHG emissions

● The sectors that will be affected

● The emissions reductions that different sectors will need to achieve

● The relationship between climate change policy goals and other policy goals such as competitiveness and energy security

To illustrate, in April 2008 (the time at which this chapter was written), companies trying to set a longer-term climate change strategy for their business were faced with trying to assess the implications of (among many other policy variables):

● The UK Climate Change Bill which was making its way through Parliament

● The implications of the US presidential elections – and the proposals from the then President, George W. Bush, that the USA would start reducing its emissions in 2025

● The national plans that are to be prepared by EU Member States detailing how they will meet the EU's target of a 20% reduction in greenhouse gas emissions by 2020

● The outcomes of international negotiations to develop a successor to the Kyoto Protocol

- The EU's proposals to include aviation in the EU ETS and to impose specific emission reduction targets on the automobile industry

- The (probably inevitable) stirrings of a business backlash against the EU's proposals, with Acelor Mittal bringing a case against the EU challenging its decision to exclude the aluminium and chemical sectors from the EU ETS, and Shell warning that refineries in the EU may need to shut if they are not granted sufficient emission permits under future phases of the EU ETS

Notwithstanding these uncertainties, climate change-related regulation and policy are increasingly recognised as proper matters for management attention and, at least in some sectors, as key drivers for future business growth. Furthermore, the strengthening scientific evidence around the urgency of climate change suggests that — if governments respond in a manner that is consistent with this evidence — the rate of policy development and implementation will accelerate.

In anticipation, companies have taken a variety of actions:

- Establishing corporate management systems

- Making public commitments to emissions reductions or carbon neutrality

- Participating in voluntary initiatives such as product labelling

- Seeking to influence their supply chains and their customers to reduce their emissions

While progress to date has been impressive, a fundamental question remains: will these regulatory and self-regulatory initiatives actually deliver the very significant reductions in greenhouse gas emissions necessary to avert the most serious consequences of climate change?

About this book

The overall aim of this book is to reflect on current business practice and performance on climate change in the light of the dramatic changes in the regulatory and policy environment over the last five years. More specifically, it examines how climate change-related policy development and implementation have influenced corporate performance, with the objective of using this information to consider how the next stage of climate change policy (regulation, incentives, voluntary initiatives) may be designed and implemented in a manner that is effective (i.e. delivers the real and substantial reductions in greenhouse gas emissions that will be required in a timely manner) while also addressing the inevitable dilemmas at the heart of climate change policy (e.g. How are concerns such as energy security to be squared with the need for drastic reductions in greenhouse gas emissions? Can economic growth be reconciled with greenhouse gas emissions? Can emissions reductions be delivered in an economically efficient manner?) The book focuses on two areas:

● **Company responses.** How have companies actually responded to the emerging regulatory framework and the growing political and broader public interest in climate change? Have companies reduced their greenhouse gas emissions and by how much? Have companies already started to position themselves for the transition to a low-carbon economy? Does corporate self-regulation — unilateral commitments and collective voluntary approaches — represent an appropriate response to the threat presented by climate change? What are the barriers to further action?

● **Drivers for action.** What have been the key drivers — regulation, stakeholder pressure, investor pressure — for corporate action on climate change? Which policy instruments have been effective, which have not, and why? How have company actions influenced the strength of these pressures?

Structure

The book is divided into five sections. First, Rory Sullivan, Rachel Crossley and Jennifer Kozak present the results of a major assessment of how 125 large UK and continental European companies are managing their greenhouse gas emissions, focusing on both the governance systems they have in place and the manner in which these are incorporated into strategy and performance management. Their analysis suggests that, while a majority of companies (though by no means all) have established emissions management systems and processes and most companies expect their emissions intensity (i.e. greenhouse gas emissions or energy consumption per unit of production or turnover) to reduce, most also expect their absolute (or total) level of greenhouse gas emissions to increase. Their research also highlights that most companies are focusing on direct emissions, emissions from electricity consumption (indirect emissions) and, to a lesser extent, business travel. It appears that companies have yet to fully engage with issues such as the embedded energy in their raw materials, the emissions associated with the use of their products or services or — notwithstanding the increasing number of companies that have made public statements on climate change — their wider role in the climate change policy debate.

The second section of the book (Chapters 3–9) examines the influence of economic instruments and government-sponsored voluntary programmes. The aim of these chapters is twofold: to examine the outcomes that have been achieved (i.e. to assess the effectiveness of these interventions) and to draw lessons/conclusions about the design and implementation of such policy instruments. The first two chapters (by William Blyth and Rory Sullivan, and by Ans Kolk and Jonatan Pinkse, respectively) focus specifically on how companies have responded to the EU ETS, examining questions such as how public policy uncertainty influences investment decision-making, how companies' responses vary depending on the specific regulatory constraints to which they are subject, and whether economic instruments function in the manner in which theory predicts. The following chapter (by Taka Miyaguchi and Rajib Shaw) considers the different motivations for private sector companies to participate in Clean Development Mechanism (CDM; one of the Kyoto Protocol flexibility mechanisms) projects in Indonesia; they then use this analysis to present recommendations on how the wider Indonesian policy framework may be strengthened to both increase private sector involvement in

CDM and to maximise the wider sustainability benefits of CDM projects. The remaining four chapters in this section focus on voluntary programmes. The chapter by Olga Fadeeva, Johannes Brezet and Yoram Krozer uses the case of the Friesland Solar Challenge to examine whether government-sponsored challenge programmes can stimulate innovation (in this case in relation to solar-powered boats) and also deliver wider business and regional development benefits. The other three chapters in this section examine the role of government-sponsored voluntary programmes — Jeffrey Apigian analyses the US Environment Protection Agency (USEPA) Climate Leaders programme, Rory Sullivan focuses on the Australian Greenhouse Challenge and Leticia Ozawa-Meida, Taryn Fransen and Rosa María Jiménez-Ambriz assess the Mexican GHG Program. These authors consider questions such as the reductions in greenhouse gas emissions delivered through these programmes, the economic implications of the programmes and the potential role such programmes may play in the context of wider climate change policy.

The third section, Chapters 10–13, examines the role that has been played by non-state actors (non-governmental organisations, investors, employees) in influencing and informing corporate responses. First, Jim Walker reviews the role that has been played by The Climate Group — an international non-profit organisation focused on convening and coordinating business and government action directed at accelerating the achievement of a low-carbon economy. The chapter presents a number of case studies where The Climate Group has catalysed significant change (developing political consensus around the need for action on climate change in Florida and California, coordinating the development of the Voluntary Carbon (Offset) Standard and supporting the 'Together' campaign) and reflects on the role that can be played by leadership initiatives in responding to the threat presented by climate change. In the following chapter, Oliver Salzmann, Ulrich Steger and Aileen Ionescu-Somers review the overall contribution of the major climate change partnerships (including The Climate Group, the Carbon Disclosure Project and the Corporate Leaders Group on Climate Change) to delivering greenhouse gas emissions reductions, to supporting public policy, and to developing the skills and competences in companies to respond effectively to climate change-related policy. The chapter by Rory Sullivan and Stephanie Pfeifer considers how the interest of institutional investors in climate change has evolved and they canvass the potential for investors to influence corporate performance on climate change (through their ability to influence the cost of capital and through their engagement with companies and policy-makers). Andrea Coulson provides a different perspective, analysing the influence of different stakeholders — investors, employees, customers, competitors — on the manner in which Lloyds TSB reports on its climate change performance.

The fourth section presents a series of company case studies, examining how and why they have responded to climate change in particular ways. The chapters cover a number of different themes. Chapter 14 by Timo Busch, Howard Klee and Volker Hoffmann reviews the (global) Cement Sustainability Initiative and assesses the potential for global initiatives of this type to play a role in the international response to climate change. The contributions by Helen Mathews and Claus-Heinrich Daub, by Ian Gill and Amanda Pitre-Hayes, and by Joan Thiesen and Arne Remmen examine the motivations for action in companies — Novartis, Vancity Credit Union and Grundfos respectively — that have sought to take a leadership position on climate change and the manner in which these motivations have translated into action. Each chapter describes the actions taken, the outcomes achieved, the barriers to action and the role of public policy in supporting or reinforcing their leadership actions. The chapter by Marlen Arnold complements these contributions by looking at one specific dimension of the discussion: namely, the role of organisational learning processes in stimulating proactive corporate responses to climate change. Finally, Christian Engau, David Sprengel and Volker Hoffmann critically examine the climate change performance of the European airline industry, focusing in particular on the relationship between emissions performance and corporate rhetoric on climate change.

The final section (Chapters 20–22) brings some of the key themes of the book together. First Rory Sullivan, Ryan Schuchard, Raj Sapru and Emma Stewart examine what moving from current good practice to best practice on climate change might look like. They argue that current good practice norms are simply not sufficient and that current conceptions of good practice need to be completely rethought if companies are to make a substantive contribution to the delivery of significant reductions in greenhouse gas emissions. Rory Sullivan then examines the potential role of voluntary approaches in the response to climate change, in particular their role as a bridging or transitional policy instrument in advance of legislation being introduced. In the final chapter, Rory Sullivan considers the question of public policy and offers some wider proposals on how policy may be designed and implemented so that it provides appropriate signals to companies, and provides stronger and clearer incentives for companies to reduce their greenhouse gas emissions.

References

IEA (International Energy Agency) (2006) World Energy Outlook 2006 (Paris: IEA).
IPCC (Intergovernmental Panel on Climate Change) (2007a) Climate Change 2007: The Physical Science Basis (Cambridge, UK: Cambridge University Press).
—— (2007b) Climate Change 2007: Impacts, Adaptation and Vulnerability (Cambridge, UK: Cambridge University Press).
—— (2007c) Climate Change 2007: Mitigation of Climate Change (Cambridge, UK: Cambridge University Press).
—— (2007d) Climate Change 2007: Synthesis Report – Summary for Policymakers (Geneva: IPCC).
Stern, N. (2006) Stern Review: The Economics of Climate Change (Cambridge, UK: Cambridge University Press).

2
Corporate greenhouse gas emissions management: the state of play*

Rory Sullivan, Rachel Crossley and Jennifer Kozak
Insight Investment, UK

As climate change has risen up the political and media agendas, companies have found themselves under increasing pressure to manage their greenhouse gas (GHG) emissions effectively. Many have responded by establishing governance and management systems, improving their reporting on greenhouse gas emissions and taking action to reduce their emissions. In this chapter we report on a benchmarking study we conducted in 2007 that sought to assess how 125 large European companies are managing their GHG emissions.

This chapter is divided into three parts. First, we provide a brief overview of the research methodology. We then present the key findings from the analysis, focusing in particular on corporate governance and management systems, emissions reporting and performance. In the third section, we discuss the wider findings and implications of our research for corporate practice and for public policy.

* This chapter is based on the report *Taking the Temperature: Assessing the Performance of Large UK and European Companies in Responding to Climate Change* (Sullivan 2008) published by Insight Investment — the asset management arm of HBOS plc. We would like to acknowledge the support of Insight Investment for this work.

About the benchmark

The principal aim of our research was to develop a comprehensive understanding of how companies manage their GHG emissions as a part of their broader approach to managing the risks and opportunities presented by climate change. The research covered 97 large UK publicly listed companies (the FTSE100, excluding investment trusts) and 28 of the largest publicly listed continental European stocks – a total of 125 companies in all.

We developed a framework against which all companies could be assessed irrespective of their sector or business model, focusing on the management processes that companies should have in place to ensure they can identify, understand and manage the risks associated with their greenhouse gas emissions. We considered corporate performance in seven areas (for more detail, see Sullivan 2008):

- **Governance.** We focused on whether companies had assigned board or senior executive responsibility for climate change issues

- **Policy.** We considered whether companies had clear climate change or environmental policies, and the quality of these policies in terms of their commitment to strong action on climate change

- **Risk assessment.** We assessed whether companies had conducted a comprehensive assessment of the risks and opportunities presented by climate change for their businesses

- **Greenhouse gas emissions inventory.** We assessed the quality (completeness, comparability, etc.) of the GHG emissions data published by companies

- **Targets.** We examined the targets set by companies: in particular, whether they expected their emissions to increase or decrease over time

- **Implementation.** We assessed whether companies had explained how they proposed to minimise or reduce their emissions

- **Leadership and performance.** We examined whether companies were playing a leadership role in the climate change debate by supporting calls for government action on reducing emissions, and whether companies had succeeded in reducing their own emissions

Within each of these areas, each company was evaluated against a number of criteria with scores awarded according to how close the company was to best practice. The categories were weighted as indicated in Table 2.1.

We initiated the project in May 2007. Between June and September 2007, we conducted a desktop review of each company's performance against the framework. Companies were assessed solely on the basis of their own non-confidential published literature including annual reports, corporate social responsibility (CSR) reports and information on their websites. The analysis took account of information published by companies up to 30 September 2007 as well as their submissions to the 2007 Carbon Disclosure Project (CDP) questionnaire.[1]

---

1 www.cdproject.net

TABLE 2.1 Benchmark weightings

Area	Points	Weighting (% of total score)
Governance	10	5.4
Policy	20	10.9
Risk assessment	26	14.1
Greenhouse gas inventory	30	16.3
Targets	30	16.3
Implementation	18	9.8
Leadership and performance	50	27.2
Total	184	100

We sent each company its summary analysis and supporting information, offering it the opportunity to review the scores and provide additional information as appropriate. Of the companies covered by the study, 64 responded within the time period allocated — a response rate of 51%. As part of this process, we had meetings and detailed discussions with companies about the benchmark and the information we were seeking.

Results

Governance, management systems and policy

Most companies have developed the management systems and processes necessary to manage their GHG emissions and related business risks effectively. The majority of the companies considered in our study have clear management accountabilities for environmental and/or climate change issues (93% publish this information), publish environmental and/or climate change policies (92%) and GHG emissions inventories (90%), and provide at least some information on their perceptions of the risks and opportunities presented by climate change (86%).

Of the companies covered by this survey, the majority (116 or 93%) have published details of how they address climate change and details of their management structures for dealing with climate change. The specific structures and approaches vary according to how the individual company manages its wider environmental issues and the relevance of climate change to the business. The majority have appointed a senior manager with explicit responsibility for managing climate change (generally as part of a broader environmental management or CSR brief), with 76% having a high-level environmental or climate change strategy group involving board and senior management personnel and

70% having a named board member with responsibility for the issue. Only 7% did not publish any information on who within the organisation (whether at board or senior management level) has responsibility for climate change or related issues.

In Western multinationals, the adoption (by the board) of a policy on a specific issue is generally seen as an essential starting point for effective management of the issue. Therefore, we sought answers to two questions in our analysis. The first was whether companies had a clear policy on environmental issues in general or on climate change in particular. The second was whether the content of these policies reflected the growing expectations that companies will take action to significantly reduce their emissions.

In relation to the former question, the findings were very encouraging. The majority of the companies surveyed (92%) have at least a published environmental policy, with many also having a stand-alone policy on climate change. While most of the policies refer to climate change-related issues such as energy efficiency or air pollution in the overall scope of the policy, ten do not explicitly mention climate change. Of the policies that do mention climate change, there is quite a wide variation in the content of the policies (see Fig. 2.1). Broadly, most companies acknowledge that climate change is a business risk and/or acknowledge that their activities contribute to GHG emissions, and many have a policy commitment to reduce their emissions. However, few companies have made explicit commitments to either achieving carbon neutrality or to making significant reductions in their total emissions over the longer term.

FIGURE 2.1 Content of company climate change/environmental policies

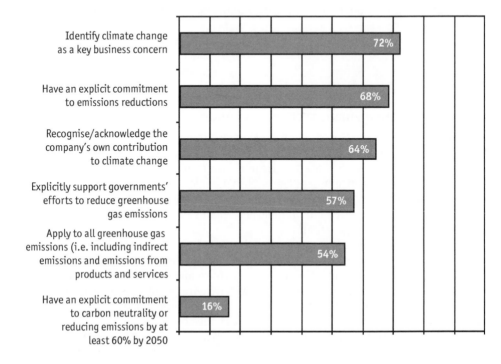

Of the 125 companies included in our study, 108 (86%) publish their views on the risks and opportunities presented by climate change, with many providing detailed accounts in their responses to the CDP. This could be seen as a positive finding as it suggests that most companies have conducted an assessment of the risks and opportunities presented by climate change. However, it is striking how few companies report on climate change risks and opportunities on their websites, or even reference their CDP responses on their websites. We noted that the majority of companies do not describe the process they follow to assess climate change-related risks. The impression created is that companies are presenting their views on these risks simply because the CDP questionnaire requested information on the physical, regulatory and other risks presented by climate change rather than because they have conducted a systematic climate change risk assessment.

Greenhouse gas emission inventories

The majority (90%) of companies covered by the survey provide some information on their emissions. But, despite the large number of companies reporting information, the quality of inventory data is mixed at best. There are number of different dimensions to this.

Most companies (91 of the 112 that report data) explain how they calculate their emissions data — generally either by referencing the *Greenhouse Gas Protocol* (WBCSD/WRI 2004) or by referencing national or industry-specific reporting guidelines. However, the degree of rigour in the application of these reporting guidelines remains unclear. Although the *Greenhouse Gas Protocol* delineates clearly between Scope 1, 2 and 3 emissions (between companies' direct and indirect [i.e. electricity] emissions, and emissions from sources not owned or controlled by the company),[2] companies present their data in different ways. For example, some include the emissions from their own vehicles in Scope 1 whereas others include these in Scope 3. Companies are similarly inconsistent in setting reporting boundaries and frequently do not state explicitly what is included or excluded from the scope of their reporting. It is often not clear whether and how subsidiaries are treated for the purposes of reporting on emissions, and it is common for corporate responsibility reports to only consider a geographic subset of operations (e.g. companies provide detailed data on UK operations but only a cursory overview of performance in non-UK operations) (Sullivan 2006: 8).[3]

---

2 Scope 1 emissions are emissions from sources owned or controlled by the company and include the generation of electricity, heat or steam, physical or chemical processing, transport in company-owned/-controlled vehicles, fugitive emissions. Scope 2 emissions are emissions from the generation of purchased electricity that is consumed in owned or controlled equipment or operations. Scope 3 emissions are emissions from other sources not owned or controlled by the company, such as business travel, external distribution, supply chain (e.g. extraction and production of purchased fuels and materials) or the use/disposal of the company's products and services (WBCSD/WRI 2004: 26-34).

3 Similar concerns about the rigour in the application of emission calculation protocols were raised in the 2007 iteration of the CDP: '. . . there were frequently discrepancies between the emissions disclosed privately to the CDP and those already disclosed elsewhere. Sometimes this was due to companies reporting to different standards other than the GHG Protocol in other documents . . . companies can use either the control or equity share methods of defining their organisational boundaries, which can make a large difference to their reported emissions . . . some FTSE350 companies included Scope 3 emissions in CDP5 reporting on Scopes 1 and 2' (Trucost 2007: 20).

Most companies (87 of the 112 that report data) have their CSR reports verified, with some having a separate verification process for their emissions data. However, it is generally not clear how data has been verified; is it simply the calculations (i.e. are the additions and subtractions correct) or are both the data calculation process and the data assessed? The verification statements provided in CSR reports are rarely clear on the verification that has been carried out or on the uncertainties in the reported data.[4]

The number of years of historical data provided by companies varies quite significantly. Thirty-two of the 112 companies that publish emissions data provide five years of data and 48 provide data for longer time periods. Among the remaining 32 (i.e. those reporting less than five years of data) the relatively short periods for which data are available reflects, in most cases, the relatively recent attention paid by many companies to this issue. It is pertinent to note that many of the companies providing five or more years' worth of data have had to restate or revise at least some of their data from earlier years. This reflects the difficulties in establishing data acquisition processes and the inevitable need to revise data as monitoring and measurement systems become more robust.

Finally, while reporting on direct and indirect (i.e. electricity) emissions is now reasonably well developed, reporting on emissions beyond this remains patchy. Even though 71 companies provide some data on emissions from sources not owned or controlled by the company, this reporting tends to be confined to business travel and, to a lesser extent, transportation and logistics. As yet, companies provide limited information on emissions from their supply chains or the use/disposal of their products and services. A number of practical issues are impeding progress on this topic at present. The first is the question of limits to responsibility: is the reporting of such emissions properly the responsibility of the company or of its suppliers or customers? The second is that calculating emissions from supply chains and product use is technically difficult and there is no consensus on the methodologies that should be used to calculate these emissions or even on the 'scope' of the supply chain or of the product life-cycle. This is an area of significant research activity at the moment. Organisations conducting work in this area include:

- The Carbon Trust – carbon auditing of supply chains

- CDP – through its Supply Chain Leadership Collaboration[5]

- British Standards Institution (BSI) with the Carbon Trust and the Department for Environment, Food and Rural Affairs (Defra) – producing a draft standard to measure greenhouse gas emissions in products and services

---

4 This issue was also raised in the analysis of the responses of companies in the FT500 to the most recent iteration of the Carbon Disclosure Project (CDP5), with Innovest noting: '. . . most responding firms are yet to have their emissions data verified by third parties. As a result, investors should recognise that the data provided to CDP may not be an accurate representation of a company's actual carbon performance. Furthermore, the lack of verification and the inconsistent use of accounting standards suggest that the disclosed data may not be comparable within and across sectors' (Innovest 2007: 38).

5 www.cdproject.net/sclc_home.asp

Targets, action and performance

We next examined whether companies had succeeded in reducing their emissions over the previous five years and how companies expect their emissions to change in coming years.

From our analysis, 65 of the 112 companies that have reported their emissions have seen these rise over the past five years. Of the remainder, 15 (out of 47) have reduced their absolute emissions by more than 3% per year, 26 have reduced their emissions by between 0.5% and 3% per year and the remaining six have stabilised their emissions.

When we look at emissions intensity (i.e. emissions per unit of production or turn-over), a more encouraging picture emerges. Sixty-one companies report that their emissions intensity has stabilised or fallen. In addition, 15 of the companies reporting that their total emissions have stabilised or reduced have not reported on changes in their emissions intensity. While we have not calculated emissions intensity for these companies, it is probably the case that all of these companies have reduced their emissions intensity. The performance outcomes achieved by many companies are impressive: 12 companies reduced their emissions intensity by more than 5% per year, 18 by 3–5% and ten by 1–3%, with the other 21 stabilising or achieving slight reductions in their emissions intensity.

Eighty companies have published emission targets, with most expressing their targets in terms of emissions intensity. Overall, only 39 companies expect their total emissions (excluding offsets) to stabilise or reduce over the next five years. This means that, despite their focus on improving their emissions intensity, most companies expect their total emissions to increase. While there is an apparent contradiction between the targets being set by companies and the expectations of government, it is important to recognise that absolute reductions in emissions present fundamental challenges for companies — particularly those that expect their businesses to grow significantly in coming years.

Two more general points need to be made about the targets being set. First, most companies' targets are relatively short-term. Of those companies with targets, 23 have set targets for the next financial or calendar year, 27 have set targets to 2010, and 20 to 2012. Only ten have set targets that extend beyond 2012 (i.e. beyond the Kyoto Protocol compliance period). Secondly, most companies use relatively recent baselines, with 40 (half of those that have set targets) using a year from 2004 to 2007 as the baseline.[6] The baseline year is particularly important when assessing corporate climate change commitments. If one assumes that a company's emissions have grown by 5% per annum, a difference of ten years in the baseline year (e.g. if 2000 rather than 1990 is taken as the starting point) could mean that the baseline against which performance is to be assessed is 62% higher than the starting point.

While setting targets for emissions reduction is important, such targets must be supported by clear delivery strategies. Even though 80 companies have published emission reduction targets (whether expressed in relative or absolute terms), most do not provide much detail on how they propose to achieve them. From our assessment, only 19 com-

---

6 There are practical reasons why more recent baselines have been chosen: many companies did not exist in 1990; the gathering of data on greenhouse gas emissions is a relatively recent activity for many companies; data gathering systems may have improved over time (i.e. historic data may no longer be valid); and more recent data may be seen as a more relevant frame of reference for assessing the performance of the business.

panies provide a good explanation of the actions they propose to take and only 32 provide information on the total costs associated with their emissions reduction efforts. Care should be taken with drawing firm conclusions on the basis of these data as it may be that:

- Some companies have yet to develop a comprehensive strategy
- Certain of the data are commercially sensitive
- Climate change is a lower priority for the business (and so for the company's reporting)

Beyond their own performance, we also examined whether companies are playing a constructive role in the policy discussions on climate change. We examined corporate support for climate change policy in two ways: through their participation in leadership initiatives and through their support for market-based (e.g. emissions trading) or regulatory instruments (e.g. mandatory efficiency standards) directed at reducing greenhouse gas emissions.

In relation to leadership initiatives, we divided these into two categories. The first comprises initiatives such as The Climate Group,[7] the Corporate Leaders Group on Climate Change[8] and the EU Corporate Leaders Group on Climate Change,[9] where the participants express clear support for emission reduction targets of 60–80% by 2050. The second comprises initiatives that involve research or action on a specific aspect of climate change (e.g. improving corporate disclosures, research into carbon capture and storage).

Sixty per cent (75 companies) report that they are involved in one or more leadership initiatives. Thirty-four companies are members of at least one of the former category of leadership initiatives (i.e. those that are calling for a strong, long-term, public policy framework) with many also participating in initiatives that focus on addressing specific aspects of implementation. Furthermore, almost half of the companies surveyed (62) have expressed some support for market-based instruments such as emissions trading as part of governments' response to climate change – though in many cases this support is qualified by comments about not harming the company's competitive position.

Leading companies

Four UK companies – BSkyB, BT, Marks & Spencer and National Grid – stand out because of the way they have integrated climate change into their core business activities, their track record of performance, their commitment to significantly reducing their own greenhouse gas emissions, and their efforts to reduce the greenhouse gas emissions of their suppliers and customers. More specifically, each company has:

- A clear structure, appropriate to its other management systems and structures, for managing climate change and related issues

7 www.theclimategroup.org
8 www.cpi.cam.ac.uk/programmes/energy_and_climate_change/corporate_leaders_group_on_cli.aspx
9 www.cpi.cam.ac.uk/programmes/energy_and_climate_change/clgcc/eu_clg.aspx

- A clear policy on climate change that includes commitments to significant reductions in total GHG emissions across the range of its activities and its wider footprint, and strong support for government action directed at meeting the commitments in the Kyoto Protocol and towards 60% or 80% emissions reductions by 2050

- Conducted a comprehensive assessment of the risks and opportunities presented by climate change including those from regulation and those from the physical effects of climate change

- Made a clear statement of its views on the financial significance of these risks and opportunities, and of the actions it intends to take to mitigate risks or maximise opportunities

- Published a comprehensive and robust emissions inventory that allows investors to understand historical changes in emissions and provides sufficient information to make an assessment of its financial exposure to climate change regulations

- Set clear targets to reduce emissions in both relative and absolute terms – while the specific targets set by the four companies vary, each sees its objective as aligning with or exceeding the targets set in the Kyoto Protocol (broadly, a reduction in emissions over 1990 levels) and the longer-term targets being mooted by the EU and others (i.e. 20% reductions by 2020 and 60–80% reductions by 2050 over 1990 levels)

- Provided a clear description of how it intends to reduce its emissions including details of the specific actions that will be taken, the emissions reductions that are expected for each action and the costs of taking action

- Made a public contribution to the climate change debate through participating in leadership initiatives and supporting calls for strong and effective government policy on climate change

- Made significant progress in reducing its own emissions in both relative and absolute terms

- Taken action to reduce the wider emission footprint of its business by working with its suppliers and/or customers

High-impact sectors

We paid specific attention to three 'high-impact' sectors: electricity utilities, integrated oil and gas. and mining.[10]

10 We acknowledge that this is not the only way that 'high-impact' could be defined. For example, companies could have been selected on the basis of their total greenhouse gas emissions or their emissions per unit of turnover. However, reflecting the focus of this report on emissions management, we note that, in the FT500: '70% of the total Scope 1 and 2 emissions reported in 2006 occurred in just four sectors: Integrated Oil & Gas (25%), Electric Utilities – International (23%), Metals & Mining and Steel (12%) and Electric Power – North America (10%) . . . As a result, investors should recognise that these sectors are most likely to be targeted by current and future GHG [greenhouse gas] regulations' (Innovest 2007: 35).

Twenty of the companies covered by our study fall into these sectors.[11] Sixteen of the companies were clear leaders, i.e. in the top third of the companies surveyed. However, four companies were particularly poor performers (in the bottom quintile of performance), which seems somewhat counter-intuitive given their significant greenhouse gas emissions. When we looked more closely at these companies, the reason became clear: although the companies are all listed in London, none of their operations or major offices is in Europe. The consequence is that, despite their significant emissions, these companies have less exposure to regulations directed at reducing emissions and have not yet faced the same degree of scrutiny from non-governmental organisations (NGOs) and other stakeholders as those companies with their headquarters or significant operations in Europe.

The 16 leading companies have a number of common features. First, climate change is a matter for senior management with almost all having a named board person with oversight responsibility, a named senior manager with executive responsibility and various committees responsible for oversight and implementation of the company's climate change policy.[12]

Second, all have comprehensive climate change policies that explain the importance of climate change to their businesses, explain the need to reduce greenhouse gas emissions and emphasise the important role of government in delivering emissions reductions. But, reflecting the tension between business growth and emissions reductions, none of these high-impact companies has set stretching long-term emissions reductions targets (e.g. 60% by 2050).

Third, the quality of their emission inventories is uniformly high, reflecting the financial as well as environmental significance of their emissions. All use the *Greenhouse Gas Protocol* (WBCSD/WRI 2004) and/or provide a detailed explanation of how they calculated their GHG emissions, and all have their emissions data audited or verified. Furthermore, 13 break their emissions down by greenhouse gas, 14 break their emissions down by country, 15 provide at least five years of historic emissions data,[13] and 14 provide data on their direct, indirect and product-related emissions. In addition, most of the mining and oil and gas companies provide information on the emissions from the combustion of the coal, oil and/or gas that they produce.

Finally, most of these companies (14 of the 16) are very clear on the actions they propose to take to minimise or reduce their emissions, with 11 providing at least some detail

---

11 Anglo American, Antofagasta, BG Group, BHP Billiton, BP, Centrica, Drax Group, E.ON, Electricité de France (EDF), Enel, ENI, International Power, Kazakhmys, Lonmin, Rio Tinto, Royal Dutch Shell, Scottish and Southern Energy, Total, Vedanta and Xstrata.
12 Three of the 16 companies have very devolved management structures with relatively small corporate (or head office) teams, with responsibility for climate change and related issues being assigned in a manner that aligns with their overall management structures and approach to managing environmental or CSR issues. In these companies, the management of issues such as greenhouse gas emissions is generally devolved to the site or business unit level.
13 At the time of the analysis, Xstrata had been in existence for less than five years. Xstrata has reported its emissions data for all the years for which it has existed.

of the emissions reductions they expect to achieve and eight providing details of the budgets allocated for emissions reductions.

Companies in high-impact sectors face two significant barriers to providing more comprehensive disclosures in this area. The first is that much of the information on costs and the emissions reductions achieved is commercially sensitive. The second is that, particularly for major projects, the objective of minimising emissions and/or maximising energy efficiency is likely to be integral to the project design and/or required by regulatory bodies as a prerequisite to obtaining planning permission and other consents. That is, it is often not possible to separate out the costs associated with reducing emissions from the overall costs of bringing a project to completion.

In relation to performance, three of the 16 companies have reduced their total emissions over the past five years, and a further five have reduced their emissions intensity over this period. A similar picture emerges with regard to expected future emissions. Of the 14 with published targets, four expect their emissions to reduce, one expects its emissions to stabilise and the remaining nine expect their emissions to increase. Furthermore, the companies producing fossil fuels all expect their production volumes (and, hence, the emissions from their products) to increase over time.

These results reflect the fundamental challenge faced by these companies: namely, that business growth is, almost inevitably, accompanied by significant increases in emissions. Moreover, the reality is that these companies have large sunk costs and fixed assets, and significantly altering the emissions profile of these assets is likely to be a long and expensive process.

Discussion

Emissions intensity

Given that many companies use emissions intensity measures as the basis for assessing their emissions performance, it is instructive to look at these measures more closely. While expressing targets in terms of emissions per unit of turnover — as many companies do — is intuitively attractive (it links emissions with a measure of financial value), the reality is that the two are not necessarily related.

For a power station (taking a simple example), total emissions (assuming no change in fuel type or quality) should be directly correlated with fuel consumption and energy output. Take the case of two electricity generating companies with exactly similar facilities (in terms of fuel type, combustion efficiency, etc.) in two different countries but where, due to the vagaries of their respective markets, one company can charge a higher price for its electricity than the other. In this situation, the company that charges the higher price would appear to have a lower emissions intensity than the other, even though the ecological impact of the companies is exactly the same.

It is also important to note that not all sectors have a clear correlation between emissions and materials consumption. For example, in a petroleum refinery, emissions may be relatively constant and the key determinant of total emissions may be the physical properties of the products being handled and produced.

Finally, emissions intensity metrics expressed per unit of turnover would be expected to improve in an environment of rising prices and/or business growth even if the business has taken no substantive action to improve its emissions performance.

There is also the question of whether emissions intensity targets are appropriate for all companies. For example, one company described its involvement in a number of quite distinct business areas and noted that a single intensity measure would not provide a meaningful measure of performance for the business as a whole. Another noted that a key part of its business model involves buying old facilities and upgrading these to be more efficient, thereby reducing emissions. For this latter example, managing the business to deliver a specific emissions intensity target may not be appropriate. There are two reasons for this. The first is that, by taking such facilities into its portfolio, the company will probably increase its average emissions intensity. The second is that, even when the facilities have been upgraded, their emissions performance may continue to be worse than other facilities in the company's portfolio – simply because of design and technology constraints. Again, the overall consequence is that, even though significant environmental benefits may accrue from the upgrading of such facilities, the company's overall performance – as measured by its overall emissions intensity – may have deteriorated.

While there is an argument that companies in a particular sector should all report in a standardised manner (i.e. using the same intensity measure), thereby allowing investors and other stakeholders to make direct comparisons between companies, we believe that caution is required when making comparisons on the basis of such intensity metrics because of:

● The general weaknesses in reported data

● The commonly used emission intensity metrics often do not capture the most significant sources of emissions (e.g. the situation where the primary source of emissions relates to product use but the intensity metric focuses on direct and indirect emissions only)

● The more general concerns outlined above about the relevance of emission intensity metrics normalised by turnover or other financial terms

In conclusion, emissions intensity is just one factor to be considered in assessing the overall performance of a company. Not only should attention be paid to data quality and the appropriateness of the intensity metric being used, but also to aspects such as the nature of the company's activities, whether GHG emissions are in fact a significant risk for the business, and other climate change-related risks and opportunities for the business.

Carbon neutrality and grid-based renewables – creating liabilities for companies?

A number of the companies covered by this assessment have highlighted their intention to go 'carbon neutral'.[14] In broad terms, carbon neutrality can be defined as meaning

---

14 For an interesting discussion of how one company, BT, understands the implications of carbon neutrality for its business, see Webb and Turner 2006.

that the company (or the specified part of its operations, activities, products or services covered by the carbon neutrality commitment) has zero net greenhouse gas emissions. Expressed another way, a commitment to carbon neutrality means that any greenhouse gas emissions from the company (or specified operations, activities, products or services) are offset by paying for emissions savings made elsewhere (i.e. by other organisations).

We recognise that carbon neutrality can provide a range of benefits to companies including:

- Sending a clear signal regarding the company's commitment to action on climate change to employees, stakeholders, government and competitors

- Providing a financial incentive for actions directed at reducing emissions

- Allowing companies to find economically efficient solutions to reducing their emissions (i.e. it may be cheaper to offset rather than reduce emissions)

- Providing brand or reputation benefits

However, we have three major concerns about carbon neutrality in practice. The first is that many of the companies with carbon neutrality commitments expect their GHG emissions to increase; our concern is that carbon offsetting may have the effect of distracting attention from the core issues of reducing energy consumption and reducing emissions. The second is that carbon neutrality commitments create an expectation that companies will continue to deliver on these commitments even if there is a significant increase in the price of offsets used to deliver this goal. While this would probably increase companies' focus on reducing that cost through fundamental emissions reductions, it may also have a significant impact on their cost base. The third concern is that there may be a backlash against the concept of carbon neutrality. Concerns have been expressed about the quality of the offsets being used by companies, though it is important to recognise that the majority of companies covered by this assessment have sought to ensure the credibility of their offsetting programmes (whether through only purchasing Certified Emissions Reductions certificates, through purchasing carbon credits that meet similar criteria[15] or through developing their own project portfolios). Another risk to the credibility of carbon neutrality is that the scope of the carbon neutrality commitments of many companies is either unclear or limited. For example, a number of companies have limited the geographic scope of their commitments to their UK operations only, or to just a part of their carbon footprint (e.g. business travel, but not other aspects of their supply chains or product use and distribution). While these limitations have not yet received much attention, they may have the effect of undermining the credibility of companies' wider climate change commitments.

We see similar issues with the use of grid-based renewable energy. Clearly, renewable energy has a key role to play in the move towards a low-carbon economy and, for many companies, purchasing 'green' electricity is the most effective way for them to reduce their emissions. This should, in turn, stimulate electricity companies to increase their investments in renewable sources. However, current practice raises a number of concerns. The first is that it is not clear that corporate demand for renewable energy has stimulated additional investment in this area. It appears that, in the UK at least, most renewable electricity actually comes from sources built to meet the requirement of electricity utilities under renewable energy legislation, i.e. corporate demand is not necessarily translating into additional renewables capacity being built.[16] The second is that – in a similar manner to the comments above about carbon neutrality – buying renewable energy may weaken the attention paid to energy efficiency. The third is that, again reflecting our comments regarding carbon neutrality, renewable electricity may become more expensive over time as demand increases, with consequent financial implications for companies.

Reflections on current practice

Most large companies have established the basic management systems and processes necessary for them to manage their greenhouse gas emissions and related business risks effectively. Most have developed clearly defined management accountabilities, published climate change or environmental policies, published greenhouse gas emissions inventories and set targets for reducing their greenhouse gas emissions intensity. Approximately one-third of companies (43 of the 125) have gone much further with:

- Policy commitments to reductions in GHG emissions across the range of their activities

- A published statement on the financial significance of the risks and opportunities presented by climate change

- Comprehensive and robust emissions inventories that allow us, as investors, to make an assessment of their exposure to climate change regulations

- Targets to significantly reduce their emissions in relative and, in many cases, absolute terms

- A clear strategy for managing their emissions

- Clear support for effective government action on climate change

Forty-seven companies have stabilised or reduced their total greenhouse gas emissions over the past five years and almost two-thirds have stabilised or reduced their emissions intensity.

There are, however, areas of concern. Most importantly, 21 companies – including four companies in high-impact sectors and a number with potentially significant GHG emissions – scored extremely poorly, suggesting that there may be significant weaknesses in the manner in which these companies are managing their greenhouse gas emissions. While some of the poor scores may reflect the quality of reporting rather than

---

16 For example, it has been argued that: 'more companies than ever are demanding green energy from their suppliers. Given the limited supply of green energy in the UK, however, there is little scope for the business sector to take up more without depriving households of their green energy supplies . . . A question of ethics hangs over the imbalance between the supply of energy from renewables and potential demand. With a limited supply now available on the grid, what real corporate virtue can subscribers to green energy claim?' (Proctor 2007).

actual performance, the reality is that one would expect all large companies to be reporting in a robust and comprehensive manner.

Even though most companies seem to have broadly sufficient governance arrangements in place, the quality of these governance arrangements may not be completely appropriate to the challenge presented by climate change. There are a number of areas to highlight:

- **Climate change policies are weak.** While most companies acknowledge that climate change is a business risk and/or acknowledge that their activities contribute to GHG emissions, and many have a policy commitment to achieving emissions reductions, few have made explicit commitments to achieving significant reductions in these emissions

- **The quality of inventory data is mixed at best.** Common issues include the lack of clarity around the scope of reporting (specifically, whether all greenhouse gases are covered and whether the reporting applies to all activities and operations), the quality of the emissions calculations and the limitations in data verification. It is worth noting that, while reporting on direct and indirect emissions is now reasonably well developed, reporting on emissions from supply chains or product use and disposal remains limited

- **Most companies have not, as yet, conducted thorough assessments of climate change-related risks and opportunities.** Even though 86% of the surveyed companies publish their views on the risks and opportunities presented by climate change, much of this reporting seems to have been triggered by questions in the CDP questionnaire. It is not clear what level of analysis actually underpins the responses submitted by many companies to CDP

- **Companies' targets seem weak.** While we acknowledge the difficulties faced by companies (in particular those in growth sectors) in reducing their emissions, we are of the view that the direction of public policy is clear: i.e. companies across the board will be expected to significantly reduce their GHG emissions. The regulatory drivers are presently strongest in the EU but we expect to see similar measures being introduced in major economies such as the USA, Canada and Australia in the coming years. Our analysis suggests that most companies do have emissions reductions targets but these are generally expressed in relative rather than absolute terms (i.e. companies expect to improve the efficiency or emissions intensity of their business activities, but also expect their total emissions to increase)

- **The means for reducing emissions is unclear.** Although most companies have targets (whether relative or absolute), most do not provide much detail on how they propose to achieve them. We recognise that, as with any activity, there are inevitably uncertainties about how much specific actions will cost and about the specific performance outcomes that will be achieved. We are nonetheless concerned that the weaknesses in the information provided in this area suggest that companies have set targets without fully understanding how these targets are to be achieved and, indeed, whether these targets are appropriate for the business (e.g. could the business do more)

● **Constructive engagement with public policy is limited.** While almost half of the companies surveyed have expressed support for market-based instruments such as emissions trading, in many cases this support is qualified by comments about not harming the company's competitive position. The overall impression is that corporate commitment to significant reductions in GHG emissions remains weak

Conclusions

The results of our benchmark suggest we have reached a point where the governance argument has been won: companies accept that they have responsibility for managing or reducing their GHG emissions and most have (notwithstanding weaknesses in individual companies and in specific aspects of emissions management and reporting) established the governance and policy frameworks and implementation mechanisms necessary for them to manage these emissions.

However, the arguments around strategy have yet to begin in earnest. The overwhelming impression is that — excluding a few leading companies who recognise the fundamental challenge presented by climate change to their business models — most are focusing their efforts on those actions that provide positive financial returns, while also seeking to improve their energy efficiency or emissions intensity per unit of production or turnover.

This is despite the strong messages being sent by policy-makers and the scientific community that significant reductions in global GHG emissions (perhaps 80% by 2050 over 1990 emission levels) will be required to avert the worst consequences of climate change. Notwithstanding the growing sense of urgency, our analysis suggests that most companies do not see climate change as a material risk to their business. The uncertainties around climate change policy — in particular around the nature and extent of the post-2012 international climate change policy framework — compound this problem. The 'wait and see' strategy being adopted by the majority of companies can be seen as a rational response to this policy uncertainty. The data from our analysis suggest that companies are not — other than to the extent that it is justified in financial terms — seeking to pre-empt public policy but instead are prepared to wait and respond as and when policy emerges.

References

Innovest (2007) *Carbon Disclosure Project Report 2007: GlobalFT500* (London: Carbon Disclosure Project).

Proctor, T. (2007) 'Getting the Balance Right', *Climate Change Corp.com*, November 2007: 20.

Sullivan, R. (2006) *Climate Change Disclosure Standards and Initiatives: Have They Added Value for Investors?* (London: Insight Investment).

—— (2008) *Taking the Temperature: Assessing the Performance of Large UK and European Companies in Responding to Climate Change* (London: Insight Investment).

Trucost (2007) *Carbon Disclosure Project Report 2007: UK FTSE350* (London: Carbon Disclosure Project).

WBCSD (World Business Council for Sustainable Development) and WRI (World Resources Institute) (2004) *The Greenhouse Gas Protocol: A Corporate Accounting and Reporting Standard* (Geneva: WBCSD, rev. edn).

Webb, R., and A. Turner (2006) *What Would a Genuinely Carbon Neutral BT Look Like?* (London: Carbon-Sense).

Part II
Public policy: regulation, economic incentives and voluntary programmes

3

The effectiveness of climate change policy as an investment driver in the power sector*

William Blyth
Chatham House and Oxford Energy Associates, UK

Rory Sullivan
Insight Investment, UK

Climate policy and the electricity sector: an overview

The electricity generation sector has a unique position in terms of its relevance to, and dependence on, regulatory drivers for climate change mitigation.

First, it is an obvious and necessary target for climate change policy. The sector is one of the largest sources of greenhouse gas emissions, representing 31% and 35% of UK and global energy-related CO_2 emissions respectively. What makes the sector even more attractive as a target for policy is the fact that it is dominated by relatively few large players (so the point of regulation is well defined) and there are several technology options for reducing greenhouse gases that are either already available or considered feasible for the sector. Indeed, in some long-term models of the energy system, the cost of abating CO_2 from electricity generation is considered to be so low relative to other sectors that

* This chapter builds on a workshop (sponsored by Insight Investment) held on 22 March 2006 at Chatham House entitled 'Addressing Uncertainty in Climate Change Policy: A Dialogue between Institutional Investors, Companies and Policy-makers' and a subsequent Chatham House briefing paper, *Climate Change Policy Uncertainty and the Electricity Industry: Implications and Unintended Consequences* (Sullivan and Blyth 2006).

future scenarios assume a major shift away from combustion technologies towards electricity consumption (e.g. in the transport and domestic sectors). Whether or not these long-term structural changes come to pass, there seems little doubt that, along with improving the efficiency of energy consumption, changes in the way we produce electricity will be critical in terms of responding to the need for emission reductions.

The second unique characteristic of the electricity generation sector is that, in many countries throughout Europe, the sector is facing a new investment cycle to replace plant built in its previous investment phase in the 1970s. Within the EU, some of this is happening for reasons of plant age. Some (particularly for coal plant) is being additionally induced by the Large Combustion Plant Directive, which requires controls on sulphur dioxide emissions that are not cost-effective to apply to the older plant. At the global level, the need for new investment is being driven by economic growth and the associated increases in primary energy demand. Estimates suggest that, by 2020, new plant representing up to 35% and 80% of current capacity will need to be added to the system in the UK (DTI 2007) and globally (IEA 2003a) respectively – although these numbers may be overestimates as they may not fully account for the potential for demand restraint through improved efficiency.

Nevertheless, this impending investment cycle represents both a threat and an opportunity in terms of climate change. In the EU, there is the backdrop of a policy debate regarding the need to push towards greater liberalisation of the electricity market. Policy-makers and politicians will strive to ensure that climate change policy does not threaten the functioning of the market, and that the necessary investment is delivered in a timely fashion to ensure that electricity supplies are not disrupted. On the other hand, the fact that a significant proportion of the electricity generation fleet is due to be replaced is a major opportunity to begin the process of stock turnover towards the lower-emitting generation sources that will be required throughout the sector over the course of this century. Indeed, this is the same opportunity, which is writ large in the case of power generation investment in China and India, of avoiding locking in to another generation of high-emitting plant.

The third unique feature of the electricity generation sector is its curious hybrid status between a real market and centralised government control. To differing degrees, governments have attempted to liberalise electricity markets with the expectation that more efficient decisions regarding the allocation of capital will result. However, governments retain responsibility for ensuring energy security including (at least a perceived) political pressure to ensure that 'the lights stay on'. This means that, although individual investments are in theory expected to be made on a purely commercial basis, the collective behaviour of the industry remains a public policy issue. Policy concerns therefore not only include the sufficiency and timeliness of investment in new generation capacity, but also the choice of technology, since coal, gas, nuclear and renewables all have their own particular issues regarding energy security, environmental impacts, infrastructure requirements and necessary incentive frameworks.

This chapter explores a particular aspect of the relationship between climate policy and the electricity sector: namely, the management of risks associated with policy uncertainty. Uncertainty is a significant feature of the climate change debate, arising at all steps in the chain between greenhouse gas emissions and the impacts of climate change on humans and the environment.

Companies tend to be exposed to these risks via uncertainties in climate policy, since regulation is the point at which companies' environmental performance is translated directly into economic performance. The IEA suggests that 'the management of risk and uncertainty in competitive markets is fundamentally changing the traditional investment paradigm' (IEA 2007).

If this is the case, then governments need to understand risk management for two reasons:

● Because risk affects the timing and choice of technology in companies' investment decisions, which will have consequences for the delivery of the public energy policy goals relating to cheap, clean and secure supplies of electricity

● Because climate change policy itself has become a significant source of uncertainty for electricity generation companies, and governments are inextricably linked to the management of this source of risk

The chapter provides an overview of the relationship between climate change policy and risk management, using the EU Emission Trading Scheme (EU ETS) as a case study. Preliminary conclusions are then drawn about the effectiveness of climate change policy as a driver for changing investment behaviour in the electricity generation sector.

Investment risk and climate policy

Companies face a range of risks and uncertainties when making investment decisions. Probably the most fundamental risk for power generation companies – particularly in competitive market conditions – is price risk. The primary short-term objective of power generation companies is to maximise profits by optimising the use of their generation assets given the price they can receive at any given time. Operational decisions on a day-to-day (or even hour-to-hour) basis are driven by this optimisation requirement. These decisions, in turn, determine the rate at which each plant in the company's portfolio will be run in any given period.

When it comes to longer-term investment decisions, companies have to decide whether to build new power plant, and how and when to retire old plant. Because of economies of scale and the historical development of the sector towards large centralised generating plant with very long lifetimes, these investments have tended to be capital-intensive and long-lived.

Different generating assets have different cost structures which expose them to different risks. For example, coal plant are very capital intensive, but coal prices (and, therefore, operating costs) have traditionally been low with little volatility. Renewable technologies such as wind and solar power have zero fuel cost, and hence low operating costs, and, again, the cost is mostly upfront capital. Gas-fired generation on the other hand has relatively low capital costs and relatively flexible operations, but gas prices form the largest part of the project's costs and are highly volatile. Table 3.1 presents a qualitative comparison of the characteristics of different types of generating technology.

TABLE 3.1 Qualitative comparison of different power generation technologies

Source: IEA 2003b

Technology	Unit size	Lead time	Capital cost/kW	Operating cost	Fuel cost	CO$_2$ emissions	Regulatory risk
CCGT[a]	Medium	Short	Low	Low	High	Medium	Low
Coal	Large	Long	High	Medium	Medium	High	High
Nuclear	Very large	Long	High	Medium	Low	Nil	High
Hydro	Very large	Long	Very high	Very low	Nil	Nil	High
Wind	Small	Short	High	Very low	Nil	Nil	Medium

[a] combined cycle gas turbine

However, the translation of these cost structures into risk exposure is not entirely straightforward. While high capital costs (e.g. for coal and nuclear plant) lead to exposure to associated risks such as cost overruns or project delays, it does not necessarily follow that high gas prices represent the biggest exposure for gas-fired plant. In fact, in many countries the cost of gas-fired power generation sets the wholesale price for electricity. This means that variations in gas price will be closely correlated to variations in electricity price, making the net revenue for gas-fired plant relatively stable. Counterintuitive as it may seem, coal and nuclear plant operating in this type of market may be *more* exposed to gas price volatility than gas-fired plant. Clearly, this depends on market structure and the details of how costs are passed through to electricity prices, but it illustrates the point about the need to understand risk management in order to be able to understand investment behaviour.

In very general terms, companies or projects that are exposed to greater levels of risk need to provide a higher return on investment in order to attract capital. The level of risk faced by a particular project or for a company as a whole is, therefore, reflected in its cost of capital. This creates a disadvantage for capital-intensive plant such as coal and nuclear since the financial case for these technologies is more sensitive to the cost of capital. Again, the risks faced by these technologies will vary depending on the specific market conditions (including any government guarantees, etc.), but a preliminary conclusion from an assessment of risk is that lower-capital plant (i.e. gas-fired plant) will tend to be favoured in more risky market conditions.

What are the sources of uncertainty in climate change policy?

There are many uncertainties in climate policy such as:

- The political context within which climate policy is developed (e.g. the level of government support for climate policy measures, concerns about energy security or wider competitiveness issues)

- The policy instruments chosen and the manner in which they are implemented

- Perceptions of the credibility of the different actors (e.g. is government seen as committed to action on climate change? Are companies committed to minimising greenhouse gas emissions?)[1]

Some of the specific sources of uncertainty faced by electricity utilities include:

- The degree of government support for policy action on climate change over the short and long term

- Whether there will be a post-2012 international regime and whether this will be target- or process-based

- The specific policy instruments used

- Differences in implementation between different countries

- The future price of greenhouse gas emissions permits (or allocations)

- Allocation rules

- Subsidy levels for specific technologies

- The timing of policy responses

- The response of other companies to specific policy measures

- The response of other sectors of the economy (i.e. how much burden will be borne by electricity utilities *vis-à-vis* other sectors?)

- The degree of support for climate policy measures among companies and investors

- The relationship between climate policy goals and other policy goals such as energy security

Efforts to evaluate in advance the effectiveness of policy measures relating to climate change are complicated by factors such as uncertainties in technology costs and uncertainties in the responses of the parties affected by the policy measures. The fact that climate policy costs and outcomes are uncertain creates pressure on policy-makers to maintain policy flexibility in order to allow them to respond appropriately to new information. If governments are too fixed in their approach, they risk committing themselves to policy actions that may turn out to be either too stringent or not stringent enough, with limited freedom to adapt or change policy in response to these outcomes. On the other hand, flexible approaches to policy may create an additional cost to companies that will have to make decisions based on a changing policy environment. Ultimately, a balance needs to be achieved between flexibility and certainty. A move towards greater policy clarity may need to take account of the potential for 'nasty surprises' as the flexibility of policy-makers to respond to such surprises may be constrained. That is, a con-

1 It is important to recognise that there is also scientific uncertainty about the magnitude of climate change and how changes in climate will translate into impacts on human society. This uncertainty, in turn, may affect the policy targets that are set and hence is another factor that feeds into the uncertainty assessment.

sequence of moving towards greater policy certainty may be that targets are more stringent than if a fully flexible policy approach was taken.

How do these uncertainties affect electricity companies?

Standard project appraisal uses a discounted cash flow (DCF) model (comprising elements such as electricity prices, fuel prices, environmental charges, taxes, tax credits, and other fixed and variable operating costs) to derive an estimate of the present value of the project's future income compared with the initial capital outlay. The simple investment rule is that if the present value of the cash flow is greater than the initial capital outlay for the project – such that the 'net present value' (NPV) is positive – then the project should go ahead. If it is not, then the project should not go ahead.

However research by the International Energy Agency (IEA) – see Box – suggests that, in practice, flexibility in timing can be a critical factor in the responses of electricity utilities to risk (Blyth and Yang 2007). This is because investments are more or less irreversible since the plant generally cannot be resold without losing considerable value. In these situations, a greater project payoff may be obtained by waiting until the uncertainty has been resolved (or reduced) than by investing immediately. Hence, in order to stimulate immediate investment, a project would not only need to achieve a positive NPV but would also need to achieve an additional return on investment sufficient to exceed the value of waiting caused by the uncertainty. This could mean that the prices (e.g. electricity or carbon prices) required to stimulate investment in low-carbon technology may be higher than expected based on normal DCF analysis. In this case, a company's response may take a number of different forms such as delaying investment, delaying

---

The implications of uncertainty (Blyth and Yang 2007)

- Uncertainty creates a financial incentive to delay investment in order to gain more information which would allow a more optimal investment choice

- In order to justify immediate investment in situations of uncertainty, a project would have to overcome this value of waiting – the gross margin would need to exceed not only the capital cost, but an additional threshold level above that

- The bigger the possible 'shock' to carbon prices, the greater the value of waiting

- The less time there is available before the 'shock', the higher the gross margin would have to be to overcome the value of waiting

- The risk premium would lead to an increase in electricity prices of 5–10% in order to stimulate investment

- In the case of carbon capture and storage, the risk premium would increase the carbon price required to stimulate investment by 16–37% compared with a situation of perfect certainty

- The option to retrofit carbon capture and storage acts as a hedge against high future carbon prices and could accelerate investment in coal plant

plant closure/replacement, giving greater preference to phased investment (e.g. prefer-ring flexible/modular plant over economies of scale), or requiring greater project cash flow for immediate investment (leading to higher prices). From a public policy perspec-tive, if investments in new technologies are deferred as a result of policy uncertainty, this could affect the emissions reductions path of the sector, while higher-than-expected carbon prices could have wider economic implications both for the power sector and for consumers.

The IEA's research also indicates that uncertainties a significant time into the future (e.g. ten years ahead) do not materially add to investment risk (Blyth and Yang 2007). However, in an emissions trading scheme, targets are set at periodic intervals. The less time a company has between taking its investment decision and the announcement of a new cap level in the trading scheme, the higher the risk premium it will add to the pro-ject. This means there would be a greater incentive to invest near the beginning of an emissions trading period than near the end, which could exacerbate periodic investment cycles as a response. It also suggests that longer trading periods may expose companies to less policy risk.

One option that appears particularly promising – in particular in the context of high gas prices – is the ability to retrofit carbon capture and storage to coal plant. The exis-tence of carbon capture and storage as a retrofit investment option makes investment in new coal plant less risky (reducing the investment threshold in the face of uncertainty) and could accelerate investment in coal. Investment in the carbon capture and storage plant itself, however, is very sensitive to carbon price, so there is an incentive to wait to gain more information about future prices before retrofitting the technology (see Box).

Case study: the EU Emission Trading Scheme (EU ETS)

The EU ETS, which came into force on 1 January 2005, is a 'cap and trade' regime which sets limits on carbon dioxide (CO_2) emissions from more than 12,000 installations rep-resenting almost half of the EU's greenhouse gas emissions. It currently includes:

● Combustion plants (including power generation)

● Oil refineries

● Coke ovens

● Iron and steel plants

● Factories making cement, glass, lime, brick, ceramics, and pulp and paper

The scheme has been introduced in distinct phases. The first phase (2005–2007) was, in some senses, designed to act as a pilot phase though the financial penalties for non-compliance were real and allowances traded for real money. Phase 2 runs from 2008 to 2012, and the European Commission's recent proposed package on climate change and renewable energy (EC 2008) implies that Phase 3 is likely to run up to 2020.

In the first phase, allowances were allocated free to the individual installations[2] and most facilities were granted sufficient allowances to cover most but not all of their emissions (each allowance representing a permit to emit one tonne of CO_2). Under the scheme, operators have to either reduce their emissions of CO_2 to the amount allocated or buy allowances in the market to cover the shortfall. When companies' allowances exceed total emissions, they are able to sell the surplus. The financial implications for individual companies depend on:

- The size of their allocations

- The method of allocation (in the future, allowances may be auctioned rather than allocated free)

- Their ability to reduce emissions

- Their ability to pass on the costs of meeting the requirements to their customers

- The price of the allowances

Overall, there needs to be a scarcity of allowances in order to create a market price for carbon and thereby induce reductions in greenhouse gas emissions.

Structurally, the key difference between Phases 1 and 2 was that banking of surplus allowances into subsequent phases was not allowed in Phase 1. This meant that any surplus allowances held at the end of 2007 had zero value. The price of CO_2 allowances fluctuated from the start of the scheme, reaching almost €30/tonne in June 2005, subsequently settling back to €20–22 at the end of 2005, and then gradually rising to almost €30/tonne in April 2006. In May 2006, the price dropped sharply to around €10/tonne as evidence emerged that most European countries had over-allocated EU Allowances (EUAs) for the first phase and then fell close to zero over subsequent months.

In Phases 2 and 3 (and beyond), banking of allowances is explicitly built into the EU Emissions Trading Directive 2003/87/EC. This means that any surplus allowances in one trading period will still have a value so long as there is an expectation that subsequent trading periods will be short of allowances. This should avoid any recurrence of the price collapse experienced in Phase 1.

Following a recent review of the Directive, other changes to the scheme are planned: for example, to include other sectors (e.g. aviation) and the provision of explicit rewards for facilities fitted with carbon capture and storage. The number of allowances that countries can allocate for free is also likely to change.

In Phase 2, the number of allowances that can be auctioned increased to 10% (up from 5% in Phase 1). The Commission (EC 2008) proposes that, in Phase 3, the power sector would be subject to full auctioning from 2013 and most other sectors would step up gradually to full auctioning in 2020. With covered emissions of around 2 billion tonnes CO_2, the asset value of allowances in the EU ETS is worth €2 billion for every €1/tonne of CO_2 ($teCO_2$) change in the price of allowances. With futures prices between €10 and

---

2 In relation to the question of why free permits were allocated, one participant in the March 2006 workshop noted: 'Why do governments need to provide a free allocation at all? The reason is that it is difficult for governments to take action that would be economically rational because of incumbent effects. The New Entrant Reserve is necessary so as not to disincentivise new entrants to the market and to avoid perverse effects.'

€20/teCO$_2$, the method of allocating these assets will have important distributional effects on companies. In Phase 1, allocating them for free led to significant windfall gains for some companies — particularly power generators — as discussed below.

The EU ETS: experiences to date

In the liberalised European energy markets, most/all of the CO$_2$ price has been internalised in power prices (IPA 2005).[3] As a consequence, the price of electricity on wholesale markets rose significantly during the periods of elevated CO$_2$ prices despite the fact that the majority of allowances that generators needed to cover their emissions had been granted to them for free. This is because, when generators bid their plants into the wholesale market, their pricing strategy is based on the marginal cost of generation. This includes fuel, variable operating and maintenance costs, and emission costs. Although generators may have received free emissions allowances, they still price emissions as if this allowance had been purchased from the market (since they could always forgo production and sell the allowances at the prevailing price). As a result, the introduction of emissions costs raises wholesale power prices regardless of the amount of free allowances the generator is granted. In the short term, the primary effect of emissions trading has been to boost the cash flow and profits of European generation companies that operate in countries in which power markets have been fully liberalised (IPA 2005; Kernan *et al.* 2005).

Although CO$_2$-free forms of generation (e.g. nuclear, renewables) do not receive free allowances, they benefit from the higher wholesale electricity prices. For many generators, these higher wholesale prices translate directly to improvements in their bottom lines.[4] Windfall profits have been controversial, with suggestions in some countries that retail prices should be capped or that full feed-through of costs should not be permitted (i.e. constraining the windfall profits achieved by generators). But, even in Phase 2 of the EU ETS, there will be significant allocations of free permits to the electricity industries and so the discussion about windfall profits is likely to continue.

Overall, Phase 1 of the EU ETS was effective in the sense that it:

● Established a price for pollution

● Incorporated this price signal into power prices

3 Research by UBS indicates that the forward electricity price in Germany at 1 January 2006 was €51/MWh — an increase of €20/MWh over the price at 1 January 2005. Of this increase, some €14/MWh could be attributed directly to the effects of purchasing CO$_2$ permits (Lekander and Gilles 2006). See also Wirtz and Koebernick 2006.
4 The specific level of windfall profits depends on the country the generator is operating in. Windfall profits are most likely in the deregulated and competitive markets of the UK, the Nordic countries, Germany and Italy, and less likely in the more regulated domestic markets of France and Spain. In competitive markets, there is a clear connection between the marginal generation cost, the wholesale market's clearing price and the retail price. An open and competitive market allows rational economic behaviour to translate into price signals regardless of the actual price and ultimate beneficiary. In a less liberalised market, the retail process may still be regulated so, even if a utility is able to make windfall profits from the wholesale market, it may not be able to pass on the increases to the retail market (Kernan *et al.* 2005).

- Successfully developed the necessary institutional structure and services required to support the trading market

- Resulted in a significant level of learning among all relevant players (including governments and companies) about the practicalities of operating a full-scale emissions trading market

However, the fact is that so far CO_2 prices have not even achieved the level to encourage fuel switching from coal to gas — even though this type of operational level action to reduce emissions is straightforward.

Incentivising the more significant shifts in investment that will be needed going forward will require not only higher prices than we have seen so far, but also a good deal more confidence that these prices will be maintained into the future to provide the necessary financial returns on investment in low-carbon sources of generation.

While emissions trading (or carbon pricing more generally) will be a crucial plank of climate policy, it has yet to prove that it will itself be the major mechanism by which deep cuts in emissions can be achieved. In order for this to be the case, investors will need to get to a state where the policy risks are sufficiently well understood and characterised so that these risks can be managed efficiently. Learning has proceeded rapidly in Phase 1 of the EU ETS with respect to the mechanics of emissions trading, but learning with respect to risk management requires a longer track record of market operation and will take significantly longer to achieve.

Looking ahead: the challenges for policy-makers

Policy-makers have been quick to take on board the message that investors and companies require long-term visibility of climate change policy in order to underpin their long-term investments. At the EU level, political leaders at the Spring Council in 2007 announced targets up to the year 2020 for greenhouse gas reductions, renewable energy and energy efficiency (Council of the European Union 2007).

Setting these medium targets was, in large part, a response to concerns that the previous long-term target (of limiting global average temperature rise to no more than 2°C) did not provide a sufficient signal to companies on time-scales that were important for their investment decisions — typically the first 15-year period that tends to dominate NPV analyses of most new investment projects. The European Commission's proposals in early 2008 (EC 2008) on how these targets for 2020 should be met (in terms of the division of effort between countries and sectors, and how much is expected to be delivered through the EU ETS versus other policy mechanisms) provided a step forward in terms of addressing this issue of policy uncertainty — though many of the details will become clear only once countries develop their national plans for meeting these targets.

The UK has gone a step further in its new Climate Change Bill, which proposes setting binding medium-term (15 years) and long-term (to 2050) caps for greenhouse gas emissions under a new system of 'carbon budgets' set in five-year tranches on a rolling basis 15 years ahead. Again, a key driver for introducing this mechanism was the need to increase the time horizon and predictability of policy-making.

However, the challenge that policy-makers face is maintaining commitment to these unilateral actions in the face of a shifting and uncertain international policy environment. Policy-makers are sandwiched in the middle of a three-layer decision-making process involving:

- Global climate negotiations
- National/regional policy-making
- Company-level investment decisions

Because they are located in the middle of this sandwich, policy-makers are playing two distinct games simultaneously – the first is a negotiation with the international community, the second is a signalling game with domestic companies.

The best hope that policy-makers have of establishing their credibility is to make their positions in these two separate games consistent. This can be harder than it sounds if policy-makers are faced with competing priorities in the two games. An example of this at the EU level was the announcement after the 2007 Spring Council that EU-level reductions of greenhouse gas emissions would be at least 20% by 2020, but that this would increase to 30% if other countries followed suit. This tactic was clearly part of the negotiating game that the EU is involved in with the international community aiming to increase participation and draw in other players. However, the credibility of this tactic is undermined by the fact that it is a bad move in the other game being played. That is, signalling that the medium-term target may be shifted at quite short notice from 20% to 30% is not consistent with the aim of creating long-term visibility and predictability of target setting.

Even if policy-makers do manage to establish a consistent position in terms of target setting, climate change policy will still be exposed to possible shifts due to factors outside the control of any particular national or regional government. There are many such factors but, to name a few, they include changes in the understanding of climate change impacts, changes in technology or fuel costs, and changes in the structure of the international agreements that affect the availability of emission reduction credits from abroad. Both companies and policy-makers will, therefore, have to be adaptive to such shifts. Ideally, they will need to develop strategies that are robust in a range of possible future outcomes.

Addressing policy uncertainty

Perhaps the most important conclusion from our analysis is that the longer-term direction of climate change policy is not seen as fixed or certain. This leads to companies requiring a greater return on their investment and, perhaps, opting for lower-capital-cost options such as gas-fired generation.

The various players (companies, institutional investors and even policy-makers) see uncertainties around the future of the emissions trading scheme itself, as well as more general uncertainty around whether CO_2 prices will (or will be allowed to) rise to a sufficiently high level that will result in companies changing their dispatch decisions or, more importantly, changing their future investment patterns.

This raises the prospect that more targeted policy intervention will be required to stimulate further emission reductions in the electricity sector. However, policy in this area is complicated by the fact that reducing greenhouse gas emissions is not the sole goal of energy policy. Specifically, issues such as energy security and, in certain countries, job creation/protection in areas such as mining, may create tensions that run counter to climate change policy goals. Policy-makers will also need to manage the risk that, by introducing several policies simultaneously, the different policy mechanisms may compete for emission reductions from the same sources — introducing inefficiencies and creating additional risks for investors.

Policy design and communication

It seems clear that the uncertainties in climate change policy — at the global and EU levels — are a significant barrier to encouraging companies to invest in new low-carbon generating capacity. As general recommendations, policy-makers and governments should:

- Ensure consistency between targets and policy measures implemented domestically, and those negotiated internationally both within the EU and within the broader UN Framework Convention on Climate Change (UNFCCC) framework

- Seek to avoid policy disconnects (e.g. avoiding a post-2012 hiatus in the international policy framework). While the politics around the specific links between EU ETS and the Kyoto Protocol are sensitive, there should be:
  - A clear EU commitment to ensuring the two remain linked
  - A fallback plan in the event of a hiatus in the international process which would be very damaging for momentum in private markets

- Make it clear that emissions trading is an integral part of the policy framework for responding to climate change

- Communicate clearly the post-2012 ambition even if the policy mechanisms remain unclear. This should include the establishment of clear national and international greenhouse gas emission targets for 2020 since a 10–15-year time horizon is the key in terms of company decision-making about investment

- Establish the credibility of emissions trading and other measures directed at reducing greenhouse gas emissions through considering competitiveness issues explicitly as a key part of the design and implementation of these policy measures. This may be through maximising participation in international initiatives and linking national and regional emissions trading schemes to increase their coverage

- Build the credibility of programmes at the domestic/regional level by demonstrating consistency with, and commitment to, broader global-level actions that will be necessary to meet global atmospheric concentration targets

- Signal their willingness — across the political spectrum and across different government departments — to provide public money to support action on climate change. Without that explicit support, companies will not take government commitments seriously

Policy-makers also need to recognise that uncertainty in climate change policy can alter the investment case for power technology if companies are fully exposed to the price of carbon. In order to incentivise investment in low-carbon technologies, carbon prices may need to be substantially higher than expected under a normal DCF analysis in order to compensate for policy risk. In addition, electricity prices may need to be higher than expected to incentivise investment in traditional (gas and coal) power generation.

A specific issue is that the EU ETS on its own (even if there is a high degree of confidence that it will remain as a key element of the policy framework for responding to climate change) is unlikely to stimulate major investments in lower-CO_2-emitting forms of power generation unless prices move up significantly from the level of around €20/teCO_2 prevailing in the markets in early 2008. This will require:

● Stringent emissions targets to be set for the EU ETS sectors

● A commitment to achieving these internally (i.e. restricting the level of external emissions credits from outside Europe)

● Sufficient political will to deal with the knock-on effects of higher carbon prices in terms of increased electricity prices and the consequences for competitiveness, fuel poverty and other policy objectives

In the absence of such supports to the carbon price, policy-makers may need to consider other policy instruments to sit alongside emissions trading. These may include regulations relating to maximum emissions from different types of generating plant or subsidies (e.g. for carbon capture and storage and renewables).[5] As noted previously, governments will have to manage the interactions that will occur as a result of such a multiple policy approach, including the effects on the price of electricity.

In terms of the design of the EU ETS, extending the allocation period under the EU ETS (e.g. changing from a five-year to a ten-year trading/allocation period) could significantly reduce the investment thresholds in the early years of the allocation period. However, any periodic allocation could encourage undesirable cyclical investment patterns. A possible solution might be to explore setting allocations on a rolling ten-year-ahead basis.

Company actions

Responding to climate change is not simply a matter of 'government dictating and companies acting', but requires companies to play their role in supporting effective and efficient policy in this area. As discussed above, decisions relating to climate change

---

5 For example, in its response to the UK Climate Change Programme EDF stated: '[W]e consider that the present ETS market is too fragile and fragmented to provide long term price signals or to sustain the necessary long-term investment. We think that the EU ETS will function best as a clearing market, enabling operators to balance their portfolios over a 3–5 year time horizon, and allowing the "invisible hand" of the market to seek out low cost opportunities that would not be seen by policy-makers . . . Accordingly, we are driven to the conclusion that while the EU ETS is an important component in a GHG [greenhouse gas] reduction strategy, at its present stage of development it cannot by itself drive the long term investments needed to make deep cuts in carbon. We believe that, for now, the EU ETS should stand as one of a package of measures to reduce greenhouse gas emissions' (EDF 2005).

mitigation are widely distributed from global negotiations, through national and regional policy-making and then to the company level, where ultimately the emission reductions will occur. This process is deeply interconnected, with decision-makers at each level acting according to their expectations of the likely actions and responses of decision-makers at the other levels. In this situation, signalling and early actions can have important consequences.

There are practical actions that can be taken by power companies that allow them to respond to future directions in policy – in particular to policy measures directed at reducing greenhouse gas emissions. Perhaps most importantly, companies need to consider explicitly current and future climate change and energy policy in their investment decisions, and to consider how they can effectively manage the risk of higher future CO_2 prices. This may involve investing in lower-greenhouse-gas-emitting technologies (e.g. preferring renewables over coal or gas) or identifying opportunities to hedge against higher-than-expected prices.

More generally, companies have an important role to play in the public policy process. This is not just a matter of contributing to discussions about the details of specific policy instruments but, perhaps more importantly, through communicating their support for a clear, long-term climate change policy framework – including emissions reduction targets – as an essential part of ensuring the delivery of climate policy goals in an economically efficient manner. It is important that companies:

- Indicate clearly that 'economically efficient' does not simply mean 'no cost'

- Recognise that effective action on climate change may have some negative impacts on individual companies – at least over the short term[6]

- Are prepared to accept these impacts as a necessary part of ensuring an effective policy response to climate change[7]

Unless these messages are an integral part of the messages from companies to government, the depth of corporate commitment to action on climate change will be perceived as superficial (or subject to renegotiation as soon as companies' commercial interests are challenged). Furthermore, these messages are essential to provide governments with the confidence to establish the long-term policy frameworks necessary to respond effectively to the threat presented by climate change.

6 In this context, it is relevant to note that the IEA (2006) has argued that countries could implement a range of measures directed at reducing energy demand growth and greenhouse gas emissions and increasing energy security, where the benefits of using and producing energy more efficiently significantly outweigh the costs incurred.

7 It is interesting to note that a number of the major UK electricity utilities have started to make these arguments to government with, for example, Centrica arguing that that the government should set 'bold' targets for cutting greenhouse gas emissions from 2008 onwards, and RWE emphasising that companies need greater regulatory certainty and transparency regarding EU ETS (see, for example, Bream and Harvey 2006).

References

Blyth, W., and M. Yang (2007) *Climate Policy Uncertainty and Investment Risk* (Paris: IEA).

Bream, R., and F. Harvey (2006) 'Call for More Certainty on Future Energy Policy', *Financial Times*, 13 April 2006: 3.

Council of the European Union (2007) *Brussels European Council 8/9 March 2007. Presidency Conclusions* (Brussels: Council of the European Union).

DTI (UK Department of Trade and Industry) (2007) *Energy White Paper* (London: The Stationery Office).

EC (European Commission) (2008) *20 20 by 2020: Europe's Climate Change Opportunity* (COM[2008]30; Brussels: European Commission).

EDF (2005) 'EDF Energy Position on UK Climate Change Programme' (letter to Lisa Stratford, 2 March 2005; London: EDF).

IEA (International Energy Agency) (2003a) *World Energy Investment Outlook 2003* (Paris: IEA).

— (2003b) *Power Generation Investment in Electricity Markets* (Paris: IEA).

— (2006) *World Energy Outlook 2006* (Paris: IEA).

— (2007) *Tackling Investment Challenges in Power Generation* (Paris: IEA).

IPA Energy Consulting (2005) *Implications of the EU Emission Trading Scheme for the UK Power Generation Sector. Report to the Department of Trade and Industry* (Edinburgh: IPA Energy Consulting).

Kernan, P., A. Zsiga and T. Hseih (2005) 'Targets and Trading: The EU's Response to Climate Change', in *Climate Change Credit Survey. A Study of Emissions Trading, Nuclear Power and Renewable Energy* (London: Standard & Poor's): 3-5.

Lekander, P., and V. Gilles (2006) *ETS Update: EU Phase II Guidance – Another 20% to go on the Power Price?* (London: UBS Investment Research).

Sullivan, R., and W. Blyth (2006) *Climate Change Policy Uncertainty and the Electricity Industry: Implications and Unintended Consequences* (Briefing Paper EEDP BP 06/02; London: Chatham House).

Wirtz, P., and M. Koebernick (2006) *RWE & E.ON* (Düsseldorf: WestLB Equity Research).

4
The influence of climate change regulation on corporate responses: the case of emissions trading*

Ans Kolk and Jonatan Pinkse
University of Amsterdam Business School, The Netherlands

Although the Kyoto Protocol envisaged the implementation of a global emissions trading regime, such a regime has not yet come about. However, a variety of emissions trading initiatives have emerged. The European Union Emission Trading Scheme (EU ETS) is the only mandatory scheme, albeit for a limited number of companies and sectors. However, a number of voluntary schemes have emerged — in particular in the USA and Australia — and there is a growing likelihood that other mandatory schemes will be introduced. As a result, multinational corporations (MNCs) now face a wide variety of emissions trading schemes that differ in scope and in maturity of implementation and stringency. This creates divergent levels of institutional constraints across locations and a high degree of uncertainty. This chapter explores how MNCs have reacted to this diversity of (emergent) emissions trading schemes and policies, and the implications of these responses for public policy on climate change.

This chapter is based on an analysis of the *Financial Times* Global 500 (FT500) companies to the fourth questionnaire sent out by the Carbon Disclosure Project (CDP) in 2006.[1] Companies responded to this questionnaire in the months shortly before June 2006, i.e. approximately halfway through the first trading period of the EU ETS. The find-

* Support from the Netherlands Organisation for Scientific Research (NWO) is gratefully acknowledged.
1 For more details of the methodology, see Pinkse and Kolk 2007. For a description of the Carbon Disclosure Project, see www.cdproject.net.

ings from this period thus provide a valuable insight into MNC attitudes to climate change regulation in a state of flux as the data was gathered at a point when:

● The institutional constraints were different in different markets (discussed further below)

● The European Commission (EC) was canvassing options for Phase 2 (2008–2012) of the EU ETS and initiating a discussion on the shape of the post-2012 policy framework

The chapter is divided into three main parts. First we provide a broad overview of the main global policy developments in relation to climate change. We then consider the manner in which MNCs have responded to different emissions trading schemes focusing in particular on:

● MNC responses to the EU ETS

● MNC responses to voluntary trading schemes

● The responses of MNCs exposed to different trading schemes in different countries where they operate

Finally, we consider the broader implications of our research for the design and implementation of public policy on climate change.

Emissions trading as a policy response

Background

Emissions trading is the climate change policy option that has received most attention – specifically so-called 'cap and trade' systems. The Kyoto Protocol first established emissions trading for the purpose of climate change mitigation. Under the Protocol, participating countries are allowed to exchange part of their obligations with another party (Grubb *et al.* 1999). This intergovernmental emissions trading regime, which enables *countries* to transfer greenhouse gas emissions, has stimulated interest in the creation of domestic systems to trade emissions at a *company* level; the logic is that emissions trading will enable companies to reduce their greenhouse gas emissions in the most economically efficient manner. In broad terms, emissions trading schemes envisage that companies need a permit to emit greenhouse gases, and that governments allocate or auction allowances that determine how much each company is allowed to emit ('the cap'). If individual countries launch similar 'national' emissions trading schemes, the two can be linked – at least in theory – and companies can engage in cross-border trade of emissions allowances (Blyth and Bosi 2004). However, this has not yet occurred in practice – with the obvious exception of the EU ETS, which is a supranational scheme.

Although the Kyoto Protocol established emissions trading between countries, it does not require participating countries to also implement a domestic emissions trading scheme applicable to companies (Blyth and Bosi 2004). Although each signatory country has to draw up a plan specifying how it intends to meet its Kyoto Protocol targets,

TABLE 4.1 **Overview of main emissions trading schemes/initiatives** (continued over)

Source: adapted and updated from Kolk and Hoffmann 2007: 412

Europe	
EU Emission Trading Scheme (EU ETS)	● Main instrument of the EU to reduce greenhouse gas (GHG) emissions ● Started trading in 2005, with two phases until 2012 ● EU plans beyond 2012 are to reduce GHG emissions by 20% by 2020 and 60–80% by 2050
UK Emissions Trading Scheme (UK ETS)	● First domestic economy-wide trading scheme ● Based on voluntary participation of companies with absolute reduction commitments ● Launched in 2002
North America	
Chicago Climate Exchange (CCX)	● Voluntary but legally binding commitment of member organisations to meet greenhouse gas emissions reduction targets of 6% by 2010 compared to average 1998–2001 emissions ● Started trading in 2003
California Global Warming Solutions Act	● Mandates state-wide greenhouse gas emissions cap for 2020 based on 1990 emissions ● Equals a 25% emissions reduction compared with business as usual
California Climate Exchange (CaCX)	● Launched by the Chicago Climate Exchange to develop trading instruments related to the California Global Warming Solutions Act
Regional Greenhouse Gas Initiative (RGGI)	● Cooperative effort by ten Northeast and Mid-Atlantic states in the USA to discuss the design of a regional cap-and-trade programme ● Initially planned to cover CO_2 emissions from power plants, but may be extended later
Western Regional Climate Action Initiative	● Initiative by seven Western states in the USA and two Canadian provinces to realise a regional, economy-wide reduction target of 15% below 2005 levels by 2020 using market-based systems such as a cap-and-trade programme ● Builds on two earlier initiatives: the West Coast Governors' Global Warming Initiative (2003) and the Southwest Climate Change Initiative (2006)
Mid Western Greenhouse Gas Reduction Accord	● Agreement, reached in November 2007, by five Midwestern states of the USA and one Canadian province to cap greenhouse gas emissions and set up an emissions trading scheme by 2010 ● Reduction targets and time-frames to be determined within a year

TABLE 4.1 (from previous page)

Asia–Pacific	
New South Wales Greenhouse Plan	• Australian state initiative to bring GHG emissions back to the level of 2000 by 2025
	• Envisages 60% reductions by 2050
	• Applies various initiatives in different sectors
Australia New Zealand	• In June 2007, Australia and New Zealand announced that they would join forces in the development of carbon-trading systems that would be compatible
	• Follows earlier announcement by Australia to move towards a domestic, nationwide emissions trading system, beginning no later than 2012
Australia Climate Exchange (ACX)	• Launched Australia's first emission trading platform in July 2007, providing a mechanism to trade emission allowances and the Australian Greenhouse Office (AGO) accredited Greenhouse Friendly™ approved abatement

governments are free to decide the exact actions they intend to take – where a domestic emissions trading scheme is just one of the policy options available.

As a consequence, governments across the globe have implemented a wide variety of policy instruments as part of their domestic climate policies, many of which are aimed directly at companies. Beyond voluntary approaches, emissions trading has emerged as a preferred policy option (see Table 4.1) – most notably its central role in the implementation of the EU's climate strategy. However, several non-European industrialised countries that have ratified the Kyoto Protocol, such as Japan and Canada, have not yet implemented trading schemes although companies from these countries can use the other Kyoto flexibility mechanisms – Clean Development Mechanism (CDM) and Joint Implementation (JI) – to offset their emissions via reduction projects in developing countries or economies in transition.

Emission Trading Scheme in Europe

The EU ETS started in January 2005 and is aimed primarily at energy-intensive activities as it covers only CO_2 emissions (Egenhofer 2007). The EU ETS targets industrial installations (EC 2003) including:

● Energy activities (combustion installations exceeding 20 MW, oil refineries, coke ovens)

● Production and processing of ferrous metals, mineral industry (installations for cement, glass and ceramic products)

● Pulp and paper production plants

The scheme thus primarily affects energy producers (including electric utilities), metals, cement, pulp and paper. But because large combustion installations are also covered, other industries with energy-intensive activities (e.g. car manufacture and food processing) also require permits and were allocated allowances. Interestingly, the aluminium and chemical industries are exempted from the EU ETS supposedly due to their strong lobbying activities (Butzengeiger *et al.* 2003; Markussen and Svendsen 2005). Although the EU ETS is meant to ensure harmonisation of emissions trading across the EU, the detailed plans for the allocation of allowances (National Allocation Plans) and the monitoring of participants' emissions have been left to the individual Member States in Phase 1 (2005–2007) and Phase 2 (2008–2012).[2]

Within Europe, the EU ETS was preceded by two schemes at the national level. Denmark launched a CO_2 emissions trading scheme for electricity producers in 1999 (suspended at the end of 2004), while the UK introduced a scheme in 2002 that covered more industries and greenhouse gases (Rosenzweig *et al.* 2002). In the UK, companies could participate in several ways but, for the majority, trading was linked directly to the Climate Change Levy – a tax on industrial and commercial energy consumption (Boemare and Quirion 2002). Remarkably, the Climate Change Levy did not cover electricity producers which, as a consequence, did not experience pressure to participate; they could only engage voluntarily in reduction projects (Roeser and Jackson 2002; Rosenzweig *et al.* 2002). Companies participating in the UK trading scheme were temporarily exempted (until the end of 2006) from joining the EU ETS.

Emissions trading in other countries

Other industrial countries that ratified the Kyoto Protocol have not yet created schemes equivalent to the EU ETS. In Japan, progress on climate policy and emissions trading has been slow due to a fundamental disagreement between the Ministry of Environment (MoE) and the Ministry of Economy, Trade and Industry (METI) (Schreurs 2002). In 2004, the MoE proposed a voluntary emissions trading scheme as well as a carbon tax, but Japanese companies strongly opposed such a voluntary scheme because they feared it might become obligatory, with METI underlining the harmful effect for the competitiveness of Japanese companies (Arita 2004; Watanabe 2005). There has, however, been some activity regarding emissions trading. Both MoE and METI conducted pilot projects to test the workings of an emissions trading scheme. The MoE followed this up with the launch of a voluntary emissions trading scheme in 2005; the scheme combines emissions trading with subsidies for emissions-reducing projects (Watanabe 2005). The only fundamental aspect of climate policy that the MoE and METI could agree on was the introduction of greenhouse gas emissions reporting standards in 2005, requiring large Japanese companies to publicly report their annual CO_2 emissions (Watanabe 2005).

In Canada, successive governments have worked on Kyoto implementation including a 2005 plan that proposed emissions trading (with emissions targets) that would be applied to the mining, manufacturing, oil and gas, and electricity utility sectors. However, no exact rules were developed and one commentator noted that the proposals were not intended as a real 'cap-and-trade system' but rather as a different means of granting subsidies (Harrison 2006). Under the current government, the 2005 plan seems to have been largely ignored and no progress has been made on implementation (Rabe 2007).

---

2 In Phase 3 of the EU ETS, the EU is expected to take responsibility for the allocation of allowances (as opposed to the present situation where Member States propose allocations and the EU then approves or rejects these proposals).

Interestingly, several Canadian provinces participate (as members or observers) in the regional, state-level initiatives that have been set up by US states (see Table 4.1 and below).

Despite the rejection of the Kyoto Protocol by the USA and, until recently, Australia, some trading schemes have emerged in both countries over the years – particularly at the state level, signalling a clear divergence in opinion between the various levels of government (Peterson and Rose 2006; Byrne et al. 2007; Griffiths et al. 2007). A considerable number of US states are preparing to participate in emissions trading, usually at a regional level, notably the Regional Greenhouse Gas Initiative (RGGI) of the north-eastern states, the Western Regional Climate Action Initiative and, most recently, the Mid Western Greenhouse Gas Reduction Accord.

A private trading scheme, the Chicago Climate Exchange (CCX), was created earlier (in 2003) with the involvement of international and local companies, governments and non-governmental organisations (NGOs). The CCX aims to demonstrate that climate change can be managed on a voluntary basis and that market mechanisms are viable. Participants commit to voluntary reduction targets, with trading of allowances and offsets as options. Geographically, the scheme is restricted to projects on the American continent. Although the CCX shows parallels to the EU ETS, participation is voluntary and it includes all six greenhouse gases covered by the Kyoto Protocol, not just CO_2 (Yang 2006).

In Australia, the New South Wales (NSW) Greenhouse Gas Abatement Scheme was launched early 2003 by the state government. It is mandatory for electricity generators, retailers and large market customers, and voluntary for other companies. And 2007 saw the launch of the Australia Climate Exchange, the first emissions trading platform in the country. Emissions trading has recently received more support at the federal level (Griffiths et al. 2007), with the former Prime Minister John Howard announcing his intention to establish a nationwide system by 2012, which would also link to a similar scheme in New Zealand (Minder 2007). The ratification of the Kyoto Protocol in December 2007 by the new Australian government is likely to lead to more activities in this area.

Concluding comments

While emissions trading overall has received considerable attention as a (preferred) policy option around the world, the levels of implementation and maturity vary considerably between countries. Emissions trading is still evolving and its final form – global, regional and/or national – is uncertain. This will remain so for a considerable period while negotiations on the Bali Roadmap (i.e. the post-2012 international climate change framework) are ongoing. This lack of policy certainty is a reality MNCs in particular need to deal with when responding to climate change.

How MNCs react to (emergent) emissions trading schemes

The overview presented above of the main emissions trading schemes and initiatives worldwide clearly shows that levels of constraints differ considerably. At one end, we see the 'hardest' constraint currently in existence: namely, the EU ETS (at least for those companies covered by the scheme). At the other end, we find many companies and sectors where the prospect of regulation is remote (because of the level of their greenhouse gas emissions and/or their countries of operation). In between, we see situations where there is a threat of action on greenhouse gas emissions (even though the specific actions and the timing of these actions are uncertain) and the emergence of voluntary trading schemes as part of the corporate and policy response.

In this section, we examine corporate responses in three different ways. First we examine how companies have responded when faced with a hard regulatory constraint (i.e. there is a clear limit on greenhouse gas emissions and there are sanctions for not complying with the requirements of the scheme). We then, by contrast, examine the views and responses of companies that do not have to meet this type of constraint. These include large emitters that are excluded by virtue of their geographic location or other reasons, as well as lower emitters that fall outside the scope of existing schemes but which may benefit from emissions trading. Finally, we examine the responses of companies who fall between these two situations, i.e. some of their facilities or operations already meet, or are likely to have to meet, hard constraints whereas others are unlikely to face any substantive obligations.

As noted above, our analysis is based on the data and responses provided by companies to the Carbon Disclosure Project in 2006. We recognise the limitations in relying solely on published/self-reported data (see Kolk and Pinkse 2007). However, we believe that analysing these data provides important insights into:

- Companies' perceptions of policy instruments such as emissions trading and, in particular, the lobbying positions they will adopt in public policy debates

- Whether and how corporate actions (e.g. in public policy debates) differ between countries.

Responses in the face of regulatory constraints

For an emissions trading scheme to be effective, it needs to create a high constraint (in line with the goal of reducing greenhouse gas emissions) and to provide for severe penalties in the event of non-compliance. Both of these characteristics are seen in the EU ETS — notwithstanding the problems of over-allocation of emissions in Phase 1 of the scheme. With some legally allowed exceptions, companies with eligible installations cannot circumvent the scheme.

In their CDP responses, many of the large energy consumers in sectors such as oil and gas, chemical, metals and pharmaceutical industries mentioned the importance of compliance with the first phase of the EU ETS. Enforcement was also taken seriously: a considerable number of companies mentioned that they expected to avoid paying non-compliance penalties. The views of the Swiss cement company Holcim are representative of the views of many of the large energy users:

> Our priorities for the EU-ETS for 2005–07 are compliance management — i.e. internal and external balancing of emissions and allowances — and learning to use the system as it is conceptually intended to be used. We do not engage in speculative trading.

This framing of trading as a compliance activity and not for speculation was shared by several other MNCs. For example, ExxonMobil did not consider 'trading emission allowances as a business' and Repsol remarked that its 'participation in the market is orientated to low cost compliance and not to speculation'. Many companies stressed the importance of minimising the cost of compliance with a number buying allowances when they feared a shortage at the end of the first trading period.

Although companies emphasised the importance of compliance, this did not necessarily mean that they had participated actively in the buying or selling of allowances. Some companies stated explicitly that they had a 'no-trading' strategy because they owned only a few installations. A number noted that, even if they had a surplus of allowances, they believed that the administration and verification costs of selling them were too high compared with the potential revenues. It is likely that this perception was compounded by the problems of over-allocation of permits in Phase 1 of the EU ETS (2005–2007), which saw the value of permits decline to virtually zero at around the time that companies would have been responding to the 2006 iteration of the Carbon Disclosure Project.

But, before this point, the price of emissions permits had risen as high as €30/tonne and many companies had to decide how they would engage with the EU ETS. In their responses to the Carbon Disclosure Project, only a small number of companies reported substantial trading, with most seeing trading simply as a means of 'get[ting] the allowances needed' (Volvo) and, although a number of companies reported having a surplus of allowances, only a few stated explicitly that they had sold excess allowances. Before selling their surplus, it seems that many MNCs first balanced their allowance accounts on a corporate level. In other words, the EU ETS enabled companies that operate in multiple countries to safeguard against the inter-institutional incompatibilities across the EU (for a further discussion, see Seo and Creed 2002).

These incompatibilities are partly due to the fact that emissions requirements (Kyoto targets) as well as national implementation vary within the EU. It is in the EU where companies reported some political activity; for example, a number of European companies tried to influence the design of National Allocation Plans and lobbied for optimal allocations (at least from their company's perspective) of allowances.

Although companies tend not to be very open about these kinds of activities (see Kolk and Pinkse 2007) and mostly reported indirect efforts to influence the EU ETS (e.g. through trade and industry associations), some examples can still be found. For example, the Italian oil company ENI asserted in its CDP response that it:

> has played a proactive role in the process for the definition of the Italian National Allocation Plan and it has supported rational allocation methodologies in line with the Kyoto targets.

In addition, those disadvantaged by the rules of the first phase expressed their concerns publicly, with large energy consumers in sectors such as chemicals, pharmaceuticals and metals in particular complaining openly about the fact that electricity companies had passed on the price of allowances to their customers.

Electricity producers indeed appear to have been most successful in influencing the design of the first phase of the EU ETS (Markussen and Svendsen 2005). They have gained from over-allocation of free allowances, not only because it has relieved the institutional constraint of the cap but also because it has led to windfall profits from passing on the opportunity costs to clients (Sijm *et al.* 2006). A recent estimate put net windfall profits for power generators in the EU in the first phase at €6–8 billion (Barber 2008).

In fact, these companies have seized the opportunity to change the institution of the EU ETS in a way that alleviates the pressure that they felt and improves the institution's efficiency — at least with regard to their own interests (Fligstein 1997). In their responses to the 2006 iteration of the Carbon Disclosure Project, electric utilities continued to show their 'entrepreneurial' approach to public policy (see Seo and Creed 2002), although their primary focus had shifted from the previous iteration in 2004 to what happens after 2012 when the first commitment period of the Kyoto Protocol expires rather than the period 2008–2012 (which would have been under discussion at the time that the CDP4 questionnaires were being completed). For example, E.ON argued for a continuation of the EU ETS in its current form to create more certainty for its long-term investments and, in a position supported by RWE, expressed its preference for a global framework to minimise the costs of reducing emissions.

Responses in the absence of (high) regulatory constraints

There are currently many situations in which companies face only limited (or no) constraints on their greenhouse gas emissions. The majority of companies in the EU, for example, are not affected by the EU ETS. This is either because their activities are not energy-intensive (i.e. increases in the cost of electricity are minimal in the context of the company's overall cost base) or they have no production sites in the EU.

This scenario applies even more to those companies with no operations at all within the EU. For these companies, the compliance requirements that apply to companies covered by the EU ETS are clearly not relevant. The manner in which these companies engage with public policy seems quite different to those covered by the EU ETS. For these companies, the drivers for their policy lobbying are the potential business opportunities that emissions trading may present or the desire to avoid costs or mandatory regulation in the future. Many of these companies are taking action to profit from such (emerging) schemes through voluntary participation in schemes, either for financial benefit or because participation in such programmes helps develop organisational capabilities or corporate image (Kolk and Pinkse 2008). In relation to the burnishing of corporate image, a number of these companies have highlighted their positive role in promoting emissions reductions or, in the case of the EU ETS, developing the scheme.

The fact that emissions trading has created an open market for emissions reductions distinguishes it from other forms of environmental regulation, opening up the possibility of involving parties not affected by the regulation itself. The financial sector is an important example here as it profits from the lack of knowledge by other companies of emissions trading, while it can use its experience from trading in other areas. Many banks — mostly European but some US banks as well — provide risk management and other services to facilitate trading by their clients or to trade allowances on their clients' behalf. In doing so, they help the further development of the EU ETS because, as argued by Fortis in its CDP response, trading services have:

> the effect of increasing liquidity by allowing many companies to trade small volumes while avoiding the administratively cumbersome of setting up of an in-house trading desk.

In its response, UK bank Barclays illustrates this by stressing the impact of its trading activities on the evolution of the EU ETS:

> Barclays was the first UK Bank to set-up a carbon-trading desk and we helped shape the development of the EU ETS market (for example in helping create standard contracts and in sharing our own trading experiences with new players).

The Slovakian subsidiary of Belgian bank, Dexia, goes one step further as it claims in its CDP response to be the only private actor administering a national allowance registry, thereby taking up a public role.

Banks play this role of financial intermediary not only for trading in the EU ETS, but also by creating and trading emissions credits from the Kyoto flexibility mechanisms (CDM and JI). By embarking upon − or otherwise supporting − particular projects which fit into their regular business activities and at the same time lead to the production of emissions credits, they are able to influence what constitute legitimate CDM/JI projects. The flexibility mechanisms are particularly attractive for MNCs because they enable them to exploit their cross-border activities. Activities with regard to the Kyoto mechanisms are not necessarily unrelated to compliance; several MNCs that currently face low constraints on their greenhouse gas emissions are building a portfolio of credits to help them meet their likely compliance requirements in future periods of the EU ETS.

In their Carbon Disclosure Project responses in 2006, companies also mentioned their role in helping design allocation plans for the second phase of the EU ETS as well as their role in influencing potential schemes in Canada, Japan and the USA. For example, carbon traders (including major banks involved in trading) have made recommendations for future phases within the EU ETS, including the introduction of auctioning, to prevent windfall profits and create a healthier, more liquid market.

Although several companies reported that they have contributed to the design of new trading schemes, most stated that they were waiting for more clarity about the exact rules for trading before taking concrete action with regard to potentially upcoming schemes. Companies' attitudes seemed to vary as emissions trading schemes evolved from initial conception through to voluntary initiatives and then to harder regulatory requirements; at which point their support in principle for emissions trading tended to be heavily qualified by concern about the financial implications for their business.

For example, the RGGI covering north-eastern states in the USA seemed to attract companies' concerns about the potential adverse effects of the regional scheme. In its CDP response, PSEG stated it was:

> very concerned about 'leakage'. Leakage refers to the market imbalance created by requiring generators within the RGGI area to internalise costs of emitting CO_2, whereas generators located outside of the region, but connected on the same electric grid, are not burdened with the same costs.

MNCs also reported that they engaged in alternative trading schemes to indirectly prepare for larger schemes expected to emerge in coming years. Participation in the CCX illustrates this. As it is purely private and voluntary, a positive selection effect appears to be at play: the CCX seems to attract those companies that can achieve their voluntary binding targets rather easily. Participants aim to influence and prepare for the development of a federal/public US emissions trading scheme. However, the political undertone of the CCX also appears to deter some companies. In its CDP response, electricity company FPL, for example, stated that the CCX is 'not yet representative of what a real regulatory driven greenhouse gas market program will be like', and Occidental Petroleum argued that schemes other than the EU ETS 'offer little business reason for most companies to participate'.

Involvement in potential future schemes in Japan, Korea, Australia and Canada was also mentioned by MNCs — albeit to a lesser extent than the USA. Looking at the corporate responses, companies appeared to be anticipating the introduction of emissions trading schemes in those countries that had ratified the Kyoto Protocol at the time, in particular in Japan and Canada. General Electric, for example, stated that it was:

> monitoring and in some cases participating in the process that other Annex B countries, such as Japan and Canada, are undertaking to ensure that they meet their Kyoto commitments.

Some European companies were following developments in the USA and elsewhere via their subsidiaries. An example is Suez North America (SENA), which mentioned in its CDP response that it was:

> actively tracking and participating in the development of US climate change legislation, such as the RGGI. While SENA supports linkage with international programs such as Kyoto, it appears that offset programs for the next few years will be limited to the US.

Responses when faced with different regulatory settings

The third situation that we examine is those companies that have to meet — now or in the future — hard constraints in certain countries but also have operations in countries or regions where emissions trading schemes have not been established. From responses to the Carbon Disclosure Project, it appears that the nature of corporate responses is related to the specific constraints that the company needs to meet. For example, MNCs in high-emission sectors face constraints in the EU at the moment, which are likely to become more stringent in the future, and can reasonably expect to be subject to regulation in other countries as well. This explains why, for example, US and Australian companies in sectors that fall under EU ETS have been relatively active in engaging in voluntary emissions trading schemes in their home countries. Interestingly, MNCs also get involved in policy discussions around emissions trading even if they are not very active in the EU and/or not covered by the EU ETS. For example, the Mexican cement company Cemex warned in its response of the consequences of 'leaking effects' (i.e. that energy-intensive industries could move their production facilities to countries that do not have an emissions target under the Kyoto Protocol). To prevent this from happening, Cemex called for a change in the EU ETS to become 'a more efficient emission trading scheme' and hopes 'that the current design will be improved in the near future'.

As noted by Kolk and Pinkse (2007), cross-border political engagement is not widely reported. The most explicit in this regard are US MNCs that have responded to (upcoming) European regulations. In addition, there are a few examples where MNCs from one EU Member State cooperate with, or lobby, a government in another country (Kolk and Pinkse 2007). In most cases, however, companies have tended to focus primarily on their home countries and to refrain from too much interference with host-country governments. This reflects the reality that MNCs' bargaining power is generally not as strong in a host country. It requires much flexibility and bargaining power to persuade host-country governments to take an MNC's interests into account (Baron 1995). In contrast, in its home country, a company typically has a much stronger foothold in the policy-making process – in part because the home country generally benefits more from the company's activities (Baron 1997).

With regard to public policy lobbying, the specific approach taken in the home country seems to reflect the national traditions (Kolk and Pinkse 2007). US MNCs state that they aim to influence opinions of experts whereas hardly any EU companies reported carrying out this type of activity. Moreover, US and Australian companies reported targeting a much wider range of actors. In contrast, the focus for EU companies was on providing policy input. Nearly all political activities for climate change on the part of Japanese companies are – in line with the country's corporatist structure – exclusively with the national government.

Finally, it should be noted that, precisely because most MNCs are geographically scattered organisations, they may well behave differently in different countries (Levy and Kolk 2002). This sometimes reflects divergent regulatory settings in home (headquarters) versus host (subsidiary) settings, while levels of decentralisation and subsidiary autonomy may also play a role. There are some notable examples from the 1990s (Kolk 2000) when General Motors (GM) opposed an international treaty via the Global Climate Coalition, while at the same time pleading for measures in its alliance with the World Resources Institute (WRI). The desire to stay involved in the US debate at the time and prevent unwanted measures conflicted with the more consensual approach advocated by GM's German subsidiary, Adam Opel GmbH.

Such internal organisational differences also played a role for Shell, where Shell Oil in the USA long refused – in spite of the insistence of the UK and Dutch parts of the company – to support climate measures and to leave the Global Climate Coalition. More recently, internal divergence has come to the fore in the case of BP where evidence from the BP Texas refinery explosion investigation revealed that a proactive role in climate change went hand in hand with explicit lobbying by the company against tougher emission controls in the USA (McNulty 2007).

These examples illustrate the difficulty for MNCs in developing a global climate change strategy given:

● The considerable variation in local institutional pressures in terms of the strength of the regulation

● The plausibility of the threat of regulation

● The company's ability to shape and influence policies in the range of contexts

But, with an increasing number of large MNCs advocating a global approach to climate change (within individual sectors and/or via a long-term overarching global climate change regime), companies may increasingly adopt a more unified, less diversified strategy in this respect.

Discussion and conclusions

The material presented in this chapter provides some important insights into corporate responses to emissions trading and how these responses have influenced emissions trading schemes.

First, corporate responses (to the Carbon Disclosure Project) provide important insights into why the EU ETS has, as yet, failed to deliver on the (theoretical) predictions that it would contribute to significant reductions in emissions in an economically optimal manner. There are two dimensions to consider. First, the design of the EU ETS is suboptimal. This is seen most starkly in the over-allocation of emissions permits in Phase 1 but also in the very limited use of auctioning and the generous permit allocations for Phase 2. The primary reason is the strong company influence on the set-up and rules of the scheme, which is compounded by the fact that the emissions allocation process is carried out by the Member States (which opens another arena where lobbying can take place). The second dimension is that the participants in the scheme seem to perceive it as a compliance programme rather than a trading scheme. That is, companies do not seem to be acting in a manner that maximises their economic welfare, but instead seem willing to accept suboptimal outcomes (perhaps, in part, because the transaction costs associated with realising these benefits are greater than the benefits that accrue). In this context, it is interesting that many companies stated explicitly in their responses to the CDP that they did not have a clear-cut emissions trading strategy, with a number noting that they had no intention of developing one either.

The second important insight is that company responses are influenced by the business context within which they operate. This is seen, as discussed above, in the differences in responses between companies that are and are not exposed to the EU ETS. Perhaps more interesting is the case of those companies that face different regulatory constraints in different countries of operation. For these companies, their response is influenced as much by the specific constraints that they face as by the overall corporate position on climate change. A corollary to this point is that it cannot be assumed that the responses or actions of subsidiaries can simply be inferred from corporate policies — rather that these are mediated through the specific operating conditions faced by the company/subsidiary in question.

We expect both of these insights (i.e. the suboptimal nature of emissions trading schemes and the variations in corporate responses) to remain valid for some time to come. Notwithstanding the broad direction of public policy on climate change, the reality is that companies will continue to face a variety of regulatory constraints with varying degrees of robustness. However, we can reasonably expect a number of changes in corporate approaches (and, indeed, a preliminary review of the results of the most recent iteration of CDP suggests that we are starting to see these changes).

First, as emissions trading schemes are proposed and/or implemented in an increasing number of countries, we can expect to see the level of participation in voluntary schemes to increase as companies seek to develop their organisational capabilities and to improve their corporate image.

Second, we expect that companies will continue to act to protect their commercial interests and, in many cases, this will take the form of strong opposition to emissions trading. We have seen this recently in advance of the European Commission's announcement in January 2008 of the concrete steps it is proposing to achieve its targets of a 20%

reduction of greenhouse gases by 2020 (compared to 1990 levels). The period prior to the announcement saw a flurry of lobbying activities by governments and companies to lower the level of ambition, with warnings that the proposals would lead to plant closures, relocation of industries to outside the EU and other damage to the economy.

Finally, and acknowledging the previous point, we expect that companies will progressively express more support for global emissions trading. As national or regional schemes become 'inevitable', companies will seek the most efficient scheme that allows emissions reduction requirements to be achieved at least cost. In principle and acknowledging that transaction costs may undermine some of the efficiency arguments, this will probably see business views converging on a global emissions trading system with the maximum possible number of participants – thereby maximising the opportunities for cost-effective emissions reductions and offsetting.

References

Arita, E. (2004) 'Proposed emissions trading, carbon tax set to be hard sell', *Japan Times*, 7 August 2004.

Barber, T. (2008) 'EU urged to use taxes in climate change', *Financial Times*, 21 January 2008.

Baron, D. (1995) 'Integrated Strategy: Market and Nonmarket Components', *California Management Review* 37.2: 47-65.

—— (1997) 'Integrated Strategy, Trade Policy, and Global Competition', *California Management Review* 39.2: 145-69.

Blyth, W., and M. Bosi (2004) *Linking Non-EU Domestic Emissions Trading Schemes with the EU Emissions Trading Scheme* (Paris: OECD/IEA).

Boemare, C., and P. Quirion (2002) 'Implementing Greenhouse Gas Trading in Europe: Lessons from Economic Literature and International Experiences', *Ecological Economics* 43: 213-30.

Butzengeiger, S., A. Michaelowa and S. Bode (2003) 'Europe: A Pioneer in Greenhouse Gas Emissions Trading. History of Rule Development and Major Design Elements', *Intereconomics* 38.4: 219-28.

Byrne, J., K. Hughes, W. Rickerson and L. Kurdgelashvili (2007) 'American Policy Conflict in the Greenhouse: Divergent Trends in Federal, Regional, State and Local Green Energy and Climate Change Policy', *Energy Policy* 35: 4,555-73.

EC (European Commission) (2003) 'Directive 2003/87/EC of the European Parliament and of the Council of 13 October Establishing a Scheme for Greenhouse Gas Emission Allowance Trading within the Community and Amending Council Directive 96/61/EC', *Official Journal of the European Union* L 275: 32-46.

Egenhofer, C. (2007) 'The Making of the EU Emissions Trading Scheme: Status, Prospects and Implications for Business', *European Management Journal* 25.6: 453-63.

Fligstein, N. (1997) 'Fields, Power and Social Skill: A Critical Analysis of the New Institutionalism', paper presented at the *Conference on Power and Organization of the German Sociological Association*, Hamburg, Germany, 1997.

Griffiths, A., N. Haigh and J. Rassias (2007) 'A Framework for Understanding Institutional Governance Systems and Climate Change: The Case of Australia', *European Management Journal* 25.6: 415-27.

Grubb, M., C. Vrolijk and D. Brack (1999) *The Kyoto Protocol: A Guide and Assessment* (London: Royal Institute of International Affairs/Earthscan).

Harrison, K. (2006) 'The Path not Taken: Climate Change Policy in Canada and the United States', paper presented at the *Conference on Global Commons and National Interests: Domestic Climate Policies in an International Context*, Vancouver, 2006 (retrieved on 19 October 2007 from www.politics.ubc.ca).

Kolk, A. (2000) *Economics of Environmental Management* (Harlow, UK: Financial Times Prentice Hall).

—— and V. Hoffmann (2007) 'Business, Climate Change and Emissions Trading: Taking Stock and Looking Ahead', *European Management Journal* 25.6: 411-14.

—— and J. Pinkse (2007) 'Multinationals' Political Activities on Climate Change', *Business and Society* 46.2: 201-28.

—— and J. Pinkse (2008) 'A Perspective on Multinational Enterprises and Climate Change. Learning from an "Inconvenient Truth"?', *Journal of International Business Studies* (forthcoming).

Levy, D., and A. Kolk (2002) 'Strategic Responses to Global Climate Change: Conflicting Pressures on Multinationals in the Oil Industry', *Business and Politics* 4.3: 275-300.

Markussen, P., and G. Svendsen (2005) 'Industry Lobbying and the Political Economy of GHG Trade in the European Union', *Energy Policy* 33: 245-55.

McNulty, S. (2007) 'BP documents reveal extent of Texas PR effort', *Financial Times*, 7 March 2007.

Minder, R. (2007) 'Howard promises emissions trading scheme', *Financial Times*, 4 June 2007.

Peterson, T., and A. Rose (2006) 'Reducing Conflicts between Climate Policy and Energy Policy in the US: The Important Role of the States', *Energy Policy* 34.5: 619-31.

Pinkse, J., and A. Kolk (2007) 'Multinational Corporations and Emissions Trading: Strategic Responses to New Institutional Constraints', *European Management Journal* 25.6: 441-52.

Rabe, B. (2007) 'Beyond Kyoto: Climate Change Policy in Multilevel Governance Systems', *Governance: An International Journal of Policy, Administration and Institutions* 20.3: 423-44.

Roeser, F., and T. Jackson (2002) 'Early Experiences with Emissions Trading in the UK', *Greener Management International* 39: 43-54.

Rosenzweig, R., M. Varilek and J. Janssen (2002) *The Emerging International Greenhouse Gas Market* (Arlington, VA: Pew Center on Global Climate Change).

Schreurs, M. (2002) *Environmental Politics in Japan, Germany and the United States* (Cambridge, UK: Cambridge University Press).

Seo, M., and W. Creed (2002) 'Institutional Contradictions, Praxis, and Institutional Change: A Dialectical Perspective', *Academy of Management Review* 27.2: 222-47.

Sijm, J., K. Neuhoff and Y. Chen (2006) 'CO_2 Cost Pass-through and Windfall Profits in the Power Sector', *Climate Policy* 6.1: 49-72.

Watanabe, R. (2005) *Current Japanese Climate Policy from the Perspective of Using the Kyoto Mechanisms Option Survey for Japan to Acquire Credits from Abroad* (Hayama: Institute for Global Environmental Strategies).

Yang, T. (2006) 'The Problem of Maintaining Emissions "Caps" in Carbon Trading Programs without Federal Government Involvement: A Brief Examination of the Chicago Climate Exchange and the Northeast Regional Greenhouse Gas Initiative', *Fordham Environmental Law Journal* 18 (Fall 2006): 1-17.

5
CDM and its development impact: the role and behaviour of the corporate sector in CDM projects in Indonesia

Takaaki Miyaguchi and Rajib Shaw*
Kyoto University, Japan

The Kyoto Protocol was adopted in December 1997 at the third session of the Conference of the Parties (COP3) to the United Nations Framework Convention on Climate Change (UNFCCC). The Protocol introduces three kinds of 'market mechanisms' that can support government efforts to reduce greenhouse gas emissions:

● Joint Implementation (JI)

● Clean Development Mechanism (CDM)

● Emissions trading

Through CDM, Annex I Parties to the Kyoto Protocol can reduce greenhouse gas (GHG) emissions by assisting non-Annex I Parties (i.e. developing countries) to implement project activities that reduce GHG emissions in their countries. Annex I Parties, in return, can use these emissions reductions — Certified Emissions Reductions (CERs) — in meeting their emissions reduction targets under the Kyoto Protocol. By participating in the CDM scheme, the benefits for non-Annex I Parties are expected to include technology transfer, capacity-building and investment that will enable them to reduce their green-

* The views and opinions expressed in this chapter are those of the authors and are not necessarily the views of Kyoto University.

house gas emissions beyond what would have been achieved in a 'business as usual' scenario.

Using the evidence from Indonesia's participation in the CDM,[1] this chapter reviews:

● The role of the private sector in the CDM process and the manner in which this has influenced the type of CDM projects that have been developed

● The benefits that have accrued to Indonesia from these projects

● The wider implications for sustainable development in Indonesia

CDM in Indonesia

Indonesia is one of the world's most biodiverse countries. It comprises more than 17,000 islands, has the third largest forest coverage in the world and has a population of 242 million people in a total area of 191 million hectares. These characteristics make the country vulnerable to emerging climate change impacts.

Since the Asian economic crisis in 1997–1998, Indonesia has experienced steady economic growth, with a growth of 5–6% in gross domestic product (GDP) each year since 2004, and an associated significant increase in the volume of industrial (and polluting) activities. Despite strong lobbying by coal and oil industry groups in Indonesia (as in many other countries) against climate change-related policy (Michaelowa 2005), the Government of Indonesia ratified the UNFCCC in 1994 and the Kyoto Protocol in 2004. Indonesia's focus on climate change was raised significantly as a consequence of its hosting of the 13th Conference of the Parties (COP13) in Bali in December 2007.

To address climate change issues properly, the Ministry of Environment established the National Commission on Climate Change in 1992. Following Indonesia's ratification of the UNFCCC and the Kyoto Protocol, the government established the National Commission on CDM (KN-MPB) as the Designated National Authority (DNA) of Indonesia in July 2005. Its role is to approve proposed CDM projects from Indonesia that meet the sustainable development criteria set by the government. The DNA also monitors the progress of each CDM project, including registration, implementation and transaction processes.

Status of CDM projects in Indonesia

According to a study on CDM commissioned by the Indonesia's Ministry of Environment, Indonesia has a potential 2% share of the global CDM market with a total volume of 125–300 million tonnes of CO_2-equivalent in the land-use, land-use change and

---

1 This chapter is based on research being conducted by the authors for the Graduate School of Global Environmental Studies of the Kyoto University, Japan. The data, unless indicated otherwise, are from in-depth interviews, company surveys and Indonesian government records.

forestry (LULUCF) sector (Ministry of Environment 2001). But as of November 2007, only nine of the CDM projects registered by the CDM Executive Board (CDM-EB) were from Indonesia. In contrast, China had registered 125 projects and India 283 projects.[2] In total, Indonesia accounts for just 2% of the total number of CDM projects in Asian countries, a proportion that is expected to remain relatively unchanged through to 2012.[3]

Even though Indonesia's CDM development looks minuscule compared with China and India, it has seen significant progress in 2007; in June 2007, the number of projects either under development or approved was only 27 but this number had nearly doubled to 52 by November 2007.[4]

Biomass energy and landfill gas related projects are the predominant sectors, followed by renewable energy and energy efficiency projects (see Fig. 5.1). But, despite the clear potential in Indonesia, renewable energies such as hydropower, geothermal and solar have yet to be fully exploited.

For biomass energy projects, the utilisation of palm oil solid waste is the most common approach, accounting for 64% of these projects, with forestry projects accounting

FIGURE 5.1 Types of CDM projects in Indonesia as of November 2007

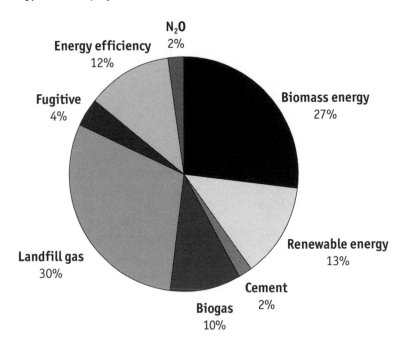

Data from UNEP Risoe Centre website, www.uneprisoe.org (accessed November 2007).

2 Data from UNFCCC website, www.unfccc.int (accessed November 2007).
3 See UNEP Risoe Centre website, www.uneprisoe.org (accessed November 2007).
4 Data from the National Commission on CDM (Ministry of Environment, Indonesia) website, dna-cdm.menlh.go.id (accessed December 2007).

FIGURE 5.2 Investment costs and expected CER returns on investment for selected CDM projects

Source: Ellis and Kamel 2007

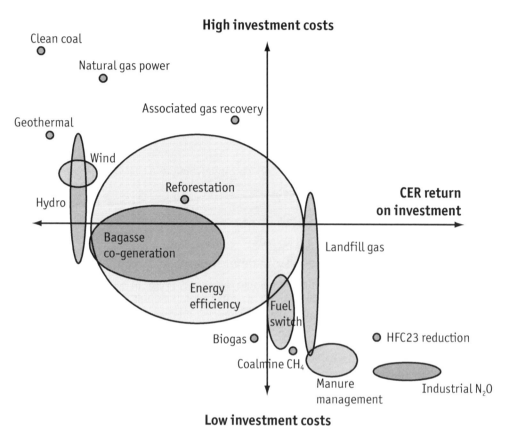

for 27% and biodiesel 9%. Indonesia is currently the world's second biggest palm oil exporter and is expected to pass Malaysia to become the largest producer in 2008, with production expected to be some 18 million tonnes of palm oil (Bhardwaj 2007).

Looking at the question of return on investment, Indonesia's CDM focus on biomass energy, landfill gas, energy efficiency and biogas is logical given the relatively low investment costs of these types of project (see Fig. 5.2). Given the higher capital costs and relatively lower return on investment on renewable energy projects, it is unsurprising there are only a couple (albeit of significant scale) of renewable energy projects in Indonesia. These projects have all been developed by partnerships between foreign carbon financing companies and the local branches of multinational corporations.

In almost all cases, the developers of Indonesia's CDM projects have established emissions reduction purchase agreements (ERPAs) with foreign credit buyers in advance of establishing the project. The majority of Indonesia's CDM projects are developed 'multilaterally', with initial capital and investment from abroad. Of the 52 CDM projects in Indonesia, the dominant credit buyer countries are from Japan (28% of projects), the UK (23%), Canada (12%) and the Netherlands 12%).[5]

With regard to project development modality, the direct involvement of the private sector (as opposed to indirectly through bilateral/multilateral carbon funds) is the dominant approach. The expected volume of CERs from those projects directly developed by the private sector companies comprises 74% of Indonesia's total expected volume of CERs.

Drivers for action

There are two major drivers for private sector involvement in CDM projects:

- Government regulations and incentives
- Motivations within the private sector

Regulations and incentives from government

In Indonesia, laws and regulations related to the energy sector in general and electricity generation and energy conservation in particular all impact on CDM projects.

For the energy sector, the national energy policy issued by the Ministry of Energy and Mineral Resources (MEMR) is the key driver for action. In 2002, the MEMR issued its *National Energy Policy 2003–2020* which, by establishing an open energy market, encourages energy ventures and industry to participate in the energy sector.[6] The policy sets a national target of 90% electrification by 2020, with a minimum of 5% of national energy to be generated from renewable energy sources.

In addition, MEMR has released the *Policy for Renewable Energy Development Energy Conservation*, whose objectives include ensuring the security of energy supply and increasing the role of renewable energy.[7] For electricity generation, the state electricity company, PLN, has historically had sole responsibility for electricity generation. PLN's monopoly position has been changed by recent legislation which allows the private sector and cooperatives to participate in electricity generation; PLN is now required to purchase the electricity generated by such parties through bidding or direct appointment.

For energy conservation, the legal framework was set by the presidential decree in 1991. The decree stipulates that energy conservation shall be implemented by all energy users in related sectors, industrial, commercial and residential areas (IGES 2006).[8]

The Indonesian government has a number of incentives for foreign direct investment (FDI), although relatively few of these relate specifically to the energy sector. The most significant of these incentives relate to:

---

5 Calculated by the authors based on data from UNEP Risoe Centre website, www.uneprisoe.org (accessed November 2007).
6 See further the MEMR website, www.djlpe.esdm.go.id (accessed November 2007).
7 See www.esdm.go.id.
8 See also the MEMR website, www.djlpe.esdm.go.id (accessed November 2007).

● Geothermal power plants – through the exemption of duty on imports

● Oil and gas facilities – through the provision of incentives for the development of marginal oil fields

● Small-scale renewable energy projects where PLN is obliged to purchase power generated by such means

● Energy efficiency – through subsidies and soft loans (see Table 5.1)

TABLE 5.1 **Incentives for FDI investment in the energy sector in Indonesia**

Source: IGES 2006

Project type	Incentive(s) available	Remarks
Geothermal	Exempted from import duty for operational goods	–
Large-scale power generation		–
Mini-hydro		MEMR determines annual capacity allocation for region/grid
Wind	Exempted from import duty for capital goods	
Co-generation		
Solar–photovoltaic and micro-hydro for rural electrification		Local government owns assets

Despite the various incentives on offer and the government's efforts to open up the energy market to the private sector, it is not clear that these are providing sufficient inducements for the private sector to be involved in CDM implementation. There appear to be three reasons for this.

First, the Indonesian government has not (at the time of writing, November 2007) implemented regulations or incentives specifically relating to CDM; the regulations and incentives mentioned above do not relate specifically to CDM, nor do they aim to promote or encourage CDM development by the private sector. Second, there is still a significant lack of transparency and accountability in many of the state-owned enterprises that continue to monopolise the energy industry in Indonesia. This makes it difficult for the private sector to compete on an equal footing with these enterprises. Third, there is an overall lack of awareness regarding CDM in Indonesia, even in industry. Even though energy efficiency and conservation projects can qualify as CDM projects, the general lack of general awareness of CDM and energy-related issues has meant that relatively few companies have explored the potential to obtain CERs for these projects.

Corporate drivers

In broad terms, there are three main drivers (motives) for companies to engage in CDM development in Indonesia. These are:

- Corporate social responsibility (CSR) and/or public relations (PR)/advertising[9]
- Meeting international obligations
- Profit

Overall, 74% (37 projects) of the CDM projects in Indonesia can described as purely motivated by the desire to make profit, 18% by CSR/PR considerations and 8% by international emissions reduction obligations.

Given the ever-increasing international attention and public scrutiny of company social and environmental performance, the private sector (particularly large companies) is under tremendous pressure to practise environmentally responsible behaviour on the ground. Many of the companies that have developed CDM projects in Indonesia have implemented a variety of CSR initiatives and, indeed, CDM projects sometimes form part of their overall CSR strategies. From the interviews conducted for this research, the companies that have participated in CDM projects for CSR or PR motives cite three main reasons, namely:

- To communicate their 'environmental conscience' as part of their investor and customer relations strategy

- To meet voluntary emissions targets that are being set by their parent companies

- To be ready to meet possible future obligations to reduce their greenhouse gas emissions

The second motive relates to the emissions reduction obligations of developed countries. One of the principles underpinning the inclusion of CDM in the Kyoto Protocol is that CDM provides a cost-effective way for developed countries to meet their emissions reduction targets. This is the main motivation underpinning all the CDM projects that are managed through bilateral (e.g. Japan Carbon Fund) and multilateral (e.g. World Bank's Prototype Carbon Fund) carbon funds, as well as those projects that involve the 'transfer' of carbon credits from the local branch to the international headquarters of multinational corporations.

The third motive relates to the profit-making potential of CDM projects. Country and CDM development risks mean that the majority of CDM projects in Indonesia are developed by carbon finance companies (e.g. EcoSecurities) acting as agents or intermediaries between the host country (Indonesia) and the credit buyer (i.e. Annex I governments). The characteristic of these carbon financing companies is that they often offer financial support for the initial investment in a CDM project in return for lower carbon credit prices than those sourced or traded in the international carbon market.

---

9 Since it is difficult to distinguish reputation-related drivers from CSR drivers, our research treated CSR and PR-/advertising-related activities as essentially the same for the purposes of assessing corporate drivers.

Trends in the type of project developer

Of the 52 CDM projects either ongoing or under development in Indonesia (as of November 2007), 18% are being developed through either bilateral or multilateral carbon funds and 61% are being developed by carbon financing companies that are primarily interested in the profit-making opportunities presented by the CDM scheme.

There are four projects being developed by non-carbon financing companies. Of these, two are being implemented by multinational corporations headquartered in Europe, with the CDM project being developed through a subsidiary firm in Indonesia and the carbon credits produced being transferred to or purchased by the corporate headquarters in Europe to enable the company to meet its obligations under the EU Emission Trading Scheme. The other two projects are being developed by Japanese companies. Given that Japanese companies do not presently have to meet emissions reduction targets from the Japanese government, these projects appear to be motivated primarily by PR/advertising considerations and by the desire to take precautionary action on climate change.

Seven projects (13% of the total) are being developed solely by domestic companies and financial institutions in Indonesia. Five of these projects are from palm oil companies (whose primary motive relates to the potential to make a profit). None of these palm oil companies actively communicates its CDM activities (e.g. through their websites) or implements comprehensive community development plans (beyond some local philanthropic or charitable activities). In contrast, the other two projects are CSR-driven: one is managed by a social entrepreneur whose motivation is the realisation of community development benefits through the CDM scheme, and the other is initiated and managed by the local government body in Bali responsible for regional waste management.

Outcomes

Indonesia overall

For the purposes of this chapter, 'outcome' is defined simply as the (projected) amount of emissions reductions from CDM projects. Compared with its potential, Indonesia has yet to realise substantial outcomes from CDM projects. Based on data from the World Resources Institute (WRI),[10] Indonesia underperforms significantly compared with other countries in South-East Asia (see Figure 5.3). This relative underperformance applies whether or not emissions from land-use, land-use change and forestry (LULUCF)[11] are included or excluded from the calculations (as can be seen by comparing Figs. 5.3 and 5.4).

---

10 Specifically, the World Resources Institute Climate Analysis Indicators Tool (CAIT), available at cait.wri.org (accessed November 2007).
11 LULUCF emissions sources (i.e. deforestation and peatland degradation) account for more than 80% of the total GHG emissions in Indonesia, making it the third largest GHG emitter in the world (PEACE 2006; Wetland International 2006).

FIGURE 5.3 Indonesia's CDM potential utilisation rate (including LULUCF)

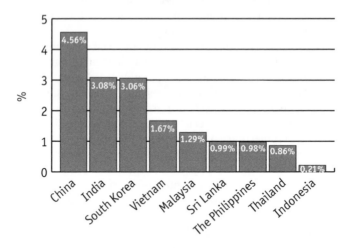

The utilisation rate is calculated by dividing the total expected CER volume from all ongoing CDM projects as well as those under development in each country as of November 2007 by the total national GHG emissions in tonnes of CO_2-equivalent, including emissions from LULUCF (calculation based on data from UNEP Risoe and WRI).

FIGURE 5.4 CDM potential utilisation rate (excluding LULUCF)

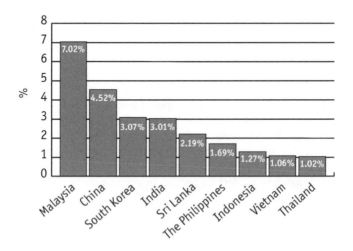

See notes for Fig. 5.3 for calculation details but note that LULUCF emissions are excluded from these calculations.

The gap becomes more apparent when Indonesia's situation is compared with that in Malaysia — a country with equally rich biodiversity and vast tropical rainforest coverage. Malaysia is in the middle rank of CDM potential utilisation rates when LULUCF sources are included, but jumps to the top when LULUCF sources are excluded from the calculation. These data suggest that, relative to Indonesia, Malaysia has more successfully exploited the opportunity presented by CDM in areas such as renewable energy, biomass energy and energy efficiency.

Industry sectors

Looking across the 52 CDM projects that have been implemented to date in Indonesia, three types of project (landfill gas, biomass energy and renewable energy) make up 72% of the total expected CER volume until 2012 (Fig. 5.5). Of these, the CER volumes expected from biomass energy projects are relatively small compared with the other two types of project. In relation to renewable energy sector, two projects (a hydroelectric

FIGURE 5.5   CER volumes (ktonnes CO_2-equivalent) and percentages of total by CDM project type

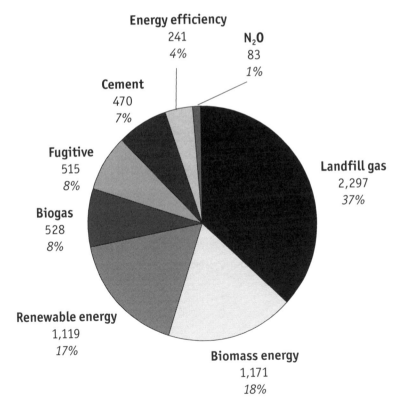

Energy efficiency
241
4%

N_2O
83
1%

Cement
470
7%

Fugitive
515
8%

Biogas
528
8%

Landfill gas
2,297
37%

Renewable energy
1,119
17%

Biomass energy
1,171
18%

Calculations based on data from UNEP Risoe.

power project and a geothermal project) will produce 90% of the expected CER volumes from the sector.

Additionality

All the projects incorporated the future revenue from the sale of CERs as part of their business and financial planning to make their projects viable. However, there are some variations between the projects with regard to the degree of additionality.

First, the technology and equipment required differs between projects. For those projects that are developed primarily for profit-making purposes (e.g. by carbon financing companies), the preference is for industry sectors that require relatively simple technology/equipment; examples include landfill gas (e.g. with gas flaring and composting technology) and biogas (e.g. with methane capture technology). However, much more effort is needed for projects in areas such as energy efficiency and renewable energy. This is because these sectors require significantly larger investment, more complicated technology or equipment and, lastly but most importantly, much consensus building and capacity-development work among political and local entities. As a result, at least to date, the motives for projects in these areas have not been dominated by profit (see Fig. 5.6).

FIGURE 5.6 Ratio of projected CERs (by volume) by motivation types

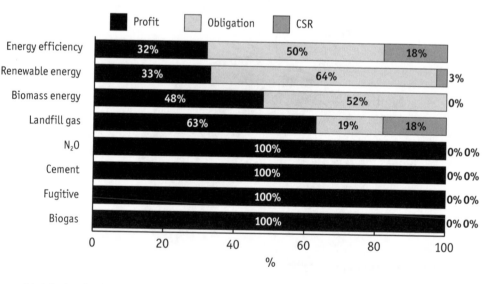

Calculation based on data from UNEP Risoe.

Second, for developers seeking to have their project registered for CERs, there are additional 'Kyoto process' costs that need to be carried by project developers — in particular the CDM registration costs and additional contract agreement and other negotiation costs. These have been estimated as adding an average €150,000 to project costs (Michaelowa *et al.* 2003). Because of the capital costs, the need to have access to specific technologies and the transaction costs, the situation has emerged where the majority of CDM projects in Indonesia are handled by either bilateral/multilateral carbon funds or carbon financing companies that provide financial and technical support, generally in return for a discounted CER purchase price.

Other outcomes: development impact

As discussed earlier, there are two 'layers' of objectives for the CDM scheme. The first is to enable Annex I countries to meet their GHG emissions reduction targets under the Kyoto Protocol. The second is to assist non-Annex I Parties to achieve clean and sustainable development. The first objective very much dictates the behaviour and modalities of CDM project development among credit buyers and foreign companies. The second objective, on the other hand, is directly related to development impact in the host country. Even though one of the key requirements for CDM projects is that projects should assist the host country in achieving sustainable development (NEDO 2006), it is the host country that has the responsibility for ensuing that this is in fact the case.

In the CDM process, each host country and its Designated National Authority are required to establish national criteria and indicators for sustainable development, and to evaluate all CDM proposals based on these criteria.

The Indonesian DNA has set four types of sustainability criteria (see Table 5.2):

● Environmental

● Economic

● Social

● Technological

The first three relate to local impacts (i.e. in the vicinity of a CDM project) whereas the technological criterion is evaluated on a national basis. In principle, all proposed CDM projects must pass all individual indicators that are applicable in order to be approved.[12]

All these indicators appear to be best described as requiring adherence to existing national/local regulations and/or not making a particular situation worse. From Table 5.2, it is difficult to find any criterion that requires improvement and/or a significant contribution to sustainable development. The consequence has been that very few CDM projects in Indonesia have incorporated activities that go beyond what is required. Only one project proposes going beyond the environmental requirements specified in legislation. In relation to social performance, only seven out of the 27 projects that had been

12 National Commission on CDM (Ministry of Environment, Indonesia) website, dna-cdm.menlh.go.id (accessed December 2007).

TABLE 5.2 Indonesia's sustainable development criteria and indicators

Source: National Commission on CDM (Ministry of Environment, Indonesia)

Category	Criteria	Indicators
Environment	Environmental sustainability by practising natural resource conservation or diversification	1 Maintain sustainability of local ecological functions 2 Not exceeding the threshold of existing national, as well as local, environmental standards (not causing air, water and/or soil pollution) 3 Maintaining genetic, species and ecosystem biodiversity and not permitting any genetic pollution 4 Complying with existing land-use planning requirements
	Local community health and safety	1 Not imposing any health risks 2 Complying with occupational health and safety regulations 3 There are documented procedures for the actions to be taken in order to prevent and manage possible accidents
Economy	Local community welfare	1 Not lowering the local community's income 2 Ensuring that there are adequate measures to overcome the possible impact of lowered income of community members 3 Not lowering local public services 4 Agreement among conflicting parties, in conformance with existing regulation, in relation to any lay-offs/redundancies
Social	Local community participation in the project	1 Local community has been consulted 2 Comments and complaints from local communities are taken into consideration and responded to
	Local community social integrity	1 Not triggering any conflicts among local communities
Technology	Technology transfer	1 Not creating dependencies on foreign parties in knowledge and appliance operation (i.e. the transfer of know-how) 2 Not using experimental or obsolete technologies 3 Enhancing the capacity and utilisation of local technology

approved by June 2007 mentioned any type of community development programme and most projects have complied with the formal requirement to consult with local communities by organising a single stakeholder consultation meeting.

A similar picture emerges with regard to economic criteria: 73% of the CDM projects either do not mention anything about local employment or provide fewer then ten jobs (see Fig. 5.7). However, even these modest levels of employment are being seen as meeting the objective of not lowering the community's income.

FIGURE 5.7 Local employment among approved CDM projects (as at June 2007)

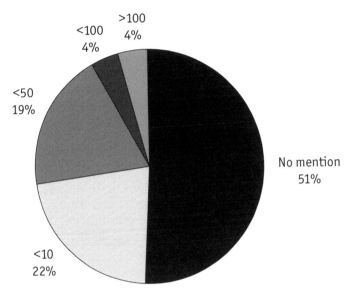

This pie chart shows the number of local jobs expected to result from CDM projects based on the Project Design Documents submitted for these projects.

Finally, since all of the CDM projects involve some kind of state-of-the-art technology, it could be argued that all of the projects deliver on the technology transfer objectives. But as many as ten of the 27 projects approved by June 2007 do not involve capacity development or skills training for local communities, raising the question of whether significant development benefits will accrue to local communities as a consequence of these projects.

Apart from direct employment, there are broadly speaking four ways in which local communities can benefit financially from CDM projects, namely:

● Charitable donations

● CER revenue reinvestment

● More (indirect) local employment

● The development of alternative income sources

However, the overall contribution of the CDM projects across these four approaches seems modest:

● Only 27% of the CDM projects implement any kind of community development programme

● Only one of the 52 projects reinvests CER revenues in the local community or local government for community development or skills training

● As discussed above, existing CDM projects do not seem to be producing much local employment

● Little attention has been paid to indirect employment or the development of alternative income sources in the CDM projects to date, nor has the government required CDM project developers to incorporate measures such as formal education, vocational training or skills training in the package of benefits provided to local communities

Despite these limitations, it must be recognised that DNAs and host countries face a fundamental dilemma: if the criteria and indicators applied are too stringent, they may not be able to attract many CDM projects.

The 'rational' market philosophy of 'the cheaper and easier the better' is ever present in the global CDM market. This pressure is compounded by the fact that there are too many buyers and too few investors; finding a buyer is relatively easy but finding investors to develop CDM projects is much more difficult. The CDM project development modalities in Indonesia confirm this point with the dominant players being carbon financing companies that also provide the capital necessary to develop these projects.

Concluding comments

Reflecting on the current status of CDM project development in Indonesia, it is clear that the Indonesian government needs to do much more to encourage the private sector to play a greater role. In this context, there are a number of important policy issues that need to be addressed.

First, the policy environment needs to be substantially overhauled. In order to generate an environment conducive to greater private sector participation in CDM, there is a need for regulations and incentives that are specific to CDM development and implementation in Indonesia. As yet, there are no specific policy measures to encourage private sector investment in CDM projects. Furthermore, there are many barriers to private sector involvement in CDM projects including:

● Lack of tax incentives

● Difficulty in attracting investment or co-financing from domestic banks

● Regulations that do not provide much support for renewable and alternative energy

● Lack of compliance with existing regulations

Second, the transparency and accountability of the state-owned enterprises and agencies that deal with conventional and alternative energy sources need to be strengthened because the behaviour of these enterprises has a significant impact on the private sector. There have already been a number of cases where entrepreneurial companies have encountered difficulties in negotiating appropriate terms and conditions for the sale of electricity from renewable and alternative energy sources. These transaction costs are a significant obstacle to greater private sector engagement in CDM project development in Indonesia.

Third, awareness of CDM needs to be increased. Knowledge and awareness of CDM is still low reflecting the sector-limited focus of government agencies and the lack of national leadership in utilising the CDM as one of the central tools to realise emissions reduction and sustainable development. Currently, among the government agencies, awareness of CDM is limited to the Ministry of Environment (where the Designated National Authority is hosted) and a handful of other agencies related to the Ministry of Energy and Mineral Resources. Unless there are concerted and coordinated efforts from national authorities, it will be difficult to move to a situation where the private sector is willing to be more involved in CDM. In addition to awareness-raising at national and ministerial levels, awareness among the local governments in Indonesia also needs to be strengthened. CDM can provide a great opportunity to be involved in emissions reduction activities while, at the same time, helping to attract foreign and domestic investment. In this context, it is interesting to note that the CDM projects that have been initiated in districts in Bali (the most popular tourist destination in Indonesia) are all being led by local government bodies.

The most important outcome from these measures in the long run will be a reduction in the perceived and the actual risks associated with CDM development in Indonesia. Overall, these risks have limited the number of CDM projects in Indonesia and have resulted in a pattern where most of the projects have been developed by bilateral/multilateral carbon funds and carbon financing companies rather than by the private sector more generally. It is thus imperative that government policy reduces CDM-related risks by:

- Deploying regulations and incentives

- Increasing the transparency and accountability of state agencies

- Building the awareness which is crucial in enlarging the private sector participation in CDM development

The effective implementation of these measures will help maximise the contribution of the Clean Development Mechanism — and the contribution of the private sector — to sustainable development in Indonesia.

References

Bhardwaj, M. (2007) 'Indonesia palm oil sales seen up in 2008 — official', Reuters India, 24 September 2007; in.reuters.com/article/businessNews/idINIndia-29689920070924 (accessed November 2007).

Ellis, J., and S. Kamel (2007) *Overcoming Barriers to Clean Development Mechanism Projects* (Paris: OECD, IEA and UNEP).

IGES (Institute for Global Environmental Strategies) (ed.) (2006) *CDM Country Guide for Indonesia* (Hayama, Japan: IGES, 2nd edn).

Michaelowa, A. (2005) *Creating the Foundations for Host Country Participation in the CDM, Climate Change and Carbon Markets: A Handbook of Emission Reduction Mechanisms* (London: Earthscan).

——, M. Stronzik, F. Eckermann and A. Hunt (2003) 'Transaction Costs of the Kyoto Mechanisms', *Climate Policy* 3: 261-78.

Ministry of Environment (Republic of Indonesia) (2001) *National Strategy Study on the Clean Development Mechanism in Indonesia* (Jakarta: Ministry of Environment).

NEDO (New Energy and Industrial Technology Development) (2006) *CDM Development in Indonesia: Enabling Policies, Institutions and Programmes, Issues and Challenges 2006* (Tokyo: NEDO).

PEACE (Pelangi Energi Abadi Citra Enviro) (2006) *Indonesia and Climate Change: Working Paper on Current Status and Policies. A Report Commissioned by DFID and the World Bank* (Jakarta: PT. Pelangi Energi Abadi Citra Enviro).

Wetland International (2006) *Peatland Degradation Fuels Climate Change* (Wageningen, Netherlands: Wetland International).

6
Encouraging innovation through government challenge programmes: a case study of PV-based boats

Olga Fadeeva and Johannes Brezet
TU Delft and Cartesius Institute, The Netherlands

Yoram Krozer
University of Twente and Cartesius Institute, The Netherlands

A number of major studies (see, for example, IEA 2006; Stern 2006) have emphasised the importance of innovation and radical change in energy systems if we are to avoid the most serious negative effects of climate change. The process of delivering innovation in energy technologies requires the accomplishment of the following two tasks:

● Development of innovative products

● The provision of appropriate incentives and other support necessary for these innovative products to be commercialised

In this chapter we examine the role that could be played by government-sponsored or -supported challenge programmes in this process of innovation using the example of the Frisian Solar Challenge and its role in stimulating the development of the photovoltaic (PV)-powered boat market in Friesland (The Netherlands).[1]

---

1 Information about the boats and partnerships participating in the Solar Challenge was collected during the competition through interviews with the participants and organisers of the race. The participating teams identified companies that played an important role in their product development process; most of these companies were interviewed to understand how the Solar Challenge influenced technology development within their businesses. Finally, data on the wider PV market were gathered from interviews with producers and users of PV boats. About 200 private boat owners from all over the world were interviewed.

The solar boats market

PV boats: a brief overview

There are four main functions on a boat where use of PV cells brings additional value to the product:

- Propulsion
- Navigation, safety and lighting
- Living (kitchen, shower, etc.)
- Keeping batteries charged

Typical power requirements for these functions are presented in Table 6.1.

TABLE 6.1 PV functions on boats and associated power requirements

PV functions	Power requirements
Propulsion	• Pedal boats: 400–600 W motor (200 Wp solar cells, 1.5 m²) • Small boats (up to 7.5 m): 1.5 kW motor (700 Wp solar cells, 5 m²) • Medium-sized boats (7.5–13 m): 4 kW motor (2 kWp solar cells, 14 m²) • Large boats (>14 m): 16 kW motor (5 kWp solar cells, 36 m²)
Navigation, safety and lighting	• Appliances: 150–250 W • Day sailing (8 hours): 70 Watt-hours/day • 24-hour sailing: 280 Watt-hours/day • 24-hour luxury sailing: 1,300-2,300 Watt-hours/day
Living (kitchen, shower, etc.)	• Appliances: 1,000–3,000 W • 24-hour luxury sailing: 3,000 Watt-hours/day
Charging batteries	• Up to 10 W

Note: Wp is the power derived from the PV panel when the light intensity is 1,000 W/m².

The use of PV for powering living applications and for battery charging is increasingly common. However, boats with a propulsion system that is fully or partly powered by PVs currently represent only a small fraction of the boats on the market. This chapter focuses mainly on opportunities for the development boats whose propulsion is fully or partly powered by PV cells; we use the term 'PV boats' to refer to these boats.

Current markets for PV boats

Solar-powered boats tend to be relatively slow, with maximum speeds typically in the range of 10–13 km/hour. While the capital cost of such boats tends to be higher than conventional fossil fuel-powered boats, running costs and maintenance are relatively low. There are three areas where these features may make solar-powered boats financially attractive:

● If the boat's batteries are charged by electricity from the grid, a PV array provides additional cruising hours. This is of particular value for relatively slow long-cruising boats

● When the boat is docked, the boat's PV array may be used to generate electricity that can then be sold to the electricity grid (i.e. providing additional income for the boat's owner)

● PV-powered boats could allow a boat owner to be completely independent of the electricity grid, though this may entail adapting sailing patterns to solar radiation patterns. For example, while the boat may be suitable for long cruises in the summer, it may be confined to short trips in the winter months

The state of the art in PV boats

About 30–40 companies worldwide are engaged in the development of solar-powered boats. Solar-powered boats are used mainly in Europe, which is where most of these companies are located (Germany, France, Switzerland, etc.). However, the solar boats market remains relatively immature and some way short of being a commercial-scale industry.

Most of the companies presently involved in the production of solar boats are focusing on larger solar boats. One of the earliest developers of solar boats, Theodor Schmidt, explains this from a technical perspective noting that, with increases in the size of the ship:

> the main source of resistance through the water, wetted surface friction, increases slightly less rapidly than the increased area available for solar cells, and resistance due to wave-making becomes less and less important.[2]

There is also a commercial reason for this focus as bigger public ships seem to be a promising area for governmental/municipal subsidies. According to interviews with producers and final customers, most large solar boats are being sold to tourist resorts for public transport and excursions. It is not uncommon for 50–70% of the capital costs of these boats to be paid by government/municipalities.

Reliance on government subsidies is a risky strategy as the reduction or elimination of such subsidies may leave companies without a market for their primary product (larger boats) and the associated risk that the technologies they have developed may not be suitable for other market niches. It therefore seems reasonable to suggest that producers should be developing products that target other areas of the market. The most obvious opportunity relates to private users (both individual boaters and rental companies) though competing in these markets will require the development of smaller boats that can go considerably faster than current PV boats.

Even if suitable smaller boats can be developed (see the discussion of technical issues below), final customers and other potential stakeholders are not currently well informed

2 www.umwelteinsatz.ch/IBS/solship2.html (accessed 20 August 2008).

about the advantages of solar-powered boats. While there is a reasonably wide uptake of PV-power equipment on boats (e.g. about 35% of our interviewees already used PVs on their boats for keeping batteries charged, and for living and navigation applications), most boat owners are not aware that PV cells can be a sufficient energy source for the propulsion system.

Interviewees also raised the issue of cost, with 45% expressing concern about the relatively high capital costs of PV boats. However, they also seemed positively disposed to the idea of PV boats with 27% expressing a wish to be more independent from fixed infrastructure and 80% stating that they would like to use PVs more (although this was primarily in relation to applications that they are already familiar with such as battery charging and the operation of living appliances). These findings suggest that there is at least the potential for the solar boats market to expand in coming years.

The other issue that may slow the development of the markets for solar boats relates to the aesthetic appearance of the majority of the solar boats that are currently on the market. Most of the boat owners we interviewed considered these boats to have an unattractive appearance. With the expansion of the target market for solar boats to include private customers, however, producers and designers are starting to focus more on boat aesthetics as an important part of the overall boat design.

Key technical developments needed for small solar boats

There are a number of areas where further technological development is required before smaller solar boats can be competitive with conventional fossil fuel-powered boats. The most significant of these are:

- **Engineering of the components and design of overall electronics.** There are a number of specific aspects that need to be addressed. First, as discussed above, boat aesthetics need to be improved. Second, the limited space available for PV cells on the boat and the heavy weight of the batteries (see the comments on batteries below) mean that maximising the energy efficiency of the boat as a whole is a priority – in particular minimising the overall boat weight, designing hulls to lower resistance and improving the energy of propulsion systems. Third, according to Hugo Grimmelius of the Marine Department at TU Delft, one of the main challenges in building solar-powered boats is engineering components and placing them together in a way that minimises energy loss. Finally, engineering of overall electronic and control systems is another key challenge in designing energy-efficient solar boats

- **Motors and propellers for small boats.** Small low-speed (around 200–600 rpm) high-efficiency motors are required for small solar boats. Such motors are not presently available on the mass market and, while the use of gearboxes could allow the use of higher-speed motors, gearbox systems are complex and could increase energy losses in the propulsion system. At present, the producers of solar boats often manufacture motors and propellers themselves. Therefore, in addition to the development of motors and propellers for small boats, there is a need to consider how the production of these could be scaled up

● **Batteries.** Most of the electric boat builders on the market, including solar boat builders, use lead–acid batteries – a well-known and proven technology. However, these batteries are heavy and have a relatively low energy density (i.e. they are relatively large in size). Lithium ion batteries offer the potential to address these problems and are presently available for applications in the 10–50 Watt-hours range although, for small-sized boats, 1 kWh is normally considered to be the minimum requirement. There are also a series of technical issues that need to be addressed with lithium ion batteries; in particular, ensuring that such batteries can be readily integrated with electrical systems on solar boats and ensuring the safety of such batteries. Research and experimentation with lithium ion batteries is ongoing in technical universities, research institutes and companies. More collaboration between these institutions might speed up the necessary breakthroughs in this area

Facilitating business innovation: a regional government perspective

There are various ways in which governments can seek to foster innovation in a region. A common strategy is to provide fiscal or other support to research and development (R&D) with the aim of attracting and concentrating R&D in the region. An example of where this strategy has been successful is in the Oresund area of Sweden and Denmark, which includes Copenhagen and Malmo (Hospers 2004).

However, regions on the periphery of major economic centres (such as Friesland in the northern part of the Netherlands) may not have a sufficient concentration of large academic and research centres for this type of strategy to work and must, therefore, search for other strategies to generate innovation. Examples of the strategies that have been successfully adopted by peripheral regions include attracting creative people to a region by developing outstanding living and working conditions – a commonly cited example is the impact of *Lord of the Rings* on New Zealand (Florida 2002, 2005) – or establishing networks to explore local opportunities for innovation and learning through the provision of exceptional meeting and information exchange facilities; an example is the activities of the Welsh Assembly Government (Cooke and Morgan 2000).

The Friesland regional government has decided to adopt a different strategy, focusing its efforts on challenging young people to create something new. This approach is based on the assumption that:

● Young people are the main source of innovation

● Young innovators will remain in the region because they get opportunities to excel

The strategy is attractive for the regional government because it is not dependent on large amounts of R&D money or on large projects. The key element of the government's challenge strategy is to generate a stream of ongoing events that create an engaging and creative atmosphere in the region. The strategy has a clear link with the concept of 'strategic niche management' in the sense that, at least in the early stages, the process involves the development of niches that are 'protected' from harsh market conditions.

The strategy involves establishing networks of actors who organise experiments with the financial support of private sponsors and local or regional government. The experimentation is initiated and later expanded through the development of the network of actors and/or through interconnections with similar experiments (Weber *et al.* 1999). The expansion, however, is difficult because the innovators will inevitably aim to become the only manufacturers of the innovation and hence maximise the returns on their investment (Douthwaite 2002: 62).

The challenge strategy in Friesland began in 2000 when a group of about ten people from within and outside the region started to generate a series of ideas for challenge programmes. This was facilitated by strong political support, government funding for start-ups and, above all, fast and flexible governmental decision-making processes. The importance of the latter should not be underestimated as the reality is that most government decision-making processes tend to be laborious, resulting in high transaction costs and potentially jeopardising the creation of market niches. In the case of Friesland, however, small amounts of money have been relatively easy to access for business start-ups and for innovative ideas. These facilitating factors have generated a stream of events and experiments for innovation in areas such as waste-water treatment, theatre and performing arts, and sport, helping create an atmosphere of 'going for innovation' in the region.

The Frisian Solar Challenge

The Frisian Solar Challenge was first run in 2006 and will be run again in 2008. The competition is a six-day race for solar-powered boats, following the route of the historical Elfstedenroute, which links the 11 Frisian cities via canals and lakes — an overall distance of 220 km.[3]

Institutes of higher education, the business community and members of the public were invited to participate in the competition. The competition was divided into three classes:

- Single-person vessels (Class A)
- Two-person vessels (Class B)
- Other vessels (Class C)

Out of the 26 participating institutions, 20 were from the Netherlands with the remainder coming from the USA, Sweden, Germany, Poland and Belgium. According to the rules of the competition, the participants in Classes A and B (21 out of the total of 26 teams) had to build their boats using standard PV panels provided by Sun Factory;[4] the oppor-

---

3 www.frisiansolarchallenge.nl (accessed 20 August 2008).
4 The dealer for PV panels in Friesland.

tunity to borrow PV panels was an important motivating factor for some of the teams to participate.

The quality of the participating boats was influenced heavily by the available resources (expertise, components, facilities, money). Some teams experimented with designing the cheapest possible solar boats, using only the team's internal resources. However, most of the teams accumulated resources through establishing relationships with other organisations such as municipalities, boat builders, designers, knowledge institutions (universities, yacht centres, research institutes, etc.), component suppliers and other actors from the supply chain.

For the companies participating in the Solar Challenge, there were a number of potential benefits, including:

● Advertising and PR

● The development of new or innovative products

● The adaptation of their existing product lines for small solar boats

Outcomes from the challenge

Addressing the barriers to PV boats

During the competition, the diversity and exotic appearance of the boats captured the attention of the companies and potential final customers. The credibility of solar boats as an alternative to conventional boats was demonstrated; indeed, there were several cases where people from the audience tried to purchase boats from the competing teams.

Apart from awareness-raising, the Solar Challenge resulted in the formation of a number of partnerships which stimulated the development of innovative components as well as developing knowledge in the partners about how these components could be engineered and applied. Experiments within different teams saw progress being made on some of the key technical barriers to PV boats. For example, a number of teams developed their own motors and propellers, and five out of the 26 teams used lithium ion batteries in their boats.

The partnerships also provided an opportunity for companies and universities/schools to cooperate on the development of solar boats. For example, the TU Delft team (made up of students from the Departments of Maritime Technology and Information Technology) partnered with the Marine Technology Naval Institute, a research organisation in the Netherlands with specialist knowledge of ship construction and industrial ship design. These partnerships allowed the university and schools participants to take their research from the laboratory to the point where they were able to test a prototype – an important intermediate step in moving towards full commercialisation.

For the teams, the six days of racing enabled them to test the capabilities of their innovative boats and to further improve the design ideas that were generated before the competition. One of the interesting outcomes from the Solar Challenge was that the teams openly shared ideas and suggestions with each other, which was another means for them to further advance the design of their boats.

Influence on businesses in Friesland

The Solar Challenge generated a number of benefits for Friesland-based companies, including:

- Allowing them to establish new partnerships

- Stimulating the development of new products or other innovations

- Leading to the formation of new businesses

New partnerships

For some of the participants, the opportunity to establish partnerships was a major objective of their participation in the Solar Challenge and an important stepping stone to the commercialisation of their ideas. For example, the Manta (which presented one of the cheapest concept-design ideas at the 2006 competition) and Andela teams intend using the next Solar Challenge as an opportunity to find more partners to further develop their designs and, hopefully, move their boats closer to commercialisation.

It is interesting to note that some of the companies did not see proceeding to the commercialisation stage as an objective for their participation in the challenge. For example, the Marine Technology Naval Institute participated because of its concerns about the environment. As a result of its partnership with TU Delft, the company increased its knowledge of the hydrodynamic design of light ships. However, the company does not intend to take this knowledge further as its commercial focus remains large industrial ships.

Product innovation

Some teams/companies developed innovative products as a result of their participation in the Solar Challenge. Examples include:

- As a result of cooperation with the NHL racing team, Combi Multi Tractje (a Frisian company that produces steering systems and outboard motors) developed a new small motor. At the time of the event, the motor was a prototype but this product has now reached the market

- Breehorn, a boat building company, helped the ROC Friese Port Sneek team with advice on boat construction and produced a propeller specifically for small light boats, which it thinks is unique to the market. However, the company does not expect this propeller will be commercialised because it is seen as being too specific to small racing boats (i.e. its wider application is seen as limited)

- ETD, located in Wanneperveen, is a dealer in outboard motors from Hungary and Austria. The company is currently helping the Manta team by making adjustments to a serial motor in order to achieve higher efficiency. The company is looking forward to the forthcoming race as an opportunity to test the new motor, which it expects to put on the market after the 2008 Solar Challenge

- Top Zeef, a family-based team from Friesland, is currently working on the development of a PV boat that would be affordable for an average buyer. At the 2006

Challenge, the company presented the cheapest concept design for a boat. It is aiming to improve the prototype through its participation in the 2008 Solar Challenge and hopes to bring the product to the market after the 2008 Challenge

● The Andela team produced a new aluminium solar boat for the 2006 competition and is currently experimenting with a prototype PV-powered displacement glider boat

Products and businesses

Some teams/companies decided to set up a business for producing or renting PV boats. One example is the company Czeers, established by two students from TU Delft, which is aiming to produce PV yachts in Friesland. The team's technical knowledge, together with the external expertise it accessed through participation in the Solar Challenge, enabled it to develop its product to a point where it was ready to go to the market. The readiness of the team to start the business was demonstrated by the successful test of the boat in the 2006 competition. The new business was partly catalysed by the start-up money from the province and this financial support, along with access with Friesland's shipbuilding industry, was an important factor in its decision to establish the business in Friesland. Czeers has already produced its first commercial prototype of the first speed solar-powered boat for wealthier buyers, offering speeds up to 50 km/hour.

Mono & Multihull Boatbuilders, a Frisian yacht building company, helped NHL to build its boat. The Solar Challenge helped the company to start developing its new solar-powered boats (Featherline). During the process of preparing for and subsequently participating in the Solar Challenge, the company identified possible business opportunities for this new product. As the company is already involved in building lightweight yachts, the expansion of its business in the PV boat construction business was not a radical change of direction for the business.

The 2006 Challenge also stimulated Kasper van der Heiden to develop a commercial prototype for a small — 6 × 2.2 m — boat (sloepen) that could travel at up to 18 km/hour. He is presently trying to develop a product that balances high technology performance with relatively low capital costs. His target market is rental companies that use boats on lakes, especially in recreational areas.

A number of companies promoted their names and gained new business opportunities throughout the competition. Perhaps the most high-profile example was Sun Factory, which received a high degree of promotional coverage through the event. Sun Factory lent solar modules to the teams for free in order to stimulate participants to participate in the first challenge. Overall, it supplied 80% of teams with solar modules and, after the event, some 70% of those teams bought PV modules from the company.

Regional policy and innovation

The Solar Challenge demonstrates the potential for challenge programmes to stimulate innovation and, in this case, help develop solutions to the problem of climate change. From the perspective of regional policy, the questions that arise are:

- Is the strategy of challenging people for innovation embedded into regional policy?

- Does this policy framework impact on business?

- Is the strategy actually stimulating innovation?

The first point to be made is that regional policy in Friesland strongly reflects the concept of sustainable development. Almost all of the major policy decisions that have been made in recent years contain elements of innovation in relation to sustainable development. Moreover, the present regional policy identifies innovation for sustainable development as a priority. At the time of writing, the portfolio of the Project Minister of Sustainable Development in the regional provincial government is held by one of the strongest and most experienced politicians. All of these facts indicate that the strategy of sustainable development is well embedded in the provincial policy-making.

The next question is whether this strategy trickles down to municipalities and companies. The answer is not a simple yes or no. Some municipalities (e.g. the capital of Friesland, Leeuwarden) strongly support sustainable development, in particular renewable energy, at the political level. One telling example was the election of the Mayor of Leeuwarden whose experience with sustainable development issues at the national level was an important factor in his election. In smaller cities in Friesland, the answer is less clear cut. Some are interested in pursuing more sustainable development whereas others do not consider sustainable development to be a policy priority. In order to create a more uniform vision, the provincial government has implemented several projects that aim at creating a broader base for sustainability. One project, Friske Fiersichten (Frisian perspectives), involved the organisation of a series of discussion sessions with the most important businesses and social and political organisations in the region, as well as with significant businesses in the Netherlands. The overwhelming majority of discussion participants agreed that sustainable development is key to the region's development.

It is difficult to provide a firm answer to the question of whether the Solar Challenge has led to innovation. The reality is that the process of innovation can take a number of years before an idea actually gets commercialised. Furthermore, challenge programmes are just one of the many influences on whether or not innovation occurs. Even if innovation occurs, it is difficult to establish precise causal relationships between innovation and specific challenge programmes or other policy measures.

Having noted these limitations, there are signs that the Solar Challenge has increased the probability of innovation in the area of PV boats. First, the Challenge allowed companies and researchers to obtain feedback on their designs and to compare their ideas with those of other teams. Second, during the competition, the teams obtained feedback from potential final customers or interested companies; this should increase the probability of commercial success for these innovations. Third, as discussed above, the programme has accelerated some aspects of technology and business development. Finally, the large variety of products/concepts launched at the competition increases the chance that new products will reach the market. This 'probabilistic approach' helps maximise the likelihood that innovations will emerge.

Conclusions

To date, the Solar Challenge seems to have been successful in achieving its objective of stimulating innovation in the PV boat market and of providing business benefits to the shipbuilding industry in Friesland. This success is likely to be reinforced with the increased international participation in the 2008 Challenge.

In the context of the Netherlands, Friesland is an example of a peripheral region that lacks significant monetary or R&D resources. In order to overcome these constraints, the province is currently testing a unique strategy where innovation is incentivised through challenge programmes. While it is too early to draw firm conclusions, the evidence to date – from the Solar Challenge and from other challenge programmes in Friesland – is that this kind of approach could lead to positive outcomes in terms of stimulating innovation, contributing to sustainable development and providing real benefits to local business.

References

Cooke, P., and K. Morgan (2000) *The Associational Economy: Firms, Regions and Innovation* (Oxford, UK: Oxford University Press).

Douthwaite, B. (2002) *Enabling Innovation: A Practical Guide to Understanding and Forecasting Technological Change* (London: Zed Books).

Florida, R. (2002) *The Rise of the Creative Class* (Cambridge, MA: Basic Books).

—— (2005) *The Flight of the Creative Class* (New York: HarperCollins).

Hospers, G. (2004) *Regional Economic Change in Europe: A Neo-Schumpeterian Vision* (Münster, Germany: LIT Verlag).

IEA (International Energy Agency) (2006) *Energy Technology Perspectives: Scenarios and Strategies to 2050* (Paris: IEA).

Stern, N. (2006) *Stern Review: The Economics of Climate Change* (Cambridge, UK: Cambridge University Press).

Weber, M., R. Hoogma, B. Lane and S. Johan (1999) *Experimenting with Sustainable Transport Innovations: A Workbook for Strategic Niche Management* (Seville: Enschede Press).

7
The role of voluntary industry–government partnerships in reducing greenhouse gas emissions: a case study of the USEPA Climate Leaders programme

Jeffrey H. Apigian
Clark University, USA

Discussions of environmental policy in the USA often focus on the merits of the two dominant regulatory approaches:

- Command and control

- Economic incentive or market-based policy

While command and control has historically been the preferred regulatory approach of the US Environmental Protection Agency (USEPA), economic incentives have gained significant traction over the past several decades as a flexible, less prescriptive alternative. The cap-and-trade scheme developed for sulphur dioxide in the 1970s is one widely cited example of an incentive-based approach (Kerr 1998; Harrington *et al.* 2004).

Yet, despite the ostensible success of market-based approaches, the relationship between government and industry remains polarised. This is particularly true in relation to the regulation of carbon dioxide (CO_2) and other greenhouse gases that are implicated in anthropogenic climate change. While some members of industry and the US Congress have begun to push for new regulation, the notion that stricter regulation of greenhouse gases is a 'zero-sum game' still prevails.

In this context, failure to examine more dynamic regulatory approaches may further entrench the view that the relationship between government and industry must be adversarial to meet environmental improvement goals, and that it may also have the effect of undermining efforts to encourage industry to reduce its greenhouse gas emissions. One of the problems with the scenario above is that it presumes that there are no alternatives to command and control and economic incentive approaches to environmental policy. However, as this chapter demonstrates, voluntary approaches seem to offer the potential for substantive action on climate change while also creating opportunities for win–win outcomes.

In the USA, voluntary programmes have become an increasingly important part of the policy mix – particularly for issues such as greenhouse gas (GHG) emissions and for persistent, non-point sources of pollution (Fiorino 2006). At least, in theory, voluntary programmes offer the USEPA a high degree of flexibility, closer collaboration with industry and a framework for meeting GHG reduction goals – particularly in the absence of tougher legislation from Congress. Participating companies may also benefit from participation through receiving technical assistance, achieving first-mover status and enhancing their 'social licence' through publicity and public interaction (Gunningham *et al.* 2004). While voluntary programmes are sometimes underpinned by the threat of stricter legislation, which in the USA has become an important driver for participation, they still represent a dramatic shift away from the traditional command-and-control approach.

This chapter examines Climate Leaders, a voluntary government–industry partnership programme from the USEPA which encourages corporations to inventory their GHG emissions and to set aggressive emissions reduction goals. The chapter seeks to answer the following questions:

● How effective is the Climate Leaders programme in driving companies towards going beyond compliance with the regulation of GHG emissions?

● What are the strengths and weaknesses of this particular voluntary approach?

● What will be the role of voluntary approaches such as Climate Leaders as US climate change policy evolves?

This chapter is divided into three main sections. The first is a brief overview of climate change policy in the USA. The second is an overview of Climate Leaders including an assessment of its effectiveness and its strengths and weaknesses. The third is a wider discussion of the role of voluntary programmes in future US regulation of greenhouse gases.

Climate change and the current US policy arena

The USA has been stalwart in its opposition to participation in international climate change agreements that mandate binding GHG emission targets and has been unable to pass unilateral federal legislation that regulates emissions domestically. The current administration's hostility towards the Kyoto Protocol in particular has drawn significant criticism internationally and, to a lesser extent, domestically. The administration's stance is to some extent due to lingering uncertainties in the science, but primarily stems from concern over the economic implications of GHG regulation in particular given the lesser restrictions on emissions from rising economies such as China, India and Brazil. Though recently surpassed by China in terms of its net CO_2 emissions, the USA is still the largest emitter of CO_2 on a per capita basis, reflecting the fact that the US economy is highly dependent on the consumption of petroleum-based fossil fuels, underpinned by a complex interaction between stakeholders within the USA, among them a regulatory agency facing numerous challenges in regulating carbon dioxide.

In a major decision in April 2007, the Supreme Court determined that the USEPA has the authority to regulate CO_2 under the Clean Air Act. Notwithstanding this decision, the current Congress is unlikely to pass tougher legislation that might harm the economy. Even if tougher legislation was enacted, it would probably be filled with the characteristically vague and general language that makes US environmental law so problematic.

The USEPA fought the Supreme Court decision because it is aware of the numerous challenges associated with GHG-related legislation. First, there is a high degree of uncertainty regarding the impacts of climate change on human health and welfare. For criteria air pollutants under the Clean Air Act, the USEPA must demonstrate a strong causal link to acute respiratory and cardiovascular health concerns. However, the reality is that the impacts of CO_2 are indirect, hard to quantify and occur more slowly via climate change. Second, the USEPA is not structured to implement or enforce regulations on diffuse sources and diverse emitters; see, for example, the case of methyl tert-butyl ether (MTBE) regulation as discussed by Franklin et al. (2000). Furthermore, given the arm's-length relationship between the USEPA and industry, new CO_2 legislation would almost inevitably open up the USEPA to a barrage of costly litigation and legal challenge. The consequence of these challenges is that it would probably take a significant amount of time before the USEPA was in a position to take even modest action to compel companies to reduce their greenhouse gas emissions.

Because they are established outside the formal legal system, voluntary programmes avoid many of these problems. They are therefore seen as the central component in meeting the Bush administration's pledge to a modest 18% reduction in US energy intensity by 2012 (Abraham 2004). The administration's approach envisages that:

- The USA will have sustained economic growth

- The use of voluntary approaches will allow a slow transition away from existing energy infrastructure in which US companies have made huge capital investments

Despite the Bush Administration's modest goals, climate change has captured the attention of US industry, environmental groups and civil society; many city governments and coalitions of states have acted independently of federal legislation, establishing fuel economy standards for automobiles and GHG emissions standards for stationary sources such as power plants. The Chicago Climate Exchange and other voluntary carbon markets are evidence that certain parts of industry are ready for stricter action on GHG emissions (Lyon and Maxwell 2004). This has not gone unnoticed in Washington, with this momentum for action leading many to suggest that it is just a matter of time before new regulations on greenhouse gases — in particular carbon dioxide — are handed down from Congress.

Background on Climate Leaders

Climate Leaders is a voluntary government–industry partnership that was initiated by the USEPA in 2002. It evolved from the Climate Wise programme, which was part of the Clinton Administration's Climate Action programme. In Climate Leaders (USEPA 2007), companies pledge to:

- Inventory their GHG emissions — specifically carbon dioxide (CO_2), methane (CH_4), nitrous oxide (N_2O), hydrofluorocarbons (HFCs), perfluorocarbons (PFCs) and sulphur hexafluoride (SF_6)

- Establish long-term reduction goals to be met over a 5–10-year time period

Partner companies are expected to set aggressive reduction goals and track their progress through annual reports submitted to the USEPA. Currently, there are 54 voluntary agreements in the USA, of which 28 address climate change (Lyon and Maxwell 2004). Given its broad-scale focus on corporate-wide GHG emissions, the USEPA has positioned Climate Leaders as an umbrella programme for its portfolio of climate-related corporate-level partnerships which include Energy Star, the Green Power Partnership and the Smartway Transport Partnership among others. The objectives of Climate Leaders align closely with the US Department of Energy's Climate VISION programme, which focuses its efforts on GHG emission reductions among trade groups. The core principle of Climate Leaders is that government and industry should work together in harmony towards environmental improvement.

At the time of its creation, Climate Leaders had 35 charter partners. As of November 2007, it had 158 partners from industries ranging from manufacturing, chemical, health services, retail and entertainment. The companies and their respective GHG reduction goals are equally diverse. For instance:

- Manufacturing company 3M pledged to reduce its total US GHG emissions by 30% between 2002 and 2007

- Lockheed Martin pledged to reduce its US GHG emissions by 30% per dollar revenue from 2001 to 2010

- Shacklee Corporation, a consumer products firm, pledged to maintain net zero US GHG emissions from 2006 to 2009 (USEPA 2007)

Even the World Bank has signed up to Climate Leaders, pledging to reduce its US GHG emissions by 7% between 2006 and 2011.[1] Companies establish their reduction goals only after completing a systematic GHG inventory protocol and management plan that

---

1 For a complete listing of the GHG emissions reduction commitments of the partners, see the Climate Leaders website, www.epa.gov/stateply (accessed 20 August 2008).

sets out the framework for measurement and reporting. The entire process can take up to two years to complete.

The Climate Leaders GHG inventory protocol has three main components:

- A set of general design principles

- A set of core modules

- A set of optional modules

The corporate GHG accounting protocol used by Climate Leaders was developed by the World Resources Institute (WRI) and the World Business Council for Sustainable Development (WBCSD) (USEPA 2007). The design principles provide companies with guidance in setting inventory parameters, establishing a reliable baseline for measurement and identifying the various sources of corporate GHG emissions at their facilities.

The core modules specify the GHG emission sources to be reported including both direct emissions (mobile and stationary fuel use, process-related emissions, refrigeration and air conditioning) and indirect emissions (primarily electricity and steam purchases). The optional modules include product transport, employee commuting and business travel.

The Climate Leaders reporting mechanism is broken down into three main components:

- Creation of an Inventory Management Plan (IMP)

- Maintenance of a comprehensive corporate-wide emissions inventory

- Annual submission to the USEPA of the partner's performance against agreed goals

The IMP provides the USEPA with a description of the company's operational boundaries, chosen methodologies quantifying emissions, data management methodologies, base-year information, management tools currently employed, and chosen auditing and verification methodologies.

Participating companies are expected to develop and begin implementing their IMP within one year of joining Climate Leaders. At the end of the first year, partners submit their IMP along with an inventory of direct and indirect emissions from their domestic and international operations. In turn, the USEPA provides each participating company with a review of its IMP, offering recommendations for improvement at both the organisational and individual facility level. In subsequent years, companies report only on current-year emission inventories and any changes they have made to their IMP. Partner companies are self-reporting and are not currently required to provide third-party verification of their inventory data, though they may choose to do so.

Once participating companies have completed their base-year inventory and IMP, the USEPA offers them a high degree of flexibility in the development of a GHG reduction goal. The goals, according to the USEPA (2007), must meet the following five requirements:

- The goal must be corporate-wide

- It must be based on the most recent base year for which data is available

- It must be achieved over a five- to ten-year time horizon

- It must be expressed as absolute GHG reduction or a decrease in GHG intensity

- It must be 'aggressive' compared with overall project performance in the sector

While companies set their own reduction goals, the USEPA evaluates these goals based on model-derived sectoral benchmarks[2] and the company's current performance. The USEPA expects that companies will choose a reduction goal that places them well ahead of the projected emissions improvement rate for their respective sector.

To facilitate this process, the USEPA provides partner companies with a range of technical assistance including:

- Support in choosing and integrating best practices

- Setting appropriate organisational boundaries

- Identifying and quantifying direct and indirect emissions

While the programme is most appealing to large companies that receive substantial public attention and have the capital to invest in abatement technologies, it has also drawn in many smaller companies seeking to benefit from information sharing with the larger companies.

The USEPA markets the programme to companies under the proposition that pollution prevention pays, suggesting that companies can make a contribution to environmental improvement while simultaneously increasing their profits by reducing waste and streamlining many aspects of their operation. The USEPA strongly recommends that partners publicise their participation and their GHG reduction pledges. In addition, Climate Leaders provides a range of promotional materials (e.g. press release templates and public service announcements) which companies are encouraged to use.

Evaluation of Climate Leaders

In this section I use a multi-criteria approach to evaluate the USEPA Climate Leaders programme adopting a framework developed by Sullivan (2005). The framework assesses programmes based on eight factors:

- Environmental effectiveness

- Efficiency

- Transaction costs

---

2 For commercial and industrial companies, the USEPA employs the US Department of Energy's National Modeling System (NEMS) and the Bureau of Labor Statistics (BLS) forecasting tables for energy intensity improvement. The USEPA uses the Integrated Planning Model developed by ICF Resources Inc to project GHG emissions from electricity generators.

- Soft effects
- Innovation
- Acceptability
- Inclusiveness and public participation
- Law and public policy issues

Environmental effectiveness

In evaluating the effectiveness of the Climate Leaders programme, there are three important points related to disparity in goals, accountability and measurement. First, there is a great disparity in the GHG reduction goals and performance indicators chosen by partner companies. While goals stated as absolute reductions allow for comparison to a base year, many companies track their progress in terms of GHG intensity, which is measured as the ratio of GHG emissions over some physical or economic normalising factor – typically a unit of production or a financial metric such as turnover. Because intensity metrics are impacted by other variables, there is the potential to obscure the actual GHG emission reductions achieved through abatement and improvements in energy efficiency. For instance, higher electricity prices or changes in the company's operations may artificially result in reductions.

There is also a great deal of variation in the time-frame over which partners seek to meet their goals and the geographic range to which they apply them. For instance Xerox Corporation pledged to reduce its total global GHG emissions by 10% below 2002 levels from 2002 to 2012, whereas General Motors pledged to reduce total GHG emissions from its North American facilities by 10% from 2000 to 2005. Although partner companies are not required to extend their GHG reduction commitments to cover international operations, some may do so in anticipation of stricter regulation abroad (USEPA 2007). Others may have large unregulated operations overseas and join Climate Leaders to deter criticism through participating in an initiative which may have little impact on overall operations.

Second, there is currently no written policy on how to deal with partner companies that are not making progress towards their stated GHG reduction goals, nor is there a policy on goal-setting. The USEPA recognises that setting goals can take up to two years to complete. A 2006 study by the Government Accountability Office (GAO) found that, as of November 2005 (i.e. three years after the programme's initiation), only 38 of the 74 participants had established their GHG reduction goals, 13 of which were charter members (GAO 2006). Currently, of the 158 total partner companies, only 81 have set a reduction goal. Weak enforcement of programme goals and failure to monitor progress can encourage free-riders; these companies may benefit from recognition without making any real progress on GHG emissions. In addition to the fairly relaxed monitoring requirements, the programme allows companies to include recorded reductions made as far back as 1990, which may result in exaggerated progress. Furthermore, these reductions may be attributable to shrinking product volume, a downsizing of operations or the moving of production overseas rather than actual GHG emission reduction actions.

Lastly, the USEPA cannot quantify accurately the total GHG emission reductions attributable to Climate Leaders at the corporate level because many of the participating companies are involved in other partnership programmes. This creates significant overlap and may lead to double-counting of reductions. The GAO study found that, in 2005, 60 out of the 74 Climate Leaders partners were in another voluntary programme and some were participating in as many as four other programmes (GAO 2006). There is also no way to measure the overall success of the programme. The USEPA estimates that the first 50 companies to join Climate Leaders account for 8% of US GHG emissions. The USEPA stated that updated performance measures would be provided in 2006, but those were still unavailable at the time of writing (October 2007).

Efficiency

Although data on the actual cost efficiency is limited to a few company-specific case studies, it is important to recognise that many of the participating companies have joined Climate Leaders in anticipation of stricter emission regulations from Congress in the future. Acting voluntarily, the partner companies are able to make the transition toward a lower carbon footprint in an atmosphere that encourages technical assistance, information and resource sharing, and relationship building. These companies see themselves as first movers, well equipped with a competitive advantage if and when new federal legislation is passed by Congress. Many act now so that they can make GHG reductions in the most cost-effective way and perhaps avoid having to adopt expensive prescribed technologies later. However, not all companies see the benefits of Climate Leaders in this way; many companies probably join solely for the positive recognition that is afforded to them through participation (Fiorino 2006). This may be why the programme, which was introduced by the Bush administration with such great optimism, has failed to entice a large number of participants even though, unlike Performance Track (another USEPA voluntary programme), there are no clearly defined prerequisites for joining.

In practice, many companies have benefited from both the recognition associated with participation in Climate Leaders and the cost savings associated with reduced energy use in their operations. For example, IBM pledged to reduce its average annual CO_2 worldwide emissions associated with energy use by 4% and to reduce its PFC emissions by 10% over the period 2000–2005. According to the USEPA (2007), IBM has met these goals and saved over $100 million since 1998 due to new energy conservation practices. SC Johnson, which pledged to reduce its US GHG emissions by 23% per pound of product from 2000 to 2005, has generated annual savings of $2.6 million in energy costs (USEPA 2007).

But how representative are the savings detailed above? Inevitably, those companies that achieve significant cost savings will tend to publicise this. However, it is not clear how systemic the benefits are because information on cost savings is released at the discretion of the companies themselves. As discussed above, there are a range of factors beyond cost savings that motivate or allow companies to join the programme.

For instance, experience with voluntary programmes indicates that larger companies are more likely to join because they have the capital to invest in abatement. They are also typically more visible and thus under more public pressure to take action. Smaller companies often join to learn from the larger companies and to gain access to technology sharing. 'Dirty' companies are also likely to join because they often have a great deal of 'low-hanging fruit' – quick and easier abatement options that can be accomplished inexpensively to yield impressive results (Videras and Alberini 2000; OECD 2003). Companies that have moderately 'clean' operations and new technology face a steeper climb to upgrade further and achieve similar reductions of a comparable percentage.

Transaction costs

While hard data on the transaction costs associated with Climate Leaders are not currently available, the transaction costs associated with participating in the programme appear to be fairly low for both the USEPA and the partner companies. Certainly the costs are lower than what could be expected under a federally legislated command-and-control regulatory approach to CO_2.

This reflects the general experience with voluntary programmes, which tend to be characterised by lower transaction costs because there is no permitting process, significantly less reporting and, as is the case with Climate Leaders, third-party verification of emissions is optional (Fiorino 2006). In addition, participating companies eliminate the costs incurred through lobbying and participating in the USEPA rulemaking sessions that typically accompany new environmental legislation. Finally, because the programme is not binding or backed by legislation, it cannot be legally challenged. As a consequence, the USEPA and partner companies avoid the costs of litigation involving national environmental groups, which has become increasingly common in command-and-control regimes in the USA.

The most significant transaction costs for companies occur within the first year of participation, during which partners conduct their base-year emissions inventory and prepare their IMP. This process demands close collaboration with the USEPA. During year one, the USEPA pledges 80 hours of technical assistance to partners that are completing their inventory and IMP (USEPA 2007). In subsequent years, the USEPA provides up to ten hours of technical assistance annually.

Soft effects

The Climate Leaders programme has been effective in providing soft benefits to both the USEPA and participating companies. The programme has the capacity to:

- Build trust and new working relationships
- Offer lessons and new ways to approach regulatory problems
- Create an atmosphere where new combinations of regulatory approaches can be tested

It also helps to dispel, or at least challenge, the commonly held belief that potential win–win situations are not feasible and that economic and environmental objectives are inevitably in conflict with each other (Fiorino 2006). Of the numerous soft benefits of this programme, the greatest may be the opportunity for industry and government to reach consensus on climate change and evolve from an adversarial relationship towards one that is more collaborative.

Partner companies also benefit from increased recognition, development of a strong relationship with the government, access to tools and information, and insider status as policy evolves (USEPA 2007). All these aspects of Climate Leaders serve to create an atmosphere in which policy formation is more inclusive and predictable. This point is important because future GHG regulation in the USA is likely resemble the structure and objectives developed through Climate Leaders. At the very least, voluntary programmes will remain an important part of the policy mix, as evidenced by the large increase in the number and scope of voluntary programmes during the Bush administration (Abraham 2004). The consequence is that participating companies may achieve first-mover status and may be able to more swiftly adapt to new regulations. In addition to domestic regulation, many companies have branches abroad where they may be facing new, stringent regulations on GHGs. Therefore having resources available to them (such as the support they receive through Climate Leaders) allows them to prepare to meet new regulations both domestically and abroad (Gillies 2003).

Innovation

One of the strengths of Climate Leaders is that it allows the USEPA to introduce and fine-tune GHG emission regulations on a voluntary basis without going through the cumbersome legal prescriptions that characterise the introduction of new environmental law. The USEPA is able to use the experience to inform and sculpt future regulations based on feedback from industry.

The importance of this benefit for the regulator should not be understated. The case of hazardous air pollutant (HAP) regulation under the Clean Air Act is a prime example: in this case, poorly delineated stipulations and unrealistic goals set by the USEPA caused a massive backlash and virtual paralysis of the USEPA's regulatory capacity due to a barrage of legal challenges (Godish 1997). This experience broke down the relationship between government and industry because the USEPA did not recognise the economic and technical reality of the regulatory situation — hazardous air pollutants were a necessary by-product of many essential products and services. This example parallels closely the economic entrenchment of GHG emissions. By fostering relationships and collaboration through Climate Leaders, future regulations may avoid the backlash that HAP regulation created in the 1980s.

At the corporate level, Climate Leaders has contributed to the advancement and dissemination of corporate climate change management practices and the development of more rigorous accounting standards, management plan tools and innovative reduction opportunities (USEPA 2007). Climate Leaders has also resulted in numerous innovations in the procedures used for conducting inventories of corporate GHG emissions.

One example is the Kaizen approach developed by Eastman Kodak. Under Kaizen, Kodak deploys small internal energy audit teams composed of technical experts, systems specialists and operations personnel to systematically assess plant operations and make recommendations to senior management on how improve efficiency in the most cost-effective way. Kodak estimates that by making many simple, investment-free changes to operations it will save over $1 million annually on electric energy usage (Climate Northeast 2006). By creating standardised accounting protocols that are easy to employ and readily available and an atmosphere conducive to innovation and information sharing, the USEPA has been able to draw in many smaller companies that are eager to make reductions but were previously hesitant because they lacked the capacity. Many GHG reporting and management tools are freely available on the Climate Leaders website,[3] increasing the transparency of the accounting methods and making them accessible to companies that are not Climate Leaders members but may be considering joining.

Acceptability

In general, industry opposes any new environmental legislation that it perceives as having the potential to increase operating costs and/or reduce managerial discretion. This is because both these changes often lead to lower revenues. As such, the traditional view is that companies would only take actions to go beyond compliance when there was some way in which they could themselves profit. This is changing though and there are now members of industry that are ambitious and genuinely committed to environmental improvement as a goal. More generally, and perhaps more significantly, others have recognised the potential cost benefits. In addition, companies emitting large amounts of GHGs have come under increasing pressure from both the local and international community.

Given the mass public attention surrounding climate change, it can be argued that the consumer is central in determining the acceptability of Climate Leaders, and the need to attain and retain their valuable 'social licence to operate' may serve to drive 'dirty' companies towards and beyond compliance (Gunningham et al. 2004). The USEPA is aware of this and has sought to influence corporate behaviour indirectly by making Climate Leaders a highly publicised voluntary programme. Accordingly, companies that are part of Climate Leaders receive support from the USEPA in marketing themselves as environmentally sound companies, potentially offering a competitive advantage over other companies in the sector (Videras and Alberini 2000; Karamanos 2001; Alberini and Segerson 2002).

The USEPA favours voluntary programmes because they allow it to collect data and fine-tune aspects of its approach, disperse information, increase communication and make modest progress on environmental improvement. As alluded to above, the USEPA has welcomed the opportunity to work outside the prescriptive and difficult to interpret confines of Congressional legislation (Lyon and Maxwell 2004).

But as Lyon and Maxwell (2004) point out, environmental NGOs dislike voluntary programmes because they do not include a mechanism for other stakeholders to get involved. As such, environmental groups lose the opportunity to influence policy through the traditional avenues of Congressional hearings and litigation. Many environmental groups view voluntary programmes as a way of pre-empting legislation with real teeth. Environmental NGOs typically favour command-and-control regulations because they are a more direct way to address pressing environmental problems. Congress, for its part, often finds voluntary programmes appealing because such programmes help to head off any criticisms that may arise due to perceived inaction on climate change or the shortcomings of environmental bills that are passed.

3 www.epa.gov/stateply

Inclusiveness and public participation

Policy becomes much more predictable through the relationships forged through Climate Leaders. But, despite attempts to establish a common ground that all stakeholders can stand on, as Lyon and Maxwell (2004) assert, voluntary approaches may actually have the effect of isolating two important actors: namely, Congress and national environmental groups. It is also important to consider the role of the American public, who through their impact on the corporate social licence to operate and through consumer choice have a strong influence on corporate practices.

While Climate Leaders has been criticised for not carrying formal legal sanctions for non-compliance, the publicity around companies' commitments and their performance against these commitments offers the potential for the public to hold these companies to account — in particular through the negative publicity that may be associated with any failure to meet these commitments.

Law and public policy issues

Climate Leaders is an inexpensive way to catalyse corporate action on climate change, particularly given that the USEPA has a constrained budget. Success in delivering on, or contributing to, the US's climate change objectives through this type of programme may:

● Serve as important means for the USEPA to legitimise its regulatory approach

● Allow the USEPA to expand its understanding of corporate GHG emissions

● Allow the USEPA to make a strong case for increased funding

But where does Climate Leaders fit into the broader mandate and regulatory programme of the USEPA? A USEPA study completed in 1999 found that voluntary programmes received less than 1% of the agency's resources (Fiorino 2006), with the majority going to Energy Star. This left a very small portion of the pie to be distributed between Climate Leaders and the numerous other voluntary programmes. Currently, Climate Leaders operates on about $2 million per year — a small portion of the USEPA's budget (USEPA 2007). Given the centrality of voluntary approaches to the current Bush administration's strategy on climate change, it is likely that funding for Climate Leaders will remain secure until through the end of President Bush's tenure. However, future funding for the programme will almost certainly depend on the USEPA's ability to demonstrate its effectiveness.

It is likely that voluntary approaches will become a more prominent component of the US policy mix over time. Because voluntary programmes do not involve the long process of Congressional legislation, they allow the USEPA to respond very rapidly to issues such as corporate GHG issues in a non-litigious way. While the reality is that soft benefits from such programmes will inevitably outweigh the hard emission reductions benefits, Climate Leaders provides an excellent foundation for stronger regulatory arrangements in the future.

Discussion and conclusions

Although a number of companies have met their pledges, it remains difficult to state conclusively whether or not Climate Leaders has been successful in actually reducing corporate GHG emissions. Despite their pledges of significant cuts, many companies remain among the largest US GHG emitters, while the metrics used by other companies to track their progress can be misleading. Even though a primary objective of the programme is to create hard impacts on GHG emissions, the USEPA to date has placed greater emphasis on rewarding, acknowledging and supporting companies that are making progress in the right direction. As such, the success of Climate Leaders may actually be linked to the various soft impacts that the programme offers.

The contentious nature of climate change means that many US politicians are reluctant to introduce stringent, new environmental regulations targeting greenhouse gas emissions. This has lead to assertions that voluntary programmes such as Climate Leaders are simply attempts to avoid enacting tougher policy. Although this may be true, it is important to recognise the role of voluntary programmes such as Climate Leaders as catalysts for future policy. While there is limited evidence that the programme has had any significant hard impacts on GHG emissions in the USA — e.g. the GAO estimates that emissions from participating companies account only for 6% of total annual GHG emissions in the USA (GAO 2006) — voluntary programmes such as Climate Leaders have an important role when there are no stronger tools available, as is currently the case in the USA (see, in general, Segerson and Miceli 1998; Sullivan 2005; Fiorino 2006).

Climate Leaders has provided real benefits by:

● Introducing greater flexibility

● Fostering a less adversarial relationship between government and industry

● Facilitating stakeholder collaboration toward common goals

These components of may provide important building blocks for US GHG regulation in the future.

References

Abraham, S. (2004) 'The Bush Administration's Approach to Climate Change', *Science* 305: 616-17.

Alberini, A., and K. Segerson (2002) 'Assessing Voluntary Programs to Improve Environmental Quality', *Environmental and Resource Economics* 22.1–2: 157-84.

Climate Northeast (2006) 'Innovation at Kodak: How GHG Reporting and New Management Practices Save Energy and Money' (Washington, DC: World Resources Institute; www.climatenortheast.org/pdfs/2006-WRI-CNE-CS-Kodak1.pdf [accessed 24 March 2008]).

Fiorino, D. (2006) *The New Environmental Regulation* (Cambridge, MA: MIT Press).

Franklin, P., C. Koshland, D. Lucas and R. Sawyer (2000) 'Clearing the Air: Using Scientific Information to Regulate Reformulated Fuels', *Environmental Science and Technology* 34.18: 3,857-63.

GAO (United States Government Accountability Office) (2006) *EPA and DOE Should Do More to Encourage Progress under Two Voluntary Programs* (Report to Congressional Requesters GAO-06-97; Washington, DC: GAO; www.gao.gov/new.items/d0697.pdf [accessed 29 April 2007]).

Gillies, A. (2003) 'Greenhouse gas soft sell', *Forbes Online*, 26 November 2003; www.forbes.com/2003/11/26/cz_ag_1126beltway.html (accessed 29 April 2007).

Godish, T. (1997) *Air Quality* (Boca Raton: Lewis Publishers).

Gunningham, N., R. Kagan and D. Thornton (2004) 'Social License and Environmental Protection: Why Businesses go Beyond Compliance', *Law and Social Inquiry* 29.2: 307-41.

Harrington, W., R. Morgenstern and T. Sterner (ed.) (2004) *Choosing Environmental Policy: Comparing Outcomes in the United States and Europe* (Washington, DC: Resources for the Future).

Karamanos, P. (2001) 'Voluntary Environmental Agreements: Evolution and Definition of a New Environmental Policy Approach', *Journal of Environmental Planning and Management* 44.1: 67-84.

Kerr, R. (1998) 'Acid Rain Control: Success on the Cheap', *Science* 282.5391: 1,024

Lyon, T., and J. Maxwell (2004) *Corporate Environmentalism and Public Policy* (Cambridge, UK: Cambridge University Press).

OECD (Organisation for Economic Cooperation and Development) (2003) *Voluntary Approaches in Environmental Policy: Environmental Effectiveness, Efficiency, and Usage in Policy Mixes* (Paris: OECD).

Segerson, K., and T. Miceli (1998) 'Voluntary Environmental Agreements: Good or Bad News for Environmental Protection?', *Journal of Environmental Economics and Management* 36: 109-30.

Sullivan, R. (2005) *Rethinking Voluntary Approaches in Environmental Policy* (Cheltenham, UK: Edward Elgar).

USEPA (US Environmental Protection Agency) (2007) EPA Climate Leaders website; www.epa.gov/stateply/index.html (accessed 30 April 2007).

Videras, J., and A. Alberini (2000) 'The Appeal of Voluntary Environmental Programs: Which Companies Participate and Why?', *Contemporary Economic Policy* 18.4: 449-61.

8
Ten years of the Australian Greenhouse Challenge: real or illusory benefits?*

Rory Sullivan
Insight Investment, UK

From 1995 to 2005, the voluntary Greenhouse Challenge formed the centrepiece of the Australian government's efforts to encourage business to take action on greenhouse gas emissions and climate change. The election of the Labour Party in 2007 promised a dramatic change in the Australian climate change debate, with Labour's campaign pledges including the ratification of the Kyoto Protocol (which it signed the day after the election results were announced) and the establishment of a national emissions trading scheme (proposed for 2010).

With the likelihood of stronger policy action on climate change in coming years, this chapter focuses on the outcomes that were achieved by the Greenhouse Challenge over the period 1995–2005 and examines whether the Greenhouse Challenge has a role to play in the new Australian policy environment.

The chapter is divided into four parts. The first is a brief overview of Australian climate change policy since 1995. The second is a description of the Greenhouse Challenge and the outcomes that it delivered. The third is a brief review of the Greenhouse Challenge Plus, which was introduced in 2005 to replace the Greenhouse Challenge. The last part considers the role of Greenhouse Challenge Plus in preparing Australian business for the stronger climate change policy measures that now seem inevitable.

---

* This chapter is based on research into the Greenhouse Challenge conducted for the author's PhD. This research was published as Chapter 6 in Sullivan 2005 and, in an updated form, in Sullivan 2006, and a subsequent research project (Sullivan 2007) for the United Nations Development Programme (UNDP) as a contribution to the 2007/2008 *Human Development Report* (UNDP 2007).

The Australian policy context 1995–2005: an overview

Over the past two decades, Australian industry has consistently argued that the Australian government should pursue only policies that are flexible and cost-effective in their own right and that have the least negative impact on competitiveness, investment, regional development and jobs (see, for example, AIGN 1999; BCA 1999; Knapp 2004; PACIA 2004).[1] This lobbying — together with the heavy dependence of the Australian economy on the mining and minerals industries, the reliance on coal-fired power generation (with over 75% of Australia's electricity generated by coal), Australia's large agriculture sector and its heavy dependence on long-haul transport — strongly influenced the Australian government's approach to international negotiations. Even though it was one of the few countries allowed to increase its greenhouse gas emissions under the Kyoto Protocol, Australia consistently refused to ratify the Protocol until agreement had been reached on issues such as the rules governing the Kyoto Protocol's flexibility mechanisms and the emissions targets to be achieved by developing countries (AGO 2004, 2006). In 2005, Australia joined the USA, China, India, Japan and South Korea to form the Asia–Pacific Partnership on Clean Development and Climate with the aim of developing global agreements on climate change based on clean technology development and deployment rather than the emissions targets approach used in the Kyoto Protocol.

Over the period 1992–2005, domestic climate change policy focused on 'no regrets measures'; a 'no regrets measure' was defined as:

> a measure that has other net benefits (or, at least, no net costs) besides limiting greenhouse gas emissions or conserving or enhancing greenhouse gas sinks (AGO 1998).

That is, the emphasis of policy was on encouraging Australian industry to contribute to reducing greenhouse gas emissions while not threatening Australia's international competitiveness. The government consistently emphasised that it would not introduce policy measures such as emissions trading in the absence of effective longer-term global action on climate change (e.g. see Commonwealth of Australia 2004). Despite this focus, the Australian government committed around two billion dollars (Australian) to climate change issues over the period 1997–2005 (AGO 2006).The government's Climate Change Strategy[2] incorporated a mix of policy measures including:

● Consumer and corporate education

● Voluntary corporate participation in emissions reduction activities

● Seed funding for renewable energy innovations

1 It is interesting to note that industry views were not homogenous (particularly in recent years) with, for example, BP calling on members of the Business Council of Australia (BCA) to adopt individual plans to cut greenhouse gas emissions (Wilson 2005) and the CEOs of a number of major Australian businesses — BP Australasia, Insurance Australia, Origin Energy, Swiss Re, Visy Industries, Westpac — calling for early action on climate change (Australian Business Roundtable on Climate Change 2006).

2 The Strategy, which consolidated previous climate change initiatives, was articulated in measures contained in the 2004/05 Federal Budget (Department of Environment and Heritage 2004) and the 2004 Energy White Paper (Commonwealth of Australia 2004).

- Mandatory standards for power generation, energy use efficiency and vehicle fuel efficiency

- The mandatory uptake of new renewable energy in power supply

- Research and policy development into sinks and emissions

- Fostering growth in plantation forestry and native vegetation

These measures were projected to deliver greenhouse emissions abatement of 87 million tonnes of carbon dioxide equivalents [MT CO_2(eq)] by 2010 (AGO 2006); for a breakdown of the expected emissions abatement from the different programmes and policy measures, see AGO 2005a: 60-66.

In parallel to the Commonwealth government's activities, each State and Territory established a greenhouse strategy to address those issues with a bearing on climate change (e.g. waste management, planning and development of power plants, land use and transport planning and vegetation management) which fell under its jurisdiction (see, for example, Government of Western Australia 2004; New South Wales Greenhouse Office 2005; State of Victoria 2005).

Australia's greenhouse gas emissions profile

In 2004, Australia's greenhouse gas emissions totalled 559.1 MT CO_2(eq) — an increase of 2.3% over the 1990 levels of 547.1 MT CO_2(eq) (AGO 2007a). Table 8.1 breaks these emissions down by sector.

TABLE 8.1 Changes in CO_2(eq) emissions and removals by sector (1990–2004)

Source: AGO 2007a

Sector	MT CO_2(eq)		Change (%)
	1990	2005	
Energy:	287.0	391.0	36.3
Stationary energy	196.0	279.4	42.6
Transport	61.9	80.4	29.9
Fugitive emissions	29.1	31.2	7.3
Industrial processes	25.3	29.5	16.5
Agriculture	87.7	87.9	0.2
Land-use, land-use change and forestry	128.9	33.7	−73.9
Waste	18.3	17.0	−6.9
Net emissions	**547.1**	**559.1**	**2.2**

The reduction in the emissions from the 'land-use, land-use change and forestry' category reflects the significant reductions in the rate at which Australian forests have been converted to agricultural or other land uses.[3] If these emissions are excluded from the inventory, Australia's greenhouse gas emissions would have increased by 20.4% (from 418.2 to 525.4.2 MT CO_2(eq)) over the period from 1990 to 2004. The inclusion of emissions from 'land-use, land-use change and forestry' in the accounting of Australia's greenhouse gas emissions has been controversial, with environmental non-governmental organisations (NGOs) arguing that this focus diverted policy attention and resources away from energy efficiency and greenhouse gas emission reduction initiatives (see, for example, ACF 1999).

Australia's greenhouse gas emissions are projected to reach 599 MT CO_2(eq) in the period 2008–2012. This figure is equal to Australia's Kyoto Protocol target (Department of Climate Change 2008) although, as with the data in Table 8.1, these numbers incorporate the reduction in emissions from 'land-use, land-use change and forestry' category. These projections incorporate the contributions from the various greenhouse gas emission reduction measures adopted by the Commonwealth, State, Territory and local governments.

The Greenhouse Challenge

Overview

Australian industry approached the Commonwealth government in 1995 with a proposal for a voluntary greenhouse gas abatement programme. The primary motivation was the threat that a carbon tax would be introduced to enable Australia to meet its commitments under the United Nations Framework Convention on Climate Change (UNFCCC) (AGO 1999; Parker 1999; Lipman and Bates 2002).

The Commonwealth government subsequently established the Greenhouse Challenge in 1995 as a voluntary programme for public and private sector organisations to undertake and report on their actions to abate greenhouse gas emissions. In line with the government's no regrets approach, the overall aim of the Greenhouse Challenge programme was to achieve the maximum practicable greenhouse gas emissions abatement while not compromising business objectives such as development and growth (Howard 1997).

Organisations wishing to participate in the Greenhouse Challenge were required to work through a six-step process, namely:

* Establish an inventory of greenhouse gas emissions with sufficient detail to identify all significant sources of emissions. Thereafter, emissions inventories were required to be prepared annually and to include an assessment of the factors that had influenced changes in emissions from previous inventories

---

3 For more information on Australia's approach to forest management, see www.greenhouse.gov.au/ncas. However, there are significant uncertainties in the contributions of carbon sequestration to Australia's total greenhouse gas emissions; see, for example, Macintosh 2007a, 2007b.

● Develop an action plan to minimise greenhouse gas emissions or enhance greenhouse sinks

● Forecast expected reductions in greenhouse gas emissions including estimates of uncertainties and an assessment of the factors that could influence changes in emissions

● Sign a Cooperative Agreement with the Australian government. Cooperative Agreements were expected to include an emissions inventory, an assessment of the opportunities available for abating greenhouse gas emissions, a greenhouse action plan and a commitment to regular monitoring and reporting of performance. The Greenhouse Challenge did not involve government imposing specific abatement targets on participating organisations (Commonwealth of Australia 2001). Rather, organisations negotiated their planned actions and expected reductions in greenhouse gas emissions with the Australian Greenhouse Office (AGO)[4]

● Monitor and report regularly on greenhouse gas emissions against targets, including an assessment of the effectiveness of policies and measures to improve energy and process efficiencies, abate emissions and enhance sinks

● Be open to independent verification. The verification process focused on those aspects that could be objectively verified (i.e. emissions inventories, actions reported as being undertaken and the accuracy of claimed greenhouse gas emission reductions). However, the verification process did not assess the adequacy of management systems nor whether all practicable actions had been undertaken. An inventory could be verified as 'materially accurate' if the verifier could confirm that the reported and actual inventories (baseline, projections and planned abatement actions) were within a 10% materiality threshold (AGO 2000a, 2000b)[5]

In addition to agreements with individual organisations, facilitative agreements (where bodies such as industry associations agreed to support and actively encourage their members to join the Challenge) and aggregate agreements (generally made by an industry sector through an industry association) could also be agreed under the Greenhouse Challenge.

The participants in the Greenhouse Challenge were allowed to use the Greenhouse Challenge logo to advertise their participation in the programme. Organisations could withdraw from the Greenhouse Challenge, without sanction, at any time (Parker 1999; AGO 2000c). The AGO stated that forecasts of emissions abatement would not be inter-

---

4 The AGO was the Australian government's lead agency with responsibility for delivering policy advice and implementing programmes relating to climate change. The Department of Climate Change, created in December 2007, is now the lead agency in these matters; www.greenhouse.gov.au.

5 There were three rounds of independent verification: a pilot programme in 1998 and two complete rounds in 2000 (AGO 2001a) and 2002 (AGO 2003a). Because of industry concerns regarding confidentiality, the report of the first complete round of verifications simply stated whether the reported inventory and actions undertaken were materially accurate or not. Greater information was provided in the report of the 2002 verification process with the report detailing the percentage error (in situations where there was a material discrepancy), but not providing information on the actual greenhouse gas emissions from the organisations involved.

preted as, or used to, set targets and that no penalty would apply where forecasts were not achieved.

The Greenhouse Challenge requirements were revised slightly in 2003 by the Joint Consultative Committee that oversaw the Greenhouse Challenge and which consisted of industry and government representatives. The main changes were that participating organisations were required to prepare an annual public statement on the undertakings contained in their agreements, including details of both absolute (i.e. bulk total) and relative (e.g. efficiency) performance. This was a change from the original Greenhouse Challenge where participating organisations were required only to report on the greenhouse gas emissions abatement achieved. Participating organisations were also expected to develop appropriate key performance indicators to allow their emissions intensity to be calculated and tracked.

Environmental effectiveness: performance against specified targets

It was envisaged that 500 companies would have signed Cooperative Agreements by the end of 2000 and that 1,000 companies would have signed by the end of 2005 (Howard 1997). This latter target was abandoned in 2004, when the AGO stated that the introduction of a mandatory requirement to join the Greenhouse Challenge Plus (discussed further below) meant that the target was no longer a useful indicator of progress (AGO 2004).

The target for the end of 2000 was met and, over 750 organisations were members as at the end of 2006.[6,7] The programme covered over 40% of Australia's greenhouse gas emissions (Department of the Environment and Water Resources 2007), with almost total coverage in a number of major industrial sectors including electricity generation and distribution, oil and gas extraction, iron and steel, aluminium and coal mining.

The original projection was that the actions taken under the Greenhouse Challenge would achieve a total emissions abatement of 15 MT CO_2(eq) by 2000 (AGO 1999). In the years leading up to 2000, however, the Australian government revised these numbers, predicting that an emissions abatement of 23.5 MT CO_2(eq) would be achieved (AGO 1999). In fact, the actual emissions abatement achieved was 19.2 MT CO_2(eq) (AGO 2002).

The reason for the discrepancy between the predicted and actual emissions abatement appears to be that many participating organisations did not implement all the actions in their Cooperative Agreements (Sullivan 2005). For example, of the 76 Greenhouse Challenge participants that had submitted progress reports by mid-2000, only eight had met their original forecasts for emissions abatement (Parliament of the Commonwealth of Australia 2000).

But, because the independent verification process did not investigate the reasons why organisations did not achieve their forecast emissions abatement, the data required to understand the reasons why certain projects or actions were not implemented were

6 This was a reduction on the 824 members at 30 June 2003 (AGO 2003b). The AGO noted that some small and medium enterprises (SMEs) had withdrawn because of problems in meeting new programme reporting requirements (AGO 2004).

7 For a current list of members, see www.environment.gov.au/settlements/challenge/members/pubs/list_of_challengers.pdf (accessed 6 June 2008).

never systematically gathered. This approach to verification was probably intended to avoid potential criticisms of the Greenhouse Challenge participants and was consistent with the strictly voluntary approach that underpinned the programme.

Despite the emissions abatement achieved, greenhouse gas emissions from participating organisations still grew by 6.4 MT CO_2(eq) (or approximately 5%) in the period to 2000. In a number of sectors (oil and gas extraction, coal mining, food processing, textiles, petroleum refining, cement manufacturing and iron and steel production), emissions actually stabilised or reduced. However, these reductions were offset by increases in emissions from the aluminium and other (non-coal) mining sectors, as well as significant increases in emissions from electricity generation (AGO 2001b).

Environmental effectiveness: beyond business as usual?

Even though the Greenhouse Challenge was effective in terms of meeting its specified targets, the critical question is whether it resulted in the abatement of greenhouse gas emissions beyond those reductions that would have been achieved anyway.

Before addressing this question, it is first necessary to consider how emissions abatement is assessed. In broad terms, emissions abatement efforts can be assessed by comparing emissions against a historical baseline or by comparing expected emissions at a given point in time with what emissions might have been at the same point in time without action to abate emissions.

Future scenarios are generally defined either in terms of static efficiency measures or 'business as usual'. The static efficiency approach assumes that there will be no changes in efficiency and, therefore, future estimates are calculated based on the assumption that greenhouse gas emissions are directly proportional to production rates or other measures of business activity (e.g. profit). In contrast, the business as usual approach takes into account the efficiency changes that would have occurred in the normal course of business; large-scale economic models typically assume a rate of improvement of 1.0–1.5% per year. Even though the business-as-usual approach is often used for economy-wide projections of emission levels, these broad assessments cannot readily be extrapolated to the level of individual facilities or specific industry sectors (Krarup and Ramesohl 2000).

The Greenhouse Challenge relied on the static efficiency approach to predict greenhouse gas emissions. As energy efficiency generally improves over time, this approach would be expected to generate higher baselines than the business as usual approach.[8] That is, the claimed emissions reductions are likely to have overestimated the outcomes achieved. Furthermore, given that participating organisations were free to define their own baselines and business-as-usual performance, there was clearly the potential for participating firms to overstate their expected emissions growth as this would allow them to claim that they had achieved even greater reductions in emissions (Sullivan and Ormerod 2002).

8 In a review of the activities of the AGO, the National Audit Office (NAO) noted that the emissions reductions claimed for the Greenhouse Challenge did not take account of (a) what would have happened in the absence of the Greenhouse Challenge, (b) the effect of corporate environmental management systems, or (c) the effect of State and Territory greenhouse programmes (NAO 2004: 43, 70).

Returning to the question of whether or not the Greenhouse Challenge resulted in the additional abatement of greenhouse gas emissions, it appears that the Greenhouse Challenge had two major effects (Sullivan and Sullivan 2005). The first was to encourage some organisations to bring forward some energy-saving or greenhouse gas emission reduction projects. The second was to help participating organisations to identify opportunities that provided clear short-term financial benefits.

However, the fact that the Greenhouse Challenge did not alter the cost or benefit side of emission reductions actions (see the discussion below on economic efficiency), the lack of accountability for delivering on Greenhouse Challenge commitments and the limited focus of the verification process on emissions inventories and the actions that have been reported as undertaken (rather than on whether all practicable actions had been undertaken or the reasons for not taking action) meant that companies had no real incentive to take emission abatement actions beyond those they would have taken anyway.

Economic efficiency

There is limited published data on the costs and benefits of energy or greenhouse expenditures by Australian organisations (public or private sector) (Sullivan 2005). The data that are available provide limited evidence that the organisations participating in the Greenhouse Challenge went beyond a narrowly defined interpretation of the costs and benefits of greenhouse gas emission reduction measures.

The majority of the projects implemented were either low-cost projects or projects that provided very short payback periods (Sullivan 2005). While Greenhouse Challenge participants also reported broader business benefits such as an improved 'social licence to operate', early capture of low-cost abatement options and the adoption of structured approaches to greenhouse gas emissions abatement (see, for example, Beresford and Waller 2000; Parker 2002), there was limited evidence that these significantly influenced the decisions on specific projects.

In this context, the Greenhouse Challenge can be said to have been economically efficient in that it did not require companies to implement measures beyond those that could be clearly justified in economic terms (Sullivan 2005). A more critical conclusion could be that the Greenhouse Challenge did not provide the strong drivers necessary to encourage companies to take advantage of all the opportunities that might be available.

The Greenhouse Challenge appeared to have had little impact on investment criteria. Apart from some isolated cases, a broader shift of investment attitude (e.g. relaxation of payback requirements) could not be observed. Indeed, it could even be argued that the Greenhouse Challenge perpetuated a 50% rate of return as an 'acceptable' target for energy investments. For example, a project conducted by the Plastics and Chemicals Industry Association (PACIA) to identify greenhouse gas emission reduction opportunities in the chemical industry involved the AGO funding a technical consultant to advise on energy-saving opportunities. The opportunities identified were those with a payback period of two years or less (Rex 2000; Sullivan 2005). This expected rate of return is significantly greater than in industries such as energy, oil, gas and mining, which typically expect large capital investments (e.g. new power generating equipment) to provide a rate of return of 15–20%. While such a comparison may not be strictly fair for SMEs (where the cost of capital may be an important barrier to investment), it suggests that opportunities for energy or environmental performance improvements that were economically viable (and relatively risk-free) were not being implemented.

Soft effects

The lack of management commitment to (or interest in) energy or greenhouse management issue has historically been a barrier to enabling Australian organisations to manage these issues effectively (Sullivan and Sullivan 2005).

One of the key contributions of the Greenhouse Challenge was to put the reduction of greenhouse gas emissions explicitly onto business decision-making agendas. Many of the participating organisations reported that greenhouse gas emissions had become an important part of their decision-making processes because of their participation in the Greenhouse Challenge (Sullivan 2005). Participating organisations also reported other changes such as (AGO 1999):

- The appointment of staff with responsibility for greenhouse issues
- The provision of training on greenhouse gas emissions for staff
- The provision of awards for staff to recognise excellence in greenhouse emissions abatement
- Skills development in relation to the development of action plans and emissions inventories

In addition, some Greenhouse Challenge participants reported that they had taken action to influence parties outside their organisation through designating other Challenge participants as preferred suppliers, including greenhouse gas emissions as a consideration in tender documents and marketing the benefits of the Challenge to suppliers (AGO 1999).

Stakeholder perspectives on the Greenhouse Challenge

Australian industry strongly supported the Greenhouse Challenge, arguing that its advantages (e.g. flexibility, lower costs, improved management practices, structured processes for measuring, reporting and forecasting emissions at the enterprise level) meant that stronger regulatory measures to reduce greenhouse gas emissions would not be required (see, for example, Beresford and Waller 2000; AIGN 2005). Notwithstanding strong public statements of support for the Greenhouse Challenge, Australian industry was actually quite sceptical of the value of the Greenhouse Challenge. For example, in interviews industry association representatives noted (Sullivan 2005):

> The Greenhouse Challenge is really a bit of a joke. Lots of people have done the easy stuff. The only effect of the Challenge is to bring these forwards a bit. Our industry is struggling with the next generation of performance improvements as the easy wins are gone.

> Basically, all the Greenhouse Challenge does is to codify what companies would be doing as part of their EMSs [environmental management systems] anyway. It's a bit more paperwork but if it helps us avoid carbon taxes or more regulations, it's worthwhile.

Despite consistent support from the Commonwealth government for the Greenhouse Challenge, a Parliamentary Inquiry into Australia's greenhouse performance (Parliament of the Commonwealth of Australia 2000) was extremely critical of the effectiveness of the Greenhouse Challenge. The inquiry concluded that there were significant limitations in the Greenhouse Challenge, specifically that:

● The Greenhouse Challenge did not distinguish between the reduction of emissions from normal business improvements and emission reductions as a result of government investment in specific programmes

● There were no penalties for companies that did not meet agreed targets

● Sector-specific abatement targets or benchmarks were not specified

● Only a small number of companies appeared to be meeting their forecast emissions abatement

● There were no incentives for 'beyond no regrets' measures

However, it was recognised that the Greenhouse Challenge had provided a range of soft effects including:

● Raising expertise in emissions abatement

● Creating CEO support for improving energy efficiency

● Stimulating the development and implementation of practical efficiency measures

● Prompting the development of methodologies for greenhouse gas emissions abatement

The Commonwealth government rejected all these criticisms, arguing that the Greenhouse Challenge was an international role model for the inclusion of business in greenhouse abatement (Commonwealth of Australia 2001).

In contrast to the public support from government and industry, environmental groups were extremely critical of the Greenhouse Challenge, arguing that it was simply a public relations campaign for activities that would have happened anyway (Sullivan 2005). Concern was expressed about the 'closed shop' and 'cosy relationship between government and industry' nature of the Greenhouse Challenge, in particular the emphasis on the confidentiality of industry data (Parliament of the Commonwealth of Australia 2000) and the absence of a formal role for NGOs in the Joint Consultative Committee overseeing the Greenhouse Challenge.

Concluding comments on the Greenhouse Challenge

The Greenhouse Challenge formed the centrepiece of the Australian government's efforts to encourage business to take action on greenhouse gas emissions and climate change for approximately ten years. By its own measures, the Greenhouse Challenge was successful — achieving significant coverage of Australian greenhouse gas emissions and contributing to many companies and industry sectors stabilising their greenhouse gas emissions.

The flexibility provisions in the Greenhouse Challenge were welcomed by industry as enabling cost-effective approaches to greenhouse gas emissions abatement to be implemented. The Greenhouse Challenge also provided a range of important soft effects, in particular making climate change issues a part of companies' decision-making processes.

Although the Greenhouse Challenge did enable companies to identify 'easy wins' or projects with very short payback periods, its overall effectiveness remains open to question. In particular, the Greenhouse Challenge did not provide strong incentives for participating organisations to set ambitious greenhouse gas emission reduction targets beyond business as usual. Furthermore, the existence of the Greenhouse Challenge allowed industry to deflect calls for the introduction of stronger policy measures such as emissions trading.

Greenhouse Challenge Plus

Overview

In May 2004, the Australian government announced budget funding of A$31.6 million for the Greenhouse Plus – Enhanced Industry Partnerships measure[9] (hereafter referred to as Greenhouse Challenge Plus). Greenhouse Challenge Plus was described as a cooperative partnership between industry and the Australian government to reduce greenhouse gas emissions, accelerate the uptake of energy efficiency, integrate greenhouse issues into business decision-making and provide more consistent reporting of greenhouse gas emissions (AGO 2005b).

The actions taken under the Greenhouse Challenge Plus programme were expected to contribute more than 15 MT CO_2(eq) in greenhouse gas emissions reductions in 2010 (AGO 2005e). Greenhouse Challenge Plus built on the infrastructure and existing commitments of the Greenhouse Challenge,[10] with the Cooperative Agreements signed under the Greenhouse Challenge being carried forward into Greenhouse Challenge Plus (AGO 2005b).

Under Greenhouse Challenge Plus, four different types of agreements were envisaged, namely:

● Cooperative agreements with individual public or private enterprises or industry associations

● Aggregate agreements with industry associations or other groups

● Facilitative agreements with industry associations or other groups

● Legally binding Deeds of Agreement with fossil fuel electricity generators subject to the Generator Efficiency Standards (AGO 2005b)

---

9 This was one part of the Australian government's Climate Change Strategy.

10 Greenhouse Challenge Plus also provided the framework for Greenhouse Friendly certification (a voluntary initiative that provides Australian businesses with the opportunity to market 'greenhouse neutral' products or services) and the Generator Efficiency Standards programme (which aims to achieve best practice in the efficiency of electricity generation).

Participants' commitments were broadly similar to those under the Greenhouse Challenge (AGO 2005b). Participants were required to:

- Measure and monitor their greenhouse gas emissions
- Deliver the maximum practicable greenhouse gas abatement
- Continuously improve the management of greenhouse gas emissions and sinks
- Work towards the milestones set in individual agreements
- Provide timely annual reports to the Australian Greenhouse Office
- Make an accurate public statement about their participation in the programme including basic information about their greenhouse gas emissions
- Promote the activities of industry participants
- Participate in independent verification of annual progress reports

The annual progress reports were expected to include (AGO 2005b):

- An updated emissions inventory
- A statement of absolute changes in emissions
- A statement of progress against significant abatement actions
- Changes in emissions intensity
- Details of the calculation methodologies and assumptions used
- An indication of which elements of the report were not confidential
- A sign-off by the chief executive or authorised delegate

From 1 July 2006, participation in Greenhouse Challenge Plus was a mandatory requirement for Australian companies receiving fuel excise credits of more than A\$3 million and for the proponents of large energy projects (AGO 2005c). The AGO estimated that these new requirements would affect around 100–200 businesses (AGO 2005c), although as many of these would probably already have been members of the Greenhouse Challenge, the overall effect on the number of participants is likely to have been modest.

Greenhouse Challenge Plus allowed participants to be recognised as Greenhouse Challenge Plus Leaders by the Minister for Environment and Heritage in an annual public statement. Participants could be considered Leaders if they went beyond the minimum requirements of Greenhouse Challenge Plus, through publicly disclosing their gross emission levels, their short-term goals for greenhouse gas emissions, their greenhouse gas strategy, the expected direction of their future greenhouse gas emissions and their proposed mitigating actions (AGO 2005b). In addition, Leaders were expected to develop action plans that include quantitative actions to (AGO 2005b):

- Meet or exceed their annual greenhouse goals
- Reference best practice in the development of their greenhouse targets and key performance indicators

● Encourage their suppliers to take action to reduce greenhouse gas emissions

The effect of these changes (i.e. mandatory participation for companies receiving fuel credits and the Greenhouse Challenge Plus Leaders) was to move the Greenhouse Challenge from being a purely voluntary programme to a programme with voluntary, incentive-based and compulsory membership requirements.

Greenhouse Challenge Plus retained verification as a key part of the programme, with the scope and objectives broadly the same as under the Greenhouse Challenge. However, the first independent verification report for Greenhouse Challenge Plus (AGO 2007b) represented something of a step back from the 2003 report under the Greenhouse Challenge (AGO 2003a) as it moved away from the provision of quantitative data such as the percentage errors in organisations' emission inventories.

A consultative committee — the Industry–Government Greenhouse Partnership Committee — made up of representatives from industry associations, individual companies and government was established to ensure the effective development and implementation of the programme (AGO 2005d).

Expected outcomes

At the time of writing (early 2008), it is too early to draw firm conclusions about the manner in which Greenhouse Challenge Plus is being implemented, whether the outcomes will be substantively different from the Greenhouse Challenge or what the role of Greenhouse Challenge Plus will be in the radically different policy environment being promised by the new Labour government. However, a number of comments can be made about the design of Greenhouse Challenge Plus.

On initial inspection, Greenhouse Challenge Plus appears to address some of the weaknesses of the Greenhouse Challenge.

First, Greenhouse Challenge Plus is no longer a purely voluntary programme but now has incentives and mandatory requirements for certain companies to participate. Though, in theory, the effect should be to increase participation, the reality is that many of these companies are already likely to be members of the programme and so the changes may not result in the significant increase in membership that has been predicted by the AGO.

Second, the Greenhouse Challenge now differentiates between participating companies. The incentives associated with Greenhouse Leaders should, *prima facie,* encourage companies to go beyond — and stay beyond — minimum compliance with the requirements of the Challenge. It remains to be seen how many companies will actually decide to become Greenhouse Leaders, i.e. whether the PR benefits outweigh the additional transaction costs.

Third, the improved disclosure requirements (in terms of the information that companies are required to put into the public domain) should address at least some of the NGO concerns about the lack of transparency. However, Greenhouse Challenge Plus retains its strong emphasis on the protection of commercial information (AGO 2005b) and it is therefore likely that NGOs will continue to be critical of the programme in this respect.

Despite the changes introduced in Greenhouse Challenge Plus, many of the weaknesses of the Greenhouse Challenge have not been addressed. Perhaps most importantly,

Greenhouse Challenge Plus does not impose specific performance targets on participating companies or provide strong incentives for companies to go beyond business as usual. The consequence is that organisations are likely to continue to make economically suboptimal decisions on investments in energy efficiency or greenhouse gas abatement.

A further issue is that the oversight structure (the Industry–Government Partnership Committee) for Greenhouse Challenge Plus continues to exclude key stakeholders – in particular environmental NGOs. Greenhouse Challenge Plus may also be less acceptable to Australian industry, which has expressed concern at the move away from the strictly voluntary approach that characterised the Greenhouse Challenge and at the proposal to highlight leaders for specific praise (e.g. AIGN 2005).

Conclusions

The first conclusion to be drawn is that the voluntary approach underpinning the Greenhouse Challenge has been supported by industry as enabling cost-effective approaches to greenhouse gas emissions abatement to be implemented by Australian business while also making greenhouse and climate change issues a part of management decision-making processes.

But, despite these positive outcomes, the overall contribution of the Greenhouse Challenge to reducing greenhouse gas emissions from Australian business appears to have been relatively modest. The Greenhouse Challenge did not provide strong incentives for participating organisations to set greenhouse gas emission reduction targets beyond business as usual, and the existence of the Greenhouse Challenge was used by industry to deflect calls for the introduction of stronger policy measures such as emissions trading. It appears that the major contributions of the Greenhouse Challenge were to encourage some organisations to bring forward some energy-saving or greenhouse gas emission reduction projects, and to help participating organisations to identify opportunities that provided clear short-term financial benefits.

Given that the Greenhouse Challenge was in place for ten years, these outcomes are quite modest– or even inadequate – when assessed in the context of the significant emissions reductions that are likely to be required of all organisations. Even though Greenhouse Challenge Plus promises a somewhat more stringent approach, it is difficult to see it providing the incentives for the dramatic reductions in emissions that will be required from Australian businesses over the coming years.

There is also an overwhelming sense that the initial momentum and enthusiasm that characterised the Greenhouse Challenge has been lost and that it is time to start again. For example, the levelling-off and reduction in the membership of the programmes, the lack of evidence that participating companies were setting strong emission reduction targets, the fact that many cooperative agreements are significantly out of date (AGO 2007b) and the negative comments from industry about the Greenhouse Challenge Plus, all indicate that a new approach is required – not simply a repackaging of the programme in the new clothes of Greenhouse Challenge Plus.

The reality is that what is needed is a far stronger and more coherent approach to climate change policy in Australia where:

● Greenhouse gas emissions are assigned a real (and significant) financial value (whether through emissions trading or other mechanisms)

● Government action is directed towards significant reductions in absolute greenhouse gas emissions

● The challenges inherent in such a move (e.g. how to deal with the competitiveness implications for export-focused industries) are explicitly recognised and addressed as part of the policy design and implementation process

In that type of policy environment, where there are clear incentives and a clear framework for action, Greenhouse Challenge Plus (at least as presently constituted) probably does not have a significant role to play.

References

ACF (Australian Conservation Foundation) (1999) *Australian Conservation Foundation Submission to the Senate Inquiry into Global Warming* (Sydney: ACF).

AGO (Australian Greenhouse Office) (1998) *The National Greenhouse Strategy* (Canberra: AGO).

—— (1999) *Greenhouse Challenge Evaluation Report* (Canberra: Commonwealth of Australia)

—— (2000a) *Greenhouse Challenge Independent Verification Program. Verification and Reporting Guidelines 1: Verification Information* (Canberra: AGO).

—— (2000b) *Greenhouse Challenge Independent Verification Program. Verification and Reporting Guidelines 2: Verification Guidelines* (Canberra: AGO).

—— (2000c) *Guidelines for the Cooperative Agreements Program* (Canberra: AGO).

—— (2001a) *Independent Verification Under the Greenhouse Challenge – 2000. Findings and Discussion Report: February 2001* (Canberra: AGO).

—— (2001b) *National Greenhouse Gas Inventory 1999* (Canberra: AGO).

—— (2002) *Annual Report 2001/2002* (Canberra: AGO).

—— (2003a) *Report on Independent Verification of the Greenhouse Challenge Programme 2002* (Canberra: AGO).

—— (2003b) *Annual Report 2002/2003* (Canberra: AGO).

—— (2004) *AGO Annual Report 2003/2004* (Canberra: AGO).

—— (2005a) *Australia's Fourth National Communication under the United Nations Framework Convention on Climate Change* (Canberra: AGO).

—— (2005b) *Greenhouse Challenge Plus: Programme Framework 2005* (Canberra: AGO).

—— (2005c) *Greenhouse Challenge Plus: An Australian Government–Industry Partnership to Reduce Greenhouse Gas Emissions and Improve Energy Efficiency* (Canberra: AGO).

—— (2005d) *Industry–Government Greenhouse Partnership Committee* (Canberra: AGO).

—— (2005e) *Australia's Response to Climate Change* (Canberra: AGO).

—— (2006) *Tracking to the Kyoto Target: Australia's Greenhouse Emissions Trends 1990 to 2008–2012 and 2020* (Canberra: AGO).

—— (2007a) *National Greenhouse Gas Inventory 2005* (Canberra: AGO).

—— (2007b) *Greenhouse Challenge Plus: Independent Verification 2005/2006* (Canberra: AGO).

AIGN (Australian Industry Greenhouse Network) (1999) *A Submission to the Senate Inquiry into Australia's Response to Global Warming. October 1999* (Melbourne: AIGN).

—— (2005) *Submission on Greenhouse Plus: Industry Consultation Discussion Paper* (Melbourne: AIGN).

Australian Business Roundtable on Climate Change (2006) 'Joint CEO Statement'; www.businessroundtable. com.au/html/jointceo.html (accessed 6 June 2008).

BCA (Business Council of Australia) (1999) *Statement of Climate Change Policy Principles* (Canberra: BCA).

Beresford, R., and S. Waller (2000) 'The Kyoto Protocol: Threats and Opportunities', *The APPEA Journal* 40.1: 645-53.

Department of Climate Change (2008) *Tracking to the Kyoto Target: Australia's Greenhouse Emissions Trends 1990 to 2008–2012 and 2020* (Canberra: Commonwealth of Australia).

Department of Environment and Heritage (2004) *Budget 2004–05* (Canberra: Department of Environment and Heritage).

Department of the Environment and Water Resources (2007) *Annual Report 2006/2007* (Canberra: Department of the Environment, Water, Heritage and the Arts).

Government of Western Australia (2004) *Greenhouse Strategy* (Perth: Government of Western Australia).

Howard, J. (1997) 'Safeguarding the Future: Australia's Response to Climate Change. Statement by The Prime Minister of Australia, The Hon. John Howard MP, 20 November 1997'.

Knapp, R. (2004) 'Australian Aluminium Council Submission to the Senate ECITA Committee Inquiry into the Kyoto Protocol Ratification Bill 2003 (No. 2), 30 January 2004' (Canberra: Australian Aluminium Council).

Krarup, S., and S. Ramesohl (2000) *Voluntary Agreements in Energy Policy: Implementation and Efficiency* (Copenhagen: AKF Institute of Local Government Studies).

Lipman, Z., and G. Bates (2002) *Pollution Law in Australia* (Sydney: Butterworths).

Macintosh, A. (2007a) *The National Greenhouse Accounts and Land Clearing: Do the Numbers Stack Up?* (Research Paper No. 38; Canberra: The Australia Institute).

—— (2007b) 'Response to the Federal Government's Critique of *The National Greenhouse Accounts and Land Clearing: Do the Numbers Stack Up?*' (Canberra: The Australia Institute).

NAO (National Audit Office) (2004) *The Administration of Major Programs: Australian Greenhouse Office* (Canberra: National Audit Office).

New South Wales Greenhouse Office (2005) *NSW Greenhouse Plan* (Sydney: NSW Greenhouse Office).

PACIA (Plastics and Chemicals Industries Association) (2004) 'Submission to Senate Standing Committee on Environment, Communications, Information Technology and the Arts in Relation to its Inquiry into the Kyoto Protocol Ratification Bill 2003 (No 2), January 2004' (Melbourne: PACIA).

Parker, C. (1999) 'The Greenhouse Challenge: Trivial Pursuit?', *Environmental and Planning Law Journal* 16.1: 63-74.

—— (2002) *The Open Corporation: Effective Self-Regulation and Democracy* (Cambridge, UK: Cambridge University Press).

Parliament of the Commonwealth of Australia (2000) *The Heat is On: Australia's Greenhouse Future. Report of the Senate Environment, Communications, Information Technology and the Arts Committee* (Canberra: Commonwealth of Australia).

—— (2001) *Government June 2001 Response to 'The Heat is On: Australia's Greenhouse Future'* (Canberra: Commonwealth of Australia).

—— (2004) *Securing Australia's Energy Future* (Canberra: Commonwealth of Australia).

Rex, L. (2000) 'Greenhouse Gas Reduction and Energy Efficiency in the Australian Plastics and Chemicals Industries', in *Proceedings of the 15th International Clean Air and Environment Conference*, Sydney, 27–30 November 2000 (Volume 2; Mitcham, Victoria: CASANZ).

State of Victoria (2005) *Victorian Greenhouse Strategy Action Plan Update* (Melbourne: State of Victoria, Department of Sustainability and Environment).

Sullivan, R. (2005) *Rethinking Voluntary Approaches in Environmental Policy* (Cheltenham, UK: Edward Elgar).

—— (2006) 'Greenhouse Challenge Plus: A New Departure or More of the Same?', *Environmental and Planning Law Journal* 23.1: 60-73.

—— (2007) *Australia Country Case Study: Human Development Report 2008/2008* (Human Development Report Office Occasional Paper 2007/61; New York: United Nations Development Programme).

—— and R. Ormerod (2002) 'The Australian Greenhouse Challenge: Lessons Learned and Future Directions for Climate Policy', in J. Albrecht (ed.), *Instruments for Climate Policy* (Cheltenham, UK: Edward Elgar): 170-91.

—— and J. Sullivan (2005) 'Environmental Management Systems and their Influence on Corporate Responses to Climate Change', in K. Begg, F. van der Woerd and D. Levy (eds.), *The Business of Climate Change: Corporate Responses to Kyoto* (Sheffield, UK: Greenleaf Publishing): 117-30.

UNDP (United Nations Development Programme) (2007) *Human Development Report 2007/2008. Fighting Climate Change: Human Solidarity in a Divided World* (New York: UNDP).

Wilson, N. (2005) 'Business told to cut emissions', *The Australian*, 4 August 2005.

9

The Mexico Greenhouse Gas Program: corporate response to climate change initiatives in a 'non-Annex I' country*

Leticia Ozawa-Meida
SEMARNAT, Mexico

Taryn Fransen
World Resources Institute, USA

Rosa M. Jiménez-Ambriz
CESPEDES, Mexico

Over the past decade, the corporate sector has taken an increasingly proactive role in crafting strategies to respond to the threat of climate change. A body of literature has emerged that attempts to describe the motivations and drivers that lead corporations to develop strategies related to climate change, such as developing inventories of their greenhouse gas (GHG) emissions, setting voluntary targets to reduce emissions, offsetting emissions, participating in GHG markets, and developing or deploying new technologies.

* The authors would like to express their gratitude to Mexico's Ministry of Environment and Natural Resources (SEMARNAT) for its support. The case studies used in this chapter were developed under contract DGRMIS-DAC-324/2007. We are grateful to the interviewees from the companies that participated in these case studies who provided insightful comments throughout the process. Finally, we would also like to thank David Rich, Edgar del Villar and Alejandro Lorea for providing feedback on early drafts of this chapter. The accuracy of this research and the conclusions reached are the responsibility of the authors alone and not of SEMARNAT, WRI or CESPEDES.

The literature has identified a range of motivations and drivers for these actions including profit, policy engagement, risk management, ethical considerations and fiduciary responsibility. Most of this research, however, has focused on companies operating in developed countries, which are obligated to reduce their GHG emissions under the Kyoto Protocol or, in the case of the USA, are facing an evolving panorama of state, regional and prospective national regulations.

Notwithstanding the fact that many companies in the rest of the world are also beginning to tackle climate change, less research is available to explain what motivates these companies to do so. Using data gathered from interviews with six companies participating in the Mexico GHG Program, this chapter explores the drivers for companies from non-Annex 1 countries to take action on climate change (in particular the reasons for participating in voluntary initiatives such as the GHG Program) as well as examining their experience in developing GHG inventories and implementing emission reduction projects.

The chapter is divided into five sections. The first gives an overview of the literature of the drivers for corporate action on climate change. The second section describes the Mexican government's responses to climate change, in particular the Mexico GHG Program. The third section profiles the six companies interviewed for this research. This is followed by an analysis of the drivers for the companies to participate in the Mexico GHG Program and of the outcomes from their participation in terms of inventory development and the design and implementation of GHG reduction projects for the purpose of participating in international GHG markets. Finally, the concluding section outlines lessons learned from the Mexico GHG Program and makes recommendations regarding future directions for the programme.

Business response to climate change: an overview of the literature

A significant and growing number of major corporations view climate change as an issue that materially affects important aspects of their businesses and around which a corporate strategy should be developed. This is true not only of companies and sectors operating in environments where GHG emissions are regulated, but also of companies headquartered in the USA, 'non-Annex I' countries and companies from non-regulated sectors within the European Union. Even in the absence of regulation, companies are developing climate strategies on a voluntary basis in order to improve operational efficiency, access new sources of capital, improve risk management, enhance corporate reputation and identify new market opportunities by addressing GHG emissions (Hoffman 2005).

Although corporate climate change strategies are becoming on average more prevalent and more sophisticated, the specific responses of any individual company to climate change depend on a variety of driving forces. These can be broadly categorised as socio-cultural, political–institutional, market-based and internal–organisational factors (Levy and Newell 2000; Kolk and Levy 2001; Levy and Rothenberg 2002) as follows:

- Socio-cultural factors are those deriving from the external context in which businesses make decisions to secure credibility, legitimacy, the 'licence to operate' and the long-term survival of the company. The stakeholders that may influence the manner in which a company responds include environmental groups, the media, local communities, governments, financial institutions, shareholders, boards of directors, customers and employees. Membership in business and industry associations may also expose companies to a diversity of perspectives and influences that aim to drive or shift their conceptions about climate science, its economic consequences and mitigation technologies

- Political–institutional factors are those associated with the relationship between the company and the public policy sphere which determine the extent of corporate access to, and representation in, public policy decision-making. Present and future regulation is a key driver of corporate climate strategies. Businesses are increasingly encouraged to work with governments to contribute to the delivery of national greenhouse gas emission targets and goals. Voluntary initiatives play an important role in these discussions, as they may allow companies to pre-empt new regulation or influence the shape of future regulation

- Corporate market-based factors are those that affect a company's competitive advantage in the marketplace such as physical risks to assets and operations deriving from climate change, as well as indirect risks associated with the effects of climate change responses on supply chains, operations, product use, market access, brand or reputation

- Internal–organisational factors refer to the company's own culture including aspects such as its policies and structures, perceptions, beliefs, and the ethical considerations of its leadership and employees. For example, some business leaders believe that creating and maintaining a culture of sustainability, encompassing economic prosperity, environmental stewardship and social responsibility, stimulates the growth of effective and efficient businesses. These beliefs may then drive the development of a set of principles that guide the company's actions on climate change, among other relevant issues

Examples of each type of factor in the context of climate change are listed in Table 9.1. Levy and Rothenberg (2002) and Oliver (1997) have noted that, in the case of climate change, the uncertainty inherent in the development of new technologies and markets, as well as in evolving policy responses, decreases the relative importance of economic and competitive factors while increasing the influence of the institutional environment. Institutional theory generally predicts the convergence of corporate strategies over time, due to the tendency of companies to imitate the actions of their successful competitors.

However, Levy and Rothenberg (2002) observed that this convergence is limited for two reasons. First, each company responds according to its history, organisational culture and market positioning. Traditionally, there has been a distinction between companies that have approached climate change regulation as a business opportunity and those that have approached it as a threat (Pulver 2002). More recently, however, some companies from GHG-intensive sectors have adopted an organisational culture that looks to create opportunities within the context of climate change mitigation. Second, organisations often operate within multiple, complex and overlapping institutional domains. For

TABLE 9.1　Factors that influence business responses to climate change

Socio-cultural	Political–institutional	Market-based	Internal–organisational
<ul><li>Public image</li><li>Stakeholder relations (employees, consumers, media, etc.)</li><li>Membership in industry associations and NGOs</li></ul>	<ul><li>Opportunities to participate in political dialogue</li><li>Credibility in political dialogue</li><li>Baseline protection</li><li>Relationship between government and business</li><li>Management of regulatory risk</li><li>Local climate change policies</li><li>Anticipation of regulation</li></ul>	<ul><li>Adaptation to physical risks</li><li>Operational efficiency</li><li>Cost savings</li><li>Product or corporate differentiation</li><li>Development of new products or processes</li><li>Identification of market opportunities</li><li>Access to capital</li></ul>	<ul><li>Leadership's perception of climate change science, impacts and solutions</li><li>Existence of a 'culture of sustainability'</li><li>Ethical considerations</li><li>Availability and type of internal climate expertise</li><li>Degree of centralisation of decision-making process</li></ul>

example, multinational corporations can face different regulatory pressures in different countries where they operate. They have the choice of setting a minimum set of environmental performance standards for their global operations, or they can adapt to the policies, technologies and standards of each individual country (see also Chapter 4).

Even though there have been national and regional differences among sectors and locations, there are signs of convergence with businesses increasingly accommodating challenges from government and interest groups in pursuing their corporate strategies. The convergence has been attributed, at least in part, to the consolidation of companies (through mergers and joint ventures) and the globalisation of corporate production and political activities.

Mexico's response to climate change

Just as the corporate response to climate change has evolved, so has that of the Mexican government. To date, Mexico has played an active role in the international policy processes related to climate change. It signed the United Nations Framework Convention on Climate Change (UNFCCC) in 1992 and ratified it in 1993. Since then, Mexico has fulfilled its commitments as a non-Annex I country by submitting three national communications[1] and updating its national GHG inventory to 2002. While Annex I parties to the UNFCCC are required to submit national GHG inventories on an annual basis, non-Annex I parties are not held to fixed dates. With respect to its national communica-

---

1 National communications are reports that present a country's inventory of GHG emissions and removals, and outline the activities the country has undertaken to implement the UNFCCC.

tion and inventory, Mexico is in an advanced position compared with other developing countries. It is currently participating in the Consultative Group of Experts on National Communications from Non-Annex I Parties (Tudela *et al.* 2003).

With regard to its economic classification, Mexico is in a unique position. Despite having being a member of the Organisation for Economic Cooperation and Development (OECD) since 1994, Mexico is considered a non-Annex I country under the UNFCCC. In fact, Mexico and South Korea are the only OECD countries not classified as Annex I. Within the UNFCCC party groupings, Mexico withdrew from the G77 in 1994 and, since 2000, has been a member of the Environmental Integrity Group along with Switzerland, South Korea, Liechtenstein and Monaco. Mexico signed the Kyoto Protocol in 1998 and ratified it in 2000.

In terms of international climate change policy mechanisms, the Mexican government supports the negotiation of a multilateral post-Kyoto Protocol agreement and favours the development of a global carbon market to foster mitigation activities. Mexico has proposed that the Clean Development Mechanism (CDM) evolve from its current state as a case-by-case, project-specific scheme to a scheme that could involve entire programmes or sectors ('programmatic CDM') in order to scale up international investment in climate change mitigation activities. Mexico also promotes the voluntary adoption of 'no-lose' targets[2] based on the progressive strengthening of national capacities to measure, monitor and mitigate GHG emissions.

At the national level, the Mexican government published its National Strategy on Climate Change in May 2007. This Strategy establishes guidelines for action on climate change, identifying specific measures for climate change mitigation in the energy and land-use sectors, proposing a step-by-step approach for the progressive valuation of carbon in the national economy, and outlining national capacity-building requirements for adaptation to climate change. The Strategy provides the basis for the development of a set of public policies known as the Special Program on Climate Change, which was under development at the time of writing.

Mexico GHG Program

Following extensive consultations with Mexico's private sector, SEMARNAT (Mexico's Ministry of Environment and Natural Resources),[3] CESPEDES (the Mexican Business Council for Sustainable Development),[4] the World Resources Institute (WRI)[5] and the World Business Council for Sustainable Development (WBCSD)[6] formed a partnership to design and implement the Mexico Voluntary GHG Accounting and Reporting Program (Mexico GHG Program) with the objective of preparing Mexican businesses for an increasingly carbon-constrained world by building their capacity to measure and manage their GHG emissions and to participate in carbon markets.

2 'No lose' targets involve the voluntary adoption of mitigation policies and measures associated with quantitative targets, but without any penalties in the event of non-compliance.
3 SEMARNAT is the Mexican federal government agency responsible for promoting the protection, restoration and conservation of natural resources, ecosystems, and environmental goods and services. For more information see its website (www.semarnat.gob.mx [Spanish language]).
4 More information about CESPEDES is available at www.cce.org.mx/CESPEDES (Spanish language).
5 More information about WRI is available at www.wri.org.
6 More information about WBCSD is available at www.wbcsd.org.

The Mexico GHG Program provides a platform for accounting for, and reporting, GHG emissions and reductions based on the WRI/WBCSD GHG Protocol (WBCSD/WRI 2004, 2005). It also organises training and capacity-building workshops and provides technical assistance in the preparation of GHG inventories and mitigation projects.

The programme was launched in August 2004 with a pilot group of 20 companies, most of which were based in Mexico City. In 2005, the capacity-building activities were extended to the cities of Monterrey and Guadalajara.

In late 2005, the programme received its first corporate GHG inventory reports from 15 participating companies that publicly reported their 2004 GHG corporate emissions, representing 84 million tonnes of CO_2 equivalent [MT CO_2(eq)]. The following year, 30 companies reported their 2005 GHG corporate emissions, accounting for 89 MT CO_2(eq). The programme has continued growing; as of 2007, 55 companies have joined the programme (see Table 9.2), including all of the companies in Mexico's cement, oil and gas, and beer brewing sectors, as well as a significant portion of its iron and steel sector. Thirty-five of these companies reported their 2006 GHG emissions,[7] accounting for 102 MT CO_2(eq) or approximately 35% of Mexico's industrial GHG emissions.[8]

When Mexico's National Climate Change Strategy was developed in 2007, it cited the Mexico GHG Program as an important capacity-building instrument and source of information to promote climate change mitigation in the industrial sector, and adopted the programme as part of the country's portfolio of efforts to address climate change. Specifically, the Mexico GHG Program contributes to two elements of the National Strategy:

- **Identification of GHG mitigation measures in the energy sector**. The National Strategy proposes that the Mexico GHG Program builds on its existing platform and portfolio of activities to:
  - Account for (and report on) 80% of Mexico's industrial GHG emissions
  - Promote the identification and implementation of GHG reduction projects
  - Serve as a registry for voluntary GHG reductions
  - Identify best practices by sector

- **Progressive valuation of carbon in the national economy.** The National Strategy outlines a medium-term, step-by-step approach for the progressive valuation of carbon in the national economy, beginning with voluntary GHG accounting and reporting, progressing to GHG caps in the energy sector, and culminating in a national cap-and-trade scheme linked to international GHG markets. The Mexico GHG Program is seen as facilitating this strategy by building the capacity of its participants in GHG accounting and emission trading, and by contributing to the formulation of sectoral baselines and benchmarks

In addition, the Mexico GHG Program played a convening role during the development of the National Strategy, facilitating dialogue between stakeholders from different industrial sectors and government agencies, and providing a venue through which the

---

7 The annual GHG reports are available at www.geimexico.org.
8 This percentage is calculated by comparing the emissions reported under the Mexico GHG Program to those from stationary combustion and industrial processes as estimated in Mexico's national GHG inventory.

TABLE 9.2 **Current participants in the Mexico GHG Program** (continued over)

Sector	Organisation
Automotive	<ul><li>Ford Motor Company</li><li>Honda de México</li><li>Nissan Mexicana</li></ul>
Beer and brewing	<ul><li>Grupo Modelo</li><li>Cervecería Cuauhtémoc Moctezuma</li></ul>
Cement	<ul><li>CEMEX México</li><li>Cooperativa La Cruz Azul</li><li>Cementos Moctezuma</li><li>Grupo Cementos de Chihuahua</li><li>Holcim Apasco</li><li>LaFarge</li></ul>
Chemicals	<ul><li>Colgate Palmolive</li><li>NHUMO</li><li>AMANCO México</li><li>ANAJALSA Agroquímicos</li><li>Boehringer Ingelheim Vetmedica</li><li>Fábrica de Jabón La Corona</li></ul>
Commercial	<ul><li>Wal-Mart de México</li></ul>
Construction	<ul><li>Urbi Desarrollos Urbanos</li></ul>
Food and beverages	<ul><li>Grupo Bimbo</li><li>Grupo Embotellador CIMSA Coca-Cola</li><li>Grupo JUMEX</li></ul>
Forestry	<ul><li>Forestaciones Operativas</li></ul>
Glass	<ul><li>VITRO</li></ul>
Iron and steel	<ul><li>Altos Hornos</li><li>De Acero</li><li>Grupo IMSA</li><li>Mittal Steel Lázaro Cárdenas</li><li>SERSIINSA</li><li>SICARTSA</li><li>Siderúrgica Tultitlán</li><li>Hierro Recuperado</li></ul>
Machinery	<ul><li>Caterpillar de México</li><li>S&C Electrics</li><li>Industrias John Deere</li><li>Johnson Controls</li><li>Hitachi Global Storage Technologies México</li><li>Assa Abloy Occidente</li><li>Siemens VDO México</li></ul>
Mining	<ul><li>Industria Minera México</li><li>Industria Peñoles</li><li>Minera Autlán</li></ul>

TABLE 9.2   (from previous page)

Sector	Organisation
Oil and gas	● PEMEX ● Gas del Atlántico
Packaging	● Tetrapak Querétaro
Public transport	● Red de Transporte Público del Distrito Federal
Refrigerants	● Ecofreeze International
Services	● Sumitomo Corporation de México
State governments and municipalities	● Secretaría de Medio Ambiente del Gobierno del Estado de México ● Ayuntamiento del Centro, Tabasco
Swine farm	● Grupo Porcícola Mexicano
Universities	● Instituto Tecnológico de Estudios Superiores de Monterrey Campus Guadalajara
Waste management	● Cappy & Asociados ● SIMEPRODE

intersecretarial commission that developed the strategy could solicit feedback from participating companies.

Company profiles

To better understand the motivations for companies to participate in the Mexico GHG Program, we selected six members of the programme — representing a range of sectors, sizes and headquarter locations — for in-depth analysis based on their active participation in the programme and their progress in implementing GHG reduction projects. Environmental managers from each company were interviewed regarding their climate change strategies, reasons for and results from participating in the programme, and perspectives on GHG markets and policies. Key information about each company is summarised in Table 9.3; a brief profile of each company and its climate change strategy is presented below.

There are two important limitations to note about the approach we adopted for this research. First, only six companies were interviewed and, because of their active participation in the programme, their responses may not be fully representative of the 'average' participant, which may be less active and sophisticated on the climate change issue in general. Second, the research relied on interviews with representatives of the companies, who may be eager to present their corporate practices in a favourable light and who may, therefore, have emphasised ethics-based reasons for taking action on climate change over more profit-based motivations.

TABLE 9.3 GHG emissions profile of the six companies

Company	Type (headquarters)	Date joined	Total GHG emissions, 2006 in kilotonnes (kt) of CO_2(eq)			GHG/energy target
			Total	Direct (Scope 1)	Indirect (Scope 2)[a]	
CEMEX	Multinational (Mexico)	2004	16,347	14,928	1,419	Reduce CO_2 per tonne of cement from global operations by 25% from 1990 levels by 2015
Ford Motor Company Mexico	Multinational (USA)	2005	145.9	28.7	117.2	1% annual increase in energy efficiency (global operations). Reduce absolute emissions by 6% from 2000 levels by 2010 (North America)
Grupo JUMEX	National	2007	16.2	7.7	8.5	Under development
Grupo Modelo	National	2004	778.3	743.5	34.9	Under development
PEMEX	Public — state-owned	2005	46,743	46,444	298.6	Reduce absolute CO_2 emissions by 18% from 2006 levels by 2012
Wal-Mart de México	Multinational (USA)	2007	726.5	9.1	717.5	Renewable energy and energy efficiency targets; see company profile

[a] Indirect GHG emissions from purchase and consumption of electricity, heat or steam from sources not controlled or owned by the reporting company.

CEMEX Mexico

Founded in Mexico in 1906, CEMEX has grown from a small, regional cement company into a global one. CEMEX is the third largest cement producer in the world, operating in 32 countries in the Americas, Asia, Europe, Australia and the Middle East. It is the largest cement manufacturer in Mexico, with 15 cement plants (with an installed capacity of 27.2 MT of cement), 234 concrete plants, 26 aggregates facilities and eight maritime terminals (CEMEX 2007).

As a member of the WBCSD, CEMEX participates in the Cement Sustainability Initiative (CSI),[9] which launched its Agenda for Action in 2002 and includes climate change,

9 See also Chapter 14.

alternative fuels and raw materials, and stakeholder communication and reporting among its themes. CEMEX Mexico has developed a sustainability development report according to the Global Reporting Initiative and identifies itself as a socially responsible company. CEMEX cites its commitment to social responsibility, recognition of the cement industry as an important energy consumer, and concerns over energy security, stakeholder relations and competitiveness among its reasons for addressing climate change.

Within its sustainability initiative, CEMEX has developed a low-carbon strategy that includes the following actions:

- Replacing clinker with less CO_2-intensive materials

- Using biomass waste and by-products as alternative fuels

- Increasing energy efficiency to decrease specific fuel consumption

- Generating electricity from renewable sources

CEMEX has set a target for its global operations to reduce CO_2 emissions per tonne of cement product by 25% from 1990 levels by 2015. For the purpose of monitoring emissions, all its cement plants around the world are connected to an online reporting tool called the CO_2 Emissions Inventory Electronic Platform, which calculates the CO_2 emissions and other indicators based on the WBCSD CSI Protocol (WBCSD 2005).[10]

CEMEX Mexico is currently developing wind power, energy efficiency, clinker reduction and alternative raw material use projects within the CDM framework (SEMARNAT 2007).

Ford Motor Company

The Ford Motor Company is the third largest manufacturer of automobiles in the world; its vehicles are sold in 200 markets. Ford Motor Company Mexico has three facilities located in Hermosillo, Chihuahua and Cuautitlán. It was the first automotive company to join the Mexico GHG Program.

The Ford Motor Company considers itself a socially responsible company and has adopted a climate change strategy; as part of that strategy, it participates in GHG initiatives such as the Chicago Climate Exchange, the Australian Greenhouse Challenge Plus,[11] the Philippine GHG Accounting and Reporting Program, the Brazil GHG Protocol Program and the EU Emission Trading Scheme. The company has a target to improve its energy efficiency in manufacturing by 1% per year for its global operations and for reducing the absolute CO_2 emissions of its North American facilities by 6% from 2000 levels by 2010. Since 2006, Ford has used a Global Emissions Manager (GEM) database to ensure that environmental metrics, such as CO_2 emissions, are tracked consistently around the globe.

Ford cites a number of drivers for its focus on climate change including: the desire to improve efficiency; the need to develop new product lines; the importance of improv-

---

10 www.ghgprotocol.org/calculation-tools
11 See Chapter 8.

ing stakeholder relations and corporate image; and the need to prepare for potential future GHG regulations.

Grupo Jumex

Established in 1961, Grupo Jumex is a Mexican company that produces natural juices, fruit nectars and other beverages; these products are now sold in 25 countries. Grupo Jumex has eight production plants located in Estado de México, Chihuahua and Mexicali. In addition, it has a research and development centre, a master distribution centre and 33 other distribution centres.

Grupo Jumex has a sustainability policy and a number of environmental initiatives including environmental audit programmes, initiatives to measure and improve environmental performance and recycling and energy efficiency programmes. In addition, it recently adopted a programme to promote environmental improvements in its supply chain.

In November 2005, Grupo Jumex became the first company from the Mexican industrial sector to have a CDM project registered. The project generates around 4,000 tonnes of CO_2(eq) of carbon credits annually.

Climate change was part of Grupo Jumex's sustainability strategy even before it joined the Mexico GHG Program. The company acknowledges its vulnerability to global warming due to its high dependence on water and crop production, and it recognises the need to adapt to climate change. Its motivations include:

- Improving energy efficiency
- Enhancing its public image
- Preparing the company for the possibility of future GHG regulation

Grupo Modelo

Grupo Modelo is a beer production, distribution and marketing company. Its 12 brands include the internationally recognised Corona and Negra Modelo. Grupo Modelo has seven breweries producing some 60 million hectolitres per year. The company's environmental initiatives include energy-saving programmes, fuel substitution, cogeneration and the use of biogas from its water treatment plants as an alternative fuel.

Grupo Modelo cites a number of factors that have influenced its response to climate change. The most important of these are:

- The inherent importance of the goals of environmental protection and sustainable development
- The potential to improve competitiveness
- The physical risks that a changing climate could pose to the availability of water and crops

- The potential to improve its brand and reputation
- The potential for stricter climate change regulation to be implemented over time
- The potential financial opportunities from participating in the CDM

Petróleos Mexicanos (PEMEX)

PEMEX is a state-owned oil company involved in hydrocarbon exploration, production, transformation and distribution. It also sells crude oil, natural gas, liquefied gas, petrochemical and various refined products in domestic and foreign markets.

PEMEX has been publishing a sustainable development report since 1999. Its sustainability strategy covers three main areas: regulatory compliance and risk elimination, business feasibility and sustainability, and social responsibility.

Within its broader environmental management and sustainable development activities, PEMEX is responding to the issue of climate change primarily through the identification and development of CDM projects. According to its 2006 Sustainable Development Report, PEMEX has identified around 60 GHG reduction opportunities (in areas such as cogeneration, energy efficiency and methane emissions abatement) that will be promoted as CDM projects.

Nineteen of these opportunities have been translated to Project Idea Notes, but only one project has yet received the letter of approval issued by the Mexican Designated National Authority (DNA). PEMEX also participates in public–private partnerships related to climate change such as the Mexico GHG Program and the Methane to Markets Initiative. Earlier in the development of its climate strategy, PEMEX adopted an internal, virtual GHG trading system. Launched in 2000, the system operated for four years, after which PEMEX decided to focus instead on the CDM, citing the lack of economic benefits associated with its virtual trading system.

In including early action on climate change within its sustainability strategy, PEMEX was responding to four key drivers. First, PEMEX is one of the two state-owned companies charged with implementing the National Energy Policy, which includes commitments on energy efficiency and renewable energy. Second, the carbon market promoted by the Kyoto Protocol provides incentives for GHG reduction. Third, PEMEX considers itself vulnerable to the physical impacts of climate change (in particular, sea level rise and hurricanes) due to the location of its plants and platforms. Therefore, PEMEX considers the identification of adaptation-related policy measures such as the payment for environmental services for forest conservation as fundamental components of its overall policy response to climate change. Finally, PEMEX considered that early action on climate change would mitigate risks to its competitiveness and would strengthen its stakeholder relationships.

Wal-Mart de México

Wal-Mart is the world's largest retailer, operating over 7,000 stores and clubs in 14 markets around the world. In Mexico, Wal-Mart is the country's leading retailer, operating 1,033 commercial units distributed across 176 cities.

Wal-Mart de México considers sustainable investment as valuable for its business operations. It views itself as a socially responsible company with a commitment to environmental care. At the global level, Wal-Mart has established ambitious environmental goals; for example, by 2025 it expects to obtain all its energy from renewable sources and to generate zero waste and zero water discharges.

Reflecting these global environmental goals, Wal-Mart de México launched its sustainability initiative in 2007 and it is currently working towards the minimisation of its GHG emissions footprint through improving energy efficiency and investing in sustainable technologies. The company has signed a partnership with a wind power generation company to supply 34% of its electricity demand and it is currently exploring the use of solar energy for water heating in new stores.

Wal-Mart de México expects that these and other efforts directed at reducing its emissions intensity will allow it to:

- Reduce its operating costs (and thereby continue to offer low prices to its customers)

- Reduce its dependence on carbon-intensive fuels

- Enhance its corporate image

Drivers for action

Drivers for addressing climate change

In relation to the drivers for action on climate change, the six companies cited a combination of internal–organisational, socio-cultural, institutional–political and market-based factors.

A number of internal–organisational factors were identified by most or all of the companies interviewed, with two being of particular importance in driving organisational action on climate change, namely:

- **The existence of sustainability or social responsibility policies within the corporate culture.** All the companies interviewed mentioned the importance of demonstrating environmental leadership as an integral part of being socially and environmentally responsible companies. Furthermore, CEMEX, Ford, Grupo Modelo, PEMEX and Wal-Mart de México publish annual sustainability reports

- **Ethical considerations.** For example, CEMEX recognises the cement industry as an important energy consumer. Thus the company is focused on the development of low-carbon strategies to reduce the impact of its operations

It is also worth noting that, once a company has developed a certain expertise on climate change, this expertise in itself may serve as a driver of further action. For example, PEMEX views its internal GHG emissions trading scheme as having allowed the company to build its capacity for GHG accounting and to participate in carbon markets.

In relation to socio-cultural drivers, all the companies interviewed cited concerns about corporate image as a factor in their decision to address climate change. However, none identified corporate image as the most important factor and, in fact, several empha-sised that the public relations benefits of their actions were secondary to the benefits to the environment itself. The stakeholders that the companies were seeking to influence varied: Ford, Grupo Modelo, PEMEX and Wal-Mart de México all highlighted sharehold-ers and investors as a key audience, while Ford and Wal-Mart de México also mentioned the public. In relation to investors, PEMEX pointed out that the Carbon Disclosure Pro-ject (an investor disclosure initiative)[12] may request GHG information from Mexican companies in the future.

Public policy (or institutional–political) drivers — especially the anticipation of future GHG regulation — were identified by a number of the companies as key motivations for action. For example, Grupo Jumex, Grupo Modelo and Ford cited the relevance of prepar-ing their companies for the possibility of future climate change regulation and several noted that they were actively monitoring the evolution of climate change policies at national and global levels (e.g. PEMEX, a state-owned company, cited its responsibility for implementing the National Energy Policy, which contains commitments on energy efficiency and renewable energy).

Most of the companies interviewed noted the importance of preparing for potential regulation of GHG emissions. Jumex, Modelo and Ford were all of the view that the Mex-ican government should take account of early action by participants in the Mexico GHG Program. CEMEX and Ford emphasised the importance of an open and transparent dia-logue between government and industry in order to ensure that the climate change poli-cies that are formulated and implemented are appropriate to the Mexican economic and business environment.

Finally, the companies interviewed identified a range of market-based drivers — effi-ciency improvements, addressing the physical risks of climate change, new product development and international competitiveness — for their actions on climate change. All the companies cited efficiency improvements as an important means of achieving cost reductions, stabilising or decreasing their energy demand, and reducing their vul-nerability to energy price fluctuations.

Several companies mentioned the potentially serious effects of climate change on their operations, including physical risks to their plants and platforms due to sea level rise and hurricanes, and disruptions to water supply and crop production (key inputs to bev-erage production). PEMEX also highlighted its market risk as its core business is based on the production of fossil fuels.

All the companies cited the desire to develop new product lines: Ford specified higher-efficiency and alternative-fuel vehicles, CEMEX quoted cement products with less CO_2-intensive materials, while PEMEX cited biofuels.

Finally, all the companies viewed improved competitiveness and market differentia-tion as important drivers of action on climate change. Interestingly, Grupo Modelo expressed concern that consumers and competitors in export markets may perceive that producing in Mexico represents unfair competition due to the absence of national GHG regulation. In spite of this lack of regulation, Grupo Modelo stated that it adopts inter-national best practices at global level to stay competitive in domestic and foreign mar-kets.

12 www.cdproject.net

Drivers for joining the Mexico GHG Program

To some extent, the six companies had previous knowledge of or experience with GHG accounting and reporting before joining the Mexico GHG Program — whether through the preparation of their own annual sustainability reports, the provision of GHG data to their corporate headquarters or participation in CDM projects.

Socio-cultural and institutional–political factors were important drivers for companies to participate in the programme. PEMEX, CEMEX and Jumex consider the programme an innovative public–private GHG initiative that not only enhances their public image, but also institutionalises their GHG reports and their efforts to reduce GHG emissions. Modelo, Jumex and Wal-Mart de México cited the importance of establishing a credible emissions baseline, which may facilitate the receipt of credit for early action under future regulations. Furthermore, Modelo mentioned the need for a systematic, measurable and internationally credible means of identifying its GHG sources as its principal reason for joining the Mexico GHG Program; this would enable it to establish a baseline from which subsequent reductions could be measured and reported.

Consistent with the drivers for taking action on climate change, CEMEX, Ford and Jumex all expressed their interest in participating in the dialogue between companies and the government facilitated by the Mexico GHG Program regarding the formulation of climate change policy. Through its participation in the programme, Wal-Mart de México expects to stay informed about the evolution of Mexico's international climate change negotiating strategy as well as national policy and incentives, which it sees as important in order to optimise its own environmental programmes and to inform its participation in the available market mechanisms.

Two multinational companies based in the USA with global operations, Ford and Wal-Mart de México, decided to join the programme because their companies' climate strategies at the global level were aligned with national climate policies. Both companies emphasised their interest in aggregating their climate change mitigation efforts with those of other actors and aligning their stance with existing national initiatives.

Despite the importance of market-based factors for stimulating action on climate change, relatively few companies cited market-based factors as a driver for participating in the programme. Of those that did mention market-based factors, Grupo Jumex's motivation for participation related to its interest in continuous improvement, while Wal-Mart de México was specifically interested in identifying GHG reduction opportunities that it could use to participate in carbon markets (specifically in the CDM).

Analysis of participant experiences

Experiences with corporate GHG accounting and reporting

Technical capacity

The companies participating in the Mexico GHG Program brought a wide range of experience to the programme. Typically, large companies, companies from energy-intensive sectors and multinational companies entered the programme with significant knowledge of energy and GHG accounting methodologies. In contrast, smaller companies, companies from less GHG-intensive sectors and local companies generally entered the programme with less experience on the topic.

The results of the interviews suggest that, regardless of prior experience, most participants were able to improve their technical capacity to some degree by participating in the programme. For example:

- CEMEX and Jumex were able to develop quantitative indicators for tracking progress

- CEMEX's previous experience supported the programme's efforts to build capacity for GHG accounting within the Mexican cement sector

- Ford was able to enhance its data on indirect GHG emissions, thereby allowing it to initiate more aggressive energy efficiency programmes

- Grupo Modelo improved its understanding of GHG accounting, identified areas where improvements to measurement techniques were needed, improved its capacity to conduct internal benchmarking between units, and enhanced its ability to respond to the mandatory government environmental report COA-RETC[13]

- Wal-Mart de México was able to develop a GHG performance baseline and translate energy data into GHG data. As a result, when projects are presented for internal approval, GHG impacts can be shown alongside energy and cost-saving impacts, thereby improving the likelihood of the projects being approved

Only PEMEX, which has been conducting GHG accounting and reporting since 1999, did not cite improvements in technical capacity as a result of participating in the programme.

GHG reduction opportunities

Most of the companies interviewed either have already identified opportunities to reduce GHG emissions and/or improve efficiency as a result of the programme, or anticipate that they will do so after participating in the programme for a longer period of time.

In the cases of CEMEX and PEMEX — both of which entered the programme with a high degree of expertise and experience in energy and GHG accounting and, therefore, a pre-existing idea of where GHG reduction opportunities would exist — the GHG inventory proved useful in confirming and reinforcing the notion of where the major opportunities could be found.

CEMEX, Ford and Wal-Mart de México all cited the fact that the GHG inventory framework included indirect GHG emissions as well as direct GHG emissions as a useful feature that enabled them to search for GHG reduction opportunities along their value chains, while Modelo expressed an interest in doing further analysis along its value chain. CEMEX is considering looking at transportation of supplies, co-processing of

---

13 COA-RETC is the Spanish acronym for the *Annual Operation Report: Register of Emissions and Transference of Pollutants.*

waste and alternative raw materials; Ford has used its GHG inventory to make the case for more aggressive energy efficiency programmes; and Wal-Mart de México is considering promoting best practices among the companies to which it outsources transportation. Jumex considers itself too new to the programme to have yet identified concrete reduction opportunities, but anticipates that it will do so.

Political–institutional effects

Consistent with the importance of political–institutional factors as key drivers for addressing climate change and for joining the Mexico GHG Program, several programme participants cited political–institutional effects as an important outcome of their experience with the programme.

Although companies did not provide detailed accounts of their experiences in this regard, they did cite two types of political--institutional benefits of their participation. First, participants have been able to 'institutionalise' their corporate GHG inventories by reporting them through a government programme. It can be inferred that they expect that, in the event of GHG regulation in Mexico, these reports will be considered in decisions regarding baselines and reduction obligations. Second, the programme has provided the opportunity for positive industry–government interactions. Ford explained it had been able to express its opinion to government officials and provide feedback which it anticipates will improve the success of both voluntary and mandatory programmes. Jumex highlighted the importance of its dialogue with government institutions regarding the development of national climate change policy and credit for early action.

Internal–organisational effects

The experience of preparing a corporate GHG inventory has created organisational change within participating companies in two ways. First, the inventory experience has raised awareness of the issue of climate change among employees. According to Grupo Modelo, employees now think of efficiency improvements not only in terms of energy savings, but also in terms of GHG reductions. Second, as noted by CEMEX, the inventory reporting experience has promoted a culture of transparency.

In some cases, these cultural shifts have extended beyond the companies' Mexican operations. In Ford's case, participation in the programme has raised CO_2 awareness at all levels within Ford Mexico, from plant engineers to the company's Board of Directors. In addition Ford Mexico shared its experience with Ford's Philippine operations, which subsequently decided to join the Philippine GHG Accounting and Reporting Program.

Socio-cultural experiences

Most companies interviewed reported that publicising their GHG inventory reports had had a positive effect on stakeholder relations. CEMEX reported a positive experience with reporting, Ford has benefited from the opportunity to inform stakeholders of its actions to address climate change, and PEMEX, which had been reporting on GHG emissions prior to joining the Mexico GHG Program, has benefited from being able to report through a government partner. Grupo Modelo, on the other hand, experienced an initial stumbling block on issuing its first GHG inventory report when it was grouped by media reports as 'among Mexico's most polluting companies'. Grupo Modelo considers it a risk that the brewing industry will be characterised as more GHG-intensive than it actually is, but did not indicate that this experience would stop it from reporting in the future.

Experiences with developing projects for GHG markets

Several of the companies interviewed for this study expressed an interest in developing GHG reduction projects in order to participate in GHG markets, and four of the six (CEMEX, Jumex, Modelo and PEMEX) had initiated the process of developing a CDM project prior to joining the programme. Anticipating that this would lead to demand for guidance and improved capacity on GHG project accounting, the Mexico GHG Program incorporated a project accounting module in the programme design. But, as discussed above, interest in carbon markets did not appear to be a principal driver for companies to join the Mexico GHG Program.

Successes

Despite its absence as a major driver for joining the Mexico GHG Program, several companies noted that their participation in carbon markets had benefited from the exercise of conducting a corporate inventory, thereby enabling them to identify more project opportunities. Jumex and Modelo stated that they had benefited from the process of 'learning by doing' in developing their projects and related documentation. Interviewees also expressed the view that the CDM in particular (PEMEX and Ford also mentioned voluntary markets) provides them with an important opportunity to obtain additional revenue and to promote GHG emission reduction projects inside their companies. Ford also expressed that its participation in voluntary programmes had allowed it to enter productive discussions about emissions reporting and market-based approaches in different countries. At the same time, the company would like to see these programmes become harmonised to accommodate emissions trading across different regions. Multinational companies such as CEMEX, Ford Motor Company and Wal-Mart de México as well as the state-owned oil company PEMEX have already established short- or medium-term targets for GHG reduction, energy efficiency improvement or renewable energy implementation. These companies all referred to their participation in carbon markets as one approach to help them achieve their targets in a cost-effective manner.

Barriers

The companies interviewed emphasised that significant barriers to the successful design and implementation of CDM projects remain. Some barriers such as a lack of internal capacity at the outset of project design have been overcome through participation in initiatives such as the Mexico GHG Program and the Cement Sustainability Initiative, or simply through experience. Other barriers, however, have proven more intractable.

Some companies emphasised the high transaction costs (related to registration with the CDM Executive Board and validation and verification costs); in the case of most of the projects that have been identified to date by the companies, these are higher than the income provided by the commercialisation of Certified Emissions Reductions

(CERs). Others emphasised policy barriers such as the need to demonstrate additionality or technical barriers; for example, Modelo, which is trying to implement a project to recover renewable energy from biomass by-products, must first resolve other environmental issues associated with the project such as the associated increases in emissions of nitrous oxides and particulates.

Conclusions

The experience of the Mexico GHG Program calls attention to the importance of internal–organisational and political–institutional factors as drivers for the development of a corporate climate change strategy and of participation in GHG reporting initiatives. Specifically, the companies interviewed for this chapter indicated that the existence of a corporate culture of sustainability, or social responsibility, preceded the development of a climate strategy or a GHG inventory.

On the political–institutional side, several of the companies highlighted that, as an OECD country, Mexico could face pressure to adopt increased mitigation measures; indeed, regulation directed at reducing GHG emissions is increasingly seen as inevitable. Therefore, the other important motivations for participating in the programme were the opportunity it provides for:

- Privileged political dialogue

- The potential to highlight the role of corporate GHG reporting through the programme

- The ability to establish a historical emissions baseline for consideration under future regulation

The importance of these internal–organisational and political–institutional factors may have previously been underestimated in efforts to engage the private sector in non-Annex-I countries, many of which have focused on market mechanisms such as the Clean Development Mechanism.

The prospective role of the Mexico GHG Program in establishing emission baselines for consideration in future regulatory schemes illuminates ways in which the programme could evolve. Among the strengths of the programme are the requirements that:

- All companies report on direct and indirect emissions in standardised metric units

- They report on all six GHGs covered by the Kyoto Protocol

- They specify a base year for tracking progress

However, several aspects of the programme still need to be strengthened, specifically:

- Quality control of inventories — credit for early action cannot be given unless there is a high degree of confidence regarding the evidence for early action (e.g. through a verified emissions registry)

- Benchmarking to establish what action represents business as usual versus a more advanced approach

- Clarity from policy-makers regarding willingness to consider registered data under a future policy scenario

An additional question relates to the ability of the Mexico GHG Program to facilitate GHG mitigation projects. Although the companies interviewed noted the utility of the GHG inventories created under the programme for identifying opportunities to reduce emissions, they also identified the need for additional technical capacity and incentives to implement reduction projects.

The Mexico GHG Program is well positioned to address the need for technical capacity development, but the question remains to what extent additional capacity would alter the cost–benefit equation for companies considering implementing GHG mitigation projects and thereby trigger actual GHG emissions reductions at the national level. It is likely that reforms would need to occur beyond the scope of the programme — for example, in the Clean Development Mechanism or in the national or international policy that drives the price of GHG emissions — in order to significantly alter this picture.

Despite its limitations, however, the Mexico GHG Program has created institutional and technical capacity for collecting and reporting corporate GHG information, and has fostered dialogue among stakeholders on the development of Mexico's GHG policy.

As Mexico's policy framework evolves, this foundation can be strengthened to support the policy needs as well as those of participating companies. However, a complementary policy and incentive structure is needed in order to take full advantage of this foundation in order to achieve real and lasting GHG efficiencies from Mexican industry.

References

CEMEX (2007) *Construimos Juntos, Informe de Desarrollo Sustentable 2006* (México: CEMEX; www.cemexmexico.com/rs/ids06/pdf/pdf.asp [accessed November 2007]).

Grupo Modelo (2007) 'Acciones en un Modelo Sustentable, Informe de Sustentabilidad 2007' (México: Grupo Modelo; www.gmodelo.com.mx/responsabilidad_social/informe_2007.pdf [accessed May 2007]).

Hoffman, A. (2005) 'Climate Change Strategy: The Business Logic behind Voluntary Greenhouse Gas Reductions', *California Management Review* 47.3: 21-45.

Kolk, A., and D. Levy (2001) 'Winds of Change: Corporate Strategy, Climate Change and Oil Multinationals', *European Management Journal* 19.5: 501-509.

Levy, D., and P. Newell (2000) 'Oceans Apart? Business Responses to Global Environmental Issues in Europe and the United States', *Environment* 42.9: 8-20.

—— and S. Rothenberg (2002) 'Heterogeneity and Change in Environmental Strategy: Technological and Political Responses to Climate Change in the Automobile Industry', in A. Hoffman and M. Ventresca (eds.), *Organizations, Policy and the Natural Environment: Institutional and Strategic Perspectives* (Stanford, CA: Stanford University Press): 173-93.

Oliver, C. (1997) 'The Influence of Institutional and Task Environment Relationships on Organisational Performance: The Canadian Construction Industry', *Journal of Management Studies* 34.1: 99-124.

Pulver, S. (2002) 'Organising Business: Industry NGOs in the Climate Debates', *Greener Management International* 39: 55-67.

SEMARNAT (Secretaría de Medio Ambiente y Recursos Naturales) (2007) *Proyectos MDL Con Carta de Aprobación. Subsecretaría de Planeación y Política Ambiental. México, D.F., México* (Mexico: SEMARNAT; www.semarnat.gob.mx/queessemarnat/politica_ambiental/cambioclimatico/Pages/mdl.aspx [accessed 5 June 2008]).

Tudela, F., S. Gupta and V. Peeva (2003) *Institutional Capacity and Climate Actions: Case Studies on Mexico, India and Bulgaria* (COM/ENV/EPOC/IEA/SLT [2003]6; Paris: OECD/IEA).

WBCSD (World Business Council for Sustainable Development) (2005) *The Cement CO$_2$ Protocol, CO$_2$ Accounting and Reporting Standard for the Cement Industry* (Geneva: WBCSD).

—— and WRI (World Resources Institute) (2004) *The Greenhouse Gas Protocol: A Corporate Accounting and Reporting Standard* (Geneva: WBCSD, rev. edn).

—— and WRI (World Resources Institute) (2005) *The GHG Protocol for Project Accounting* (Geneva: WBCSD).

Part III
Non-state actors and their influence on corporate climate change performance

10

The Climate Group: advancing climate change leadership

Jim Walker
The Climate Group, UK

The Climate Group[1] was formed in 2004 with one aim in mind — to prevent dangerous climate change by accelerating business and government leadership towards a low-carbon economy. Now, with more than 90 professional staff based in the UK, USA, Australia, China, India and Belgium, we believe we are the only international non-profit organisation to focus exclusively on climate change solutions and to convene business and government on the issue. This chapter describes the key activities of The Climate Group, presents a number of case studies where The Climate Group has catalysed significant change and reflects on the role that can be played by leadership initiatives in responding to the threat presented by climate change.

About The Climate Group

Goals and objectives

The formation of The Climate Group was driven by insights from research carried out in 2003 by Rockefeller Brothers Fund (New York) into the multiple actions being undertaken worldwide by businesses and governments. While these actions all combined success in reducing greenhouse gas emissions with financial benefit, this evidence was

---

1 www.theclimategroup.org

being largely ignored in the wider climate change debate which was focused on scepticism about the state of the science and the feasibility of meaningful action.

A meeting was convened in The Hague in 2003 involving diverse organisations including the city of Heidelberg, the state of California, BP and the UK Met Office, to share lessons learned and to develop a plan for the initiative that was subsequently launched as The Climate Group in April 2004 with the endorsement of then UK Prime Minister, Tony Blair, and 20 CEOs and VIPs.[2]

The Climate Group aims to work in partnership with a target list of approximately 50 governments (G8+5 nations, regions and cities) and 100 multinational businesses. Our current network of members and partners stands at approximately 50% of this goal. Our research and communications are primarily directed at 1,000 influential individuals worldwide, predominantly drawn from these 150 businesses and governments (i.e. 'C-suite' executives and political appointees respectively, and their lead advisors), but also from world-leading media channels and the opinion-former, investor and philanthropic communities.

Our mission is to help these key players understand the emerging financial and economic opportunities associated with leadership on climate change and the wider links between greenhouse gas emissions reductions and increased global prosperity. These links have recently been reinforced by the *Stern Review on the Economics of Climate Change*[3] and extensive research by McKinsey & Company[4] among others.

Our core belief is that strategic and political leadership among this small but influential community can bring about the necessary shift in the world economy towards a low-carbon model of energy generation and use by 2015. In particular, given the policy, financial and technological challenges faced in this shift, it is clear that collaboration within this community will be essential to devising and demonstrating the new business models and more effective government policies required.

Going forward we aim to facilitate stronger collaboration within and across sectors on market mechanisms and policy development. By understanding and communicating successes in greenhouse gas reduction in the multiple contexts that our partners operate (by sector, by continent and link in the value chain), we believe we can encourage this critical mass of organisations to act more urgently and with greater confidence on the issue.

Our view is that the window of opportunity for the world economy to succeed on climate change is closing fast and that 'business as usual' growth in global emissions, coupled with an unwieldy international process for negotiating an appropriate framework for action, means it will soon become too late to avert the worst consequences of inaction — likely to include major humanitarian and ecological disasters. Because of this inherent urgency and the rapidly evolving agenda surrounding the issue, we are committed to maintaining a flexible strategy and to pursuing 'whatever works' to meet our objectives.

---

2 The Climate Group's funding comes from a diverse range of sources including foundations, government funding, private individuals, corporate sponsorship and membership. In 2007 the organisation was selected as one of four NGO partners in the five-year HSBC Climate Partnership (see www.theclimategroup.org/special_projects/hsbc_climate_partnership [accessed 20 August 2008]).

3 www.hm-treasury.gov.uk/independent_reviews/stern_review_economics_climate_change/sternreview_index.cfm (accessed 20 August 2008).

4 www.mckinsey.com/clientservice/ccsi (accessed 20 August 2008).

Global membership network

The Climate Group currently operates out of offices in the UK (our international head-quarters), USA, India, China, Hong Kong, Belgium and Australia. Our core membership includes over 40 global corporations and governments, including HSBC, JP Morgan, News Corporation, Dell, Nike, Virgin, Google, BP, Duke Energy, Tesco, Swire Group, New York City, London, Victoria (Australia) and California.[5] Our current business membership is drawn from sectors with high direct impact (e.g. power, oil and gas, aviation) and those with significant wider influence on the economy through investment, supply chain, consumer behaviour and media coverage.

The prerequisites for membership are built around leadership, or an aspiration to leadership, in each sector. Members are also targeted for influence, i.e. their potential to significantly drive forward wider progress by virtue of their scale and profile. Although leadership is difficult to define precisely, one key criterion is a commitment to reducing greenhouse gas emissions within members' own organisations (absolute emissions or emissions intensity where companies are growing rapidly). Commitment is tested partially through the willingness of members to sign a set of Membership Principles[6] at board level, and also through a process of research and interviewing key staff during the recruitment process.

At the outset, we work with each member to develop an annual engagement plan, identifying actions that they can take in the coming year and targets for the partnership to assist in achieving those actions. Opportunities are identified relating to members' direct business, operational and policy activities, and through wider communications. An engagement plan typically incorporates both formal and informal agreements to collaborate on specific tasks. The Climate Group may lead on specific activities (e.g. events, convening working groups, conducting research), or may advise on or support member activities (e.g. developing a climate strategy, conducting trades on the carbon market).

Leadership groups

We estimate that up to a 20% reduction in greenhouse gas emissions can be achieved relatively easily in most organisations through 'good housekeeping' and through pursuing existing best practice in emissions management in higher impact activities.

With this in mind, The Climate Group has developed a structured Corporate Leadership Programme for its corporate members, focusing on achieving best practice. This programme is built around a framework of seven strategic objectives that leading organisations consider in tackling climate change, as follows:

- **Policy & Process:** developing and implementing greenhouse gas emission management strategies and policies on communication and government engagement

- **Pounds & Dollars:** financing low-carbon solutions, and managing the greenhouse gas emission impacts of wider investment portfolios

---

5 www.theclimategroup.org/index.php/our_partners/members (accessed 20 August 2008).
6 www.theclimategroup.org/assets/The%20%c2%baClimate%20Group%20Principles.pdf (accessed 20 August 2008).

- **Power:** managing the greenhouse gas emissions intensity of energy supply

- **Products & Services:** extending greenhouse gas emissions management through the supply chain, and developing and promoting lower-impact product and service offerings

- **People:** facilitating action among consumers and employees

- **Planes, Trains, Automobiles & Ships:** managing emissions from transport

- **Property:** reducing the greenhouse gas emissions associated with new and existing building stock

To support progress on these selected themes, The Climate Group supports Leadership Groups drawn from our partners. For example, under 'Products & Services' we support a leadership group on supply chain management where members can seek guidance on measuring and managing their greenhouse gas emissions across the value chain. Under 'Policy & Process' a Carbon Neutrality Group brings together organisations pursuing commitments to carbon neutrality to develop guidance for others to follow suit. The Climate Group's role in these groups is to facilitate progress, and to gather and analyse information on the many initiatives under way.

For government partners, a strategy is under development for state and regional government members, building on partnerships with signatories to the 2006 Montreal Declaration,[7] supported by a recent €1.5 million commitment from the Netherlands Postcode Lottery. Engagement with a limited number of cities is also planned as part of the HSBC Climate Partnership.

Building partnerships

Where feasible we seek to partner with like-minded organisations to deliver on shared objectives. Most recently, The Climate Group announced a partnership with Tony Blair to support his work on international climate policy. The partnership, Breaking the Climate Deadlock,[8] seeks to broker a new global deal on climate change between the main global economies, and supports the official negotiation processes of the UNFCCC and the G8.

The Climate Group and World Resources Institute (WRI) have been working together for the last three years to extend WRI's programme on energy procurement in large companies into Europe. The project, Green Power Market Development Group Europe,[9] was launched in 2005 as a partnership dedicated to building commercial and industrial markets for renewable energy. Partners include BT, Dow, DuPont, General Motors, Holcim, IBM, IKEA, Interface, Johnson & Johnson, Michelin, Nike, Staples, TetraPak and Unilever. The network has demonstrated the business case for corporate procurement

---

7 www.theclimategroup.org/index.php/news_and_events/event/regional_governments_endorse_ a_common_declaration_on_climate_change (accessed 20 August 2008).

8 www.theclimategroup.org/index.php/news_and_events/news_and_comment/breaking_the_ climate_deadlock (accessed 20 August 2008).

9 www.thegreenpowergroup.org/eu.cfm (accessed 20 August 2008).

and/or development of on-site 'green'[10] energy, and has evaluated and deployed a variety of renewable energy technologies. The Group has already completed more than 100 MW of projects and purchases since its launch, providing more than 50 corporate facilities in 16 European countries with green power.

The Climate Group also partnered with the British Council in 2005 on a ground-breaking international book and exhibition, *NorthSouthEastWest*,[11] communicating climate threats and opportunities around the world.

In addition, we regularly partner with business and government on events; collaborators include the UK Government, Florida Governor's Office, the State of Victoria (Australia), Toronto City Government, the Mayor of London and Arup. Finally, a partnership with Duke University (North Carolina) and the Cambridge Programme for Industry delivers executive education courses on climate solutions under the umbrella of the Climate Academy.[12]

Developing effective partnerships within the non-profit sector is challenging. At best, non-profit partnerships are essential to maximise the impact of a limited pool of philanthropic capital available for the issue by avoiding overlap of funded programmes. At worst, they can be a drain on management resources, with the need for consensus and coordination between partners, potentially acting as a brake on delivery. In our experience partnerships have been most effective where there has been:

● Clear delineation of roles

● Clarity of vision and mission

● Clarity of resource deployment

● A high degree of trust between the partners at the outset (most important)

Original research

To further support our partners and to develop knowledge and understanding throughout our network, The Climate Group runs a programme of webinars[13] on different topics of interest and undertakes original research on specific topics.

In September 2006, The Climate Group commissioned independent research aimed at better understanding the 'climate change conscious consumer'.[14] The work was under-

---

10 'Green power' includes renewable energy as well as efficient generation techniques such as combined heat and power (CHP).

11 www.northsoutheastwest.org

12 www.theclimategroup.org/index.php/special_projects/climate_leadership_programme (accessed 20 August 2008).

13 Web-based seminars with moderated interactive elements.

14 www.theclimategroup.org/assets/Serving%20CCC%20consumer.pdf; www.theclimategroup.org/assets/resources/Climate_Change_UK_07.pdf (both accessed 20 August 2008).

taken on behalf of a group of retailers and high-street brands which, at that time, wanted reassurance that there was a market for 'climate-friendly' products, services and brands. They also felt there was a lack of understanding of the types of brand messages that might resonate among consumers on this issue.

The research was carried out with Lippincott Mercer and was supported by BSkyB. It revealed that 28% of consumers in the UK and 19% in the USA were 'strongly concerned' about climate change, suggesting a potentially much larger market than for organic or Fairtrade products when those markets first took off. Furthermore, this research indicated that there was a latent demand for products, services and brands that would allow this group to reflect their concern about climate change in their spending. However, significant barriers to action were also identified, particularly around convenience and fairness; consumers wanted their commitment to be 'met halfway' by government and businesses and they wanted 'low-carbon' choices to be easy and affordable.

It was proposed that brands could play a number of roles — as educator, leader, facilitator, contributor and marketer — to empower consumers and, at the same time, unlock a significant business opportunity. Against this backdrop, in April 2007 The Climate Group launched 'Together',[15] a ground-breaking new campaign backed by some of the UK's major business brands committed to offering simple, inexpensive solutions to consumers. The campaign has been endorsed by the UK Government and civil society groups, has partner campaigns in the USA and Australia, and will soon launch in China.

The Climate Consumer Research project was developed in 2007 with the launch of the UK's first-ever Climate Brand Index, tracking year-on-year perceptions of how brands are performing on climate change.[16] Phase 2 of the research showed a significant gap between what consumers want and expect from brands on climate change and what they think they are doing about it. Awareness of what companies are doing is low and most people (69% in the UK, 74% in the USA) remain unable to identify any brands as taking a lead on climate change without prompting. However, the research also showed clearly that people want brands (rather than green specialists) to play a bigger role and that they are more likely to trust those companies acting under the auspices of an umbrella scheme such as Together rather than the companies acting on their own.

Another major research project we have recently conducted in partnership with the Global e-Sustainability Initiative (GeSI) and McKinsey & Company explores the direct, indirect and enabling roles of information and communication technology (ICT) in climate change. The aim is to understand both the climate change impact of ICT and how ICT could enable the reduction of emissions across the economy. The project presents in-depth case studies around key areas where there is significant abatement potential through ICT solutions. The final analysis brings the different strands together in an over-

15 www.together.com
16 www.theclimategroup.org/index.php/news_and_events/news_and_comment/consumers_
demand_green_action_from_uks_top_brands (accessed 20 August 2008).

arching report identifying where ICT can play a role in facilitating a low-carbon economy, and at what cost.[17]

Running across these specific research projects, we also track trends in emerging low-carbon markets and conduct research into how organisations across sectors are successfully reducing greenhouse gas emissions — focusing particularly on the economic and financial consequences of these actions. This research is made available to members and disseminated through a range of case study material and annual publications *In the Black*[18] and *Carbon Down, Profits Up.*[19]

Developing our leadership programme: monitoring progress

Because The Climate Group is a young organisation with an urgent and unprecedented mission, we need to be able to learn quickly which actions are effective and why, and to act on that knowledge.

In addition to our *Carbon Down, Profits Up* research, which enables us to monitor corporate performance on climate change and, in particular, to track our members' performance, we have also implemented an 'Impacts, Planning and Improvement' programme to analyse the impact of our activities, monitor member progress, inform strategic decisions and strengthen our core programmes. The information captured as part of this process is also a powerful communications resource, both for external communications and for reporting to funders and other partners.

Annual independent surveying of our members and of other businesses and government leaders began in 2007/08. Having established a baseline for business and government responses to climate change in seven world cities, future surveys will help us to monitor change and the extent to which progress is attributable to The Climate Group's activities.

We are also piloting an integrated planning approach in two core programmes: Corporate Leadership and External Communications. Programme directors and managers determine which outputs and impacts are worth measuring and how best to measure them, helping to ensure that the information gathered has genuine practical value. There is also group-level oversight to see that metrics that cut across programmes and regions (media coverage, reduced emissions, stakeholder engagement, etc.) are included in all relevant programme plans and that this data is recorded consistently.

Our leadership programme will continue to evolve as the findings from this monitoring process feed through, and as our group of core members grows and moves ahead with their own programmes. We are also extending and adapting the leadership programme as our activities and influence in China and India develop. Our strategy in both of these countries is to work with local leading enterprises and with the regional contacts of our European and North American members. This regional cross-sharing of information is something that our members particularly value and which will become an even greater priority over the next 2–3 years.

---

17 For the latest details, see www.smart2020.org.
18 www.theclimategroup.org/assets/resources/TCG_ITB_SR_FINAL_COPY.pdf (accessed 20 August 2008).
19 www.theclimategroup.org/assets/resources/cdpu_newedition.pdf (accessed 20 August 2008).

Case studies

Ultimately, all our work with our members and partners is focused on demonstrating best practice on action to reduce greenhouse gas emissions and accelerating action among the most influential organisations globally on this basis. To dig deeper into how this works in practice and to highlight the successes and challenges associated with our unique approach, the following section presents three case studies:

● Growing political momentum: building confidence in Florida and California

● Game-changing: the Voluntary Carbon Standard

● Building coalitions: the 'Together' campaign

Facilitating change: building confidence in Florida and California

One of the most important elements of The Climate Group's work is supporting political leaders who are prepared to show courage in setting the regulatory frameworks required to combat climate change. Although government action at national and international level – setting mandatory standards and making global commitments – is vitally important, regional policy and regulation can be effective in their own right as well as acting as a vital test bed for national climate change policy. Over the last three years, we have seen a range of ground-breaking targets and standards introduced at regional and state level, with clear policy strategies implemented to support their delivery.

The Climate and Energy Roundtable in California, convened by The Climate Group, brought together 14 CEOs, the UK's then Prime Minister Tony Blair and California's Governor Arnold Schwarzenegger to discuss how businesses and government could work together to address the global problem of climate change. Business participants spoke of wanting a long-term policy framework within which they could operate with confidence; the coalition of the willing which was formed played an important part in generating political momentum towards legislative change. Two months after the meeting, on 27 September 2006, Governor Schwarzenegger signed California's Global Warming Solutions Act (AB-32) into law, introducing aggressive targets to reduce greenhouse gas emissions to 2000 levels by 2010 and by 80% on 1990 levels by 2050: 'When you get the private sector and the public sector together, that is when you create real action', Governor Schwarzenegger noted.

A year later, at an event organised by The Climate Group, Florida Governor Charlie Crist met numerous corporate executives as part of the Serve to Preserve Florida Summit on Global Climate Change initiative. They explored how business and government could cooperate to develop an effective climate strategy. Governor Crist has now signed into law three executive orders on climate change, one of which aims to reduce Florida's greenhouse gas emissions by 80% from 1990 levels by 2050.

While The Climate Group does not claim responsibility for these policy breakthroughs, California and Florida's success in adopting two of the most ambitious climate change policy frameworks in the USA is a testament to the importance of consensus-building in developing successful public policy.

The Climate Group is convinced that partnerships and dialogues at the regional level, such as those facilitated in California and Florida, are a key avenue for influence. We also recognise the need to establish parallel activities at the global level encouraging:

● The sharing of best practice, knowledge and resources across regions internationally

● The building of dialogue between the business and policy sectors within this regional–international framework

This will help ensure that regional successes can be replicated and opportunities for public–private partnerships identified. With this in mind, we currently hold regular state and regions meetings at the UN Conference of the Parties (COP) meetings.

Game-changing: the Voluntary Carbon Standard

As mentioned above, The Climate Group is committed to maintaining a flexible strategy and to pursuing 'whatever works' to meet our objectives. This approach enables us to actively look for game-changing opportunities, i.e. projects we can undertake, or at least facilitate, that once delivered will significantly alter the playing field for our members in terms of reducing their negative impact on the world's climate.

One area we have focused on is voluntary carbon offsets. Although direct cuts in greenhouse gas emissions should be the first step for any organisation looking to minimise its impact on climate change, the voluntary carbon market — where buyers can purchase emission reductions generated by projects outside the compliance system to 'offset' or compensate for their own greenhouse gas emissions footprint — has an important role to play. Voluntary offsetting is a valuable transition solution because it allows additional investment in emission reductions over and above that which government regulations have achieved, and makes those reductions wherever they can be made fastest and at the lowest cost.

Over the last four years, many of the companies with which The Climate Group works have expressed an interest in voluntary carbon offsetting as part of a suite of solutions to reduce their greenhouse gas emissions footprint. However, soon after our launch, we noticed that concerns about a lack of transparency and accountability in the market were creating a significant barrier to action. This is why, in late 2005, working in partnership with the International Emissions Trading Association and the World Economic Forum, The Climate Group started work on the Voluntary Carbon Standard (VCS) — believing that global standardisation would make the market more robust and more appealing for our members.

Version 1 of the VCS was released on 28 March 2006 as both a consultation document and a pilot standard for use in the market. Version 2 was released in October 2006; 150 written submissions were received from carbon market stakeholders on VCS versions 1 and 2, and a Steering Committee with 19 members was established to consider these and develop the final standard.

2007 saw the launch of the final VCS[20] and the World Business Council for Sustainable Development (WBCSD) joined the initiative as a partner. VCS provides a robust,

20 www.v-c-s.org

new global standard and programme for approval of credible voluntary offsets. VCS offsets must be:

- Real (have happened)
- Additional (beyond business as usual activities)
- Measurable
- Permanent (i.e. not just temporarily displace emissions)
- Independently verified
- Unique (i.e. not used more than once to offset emissions)

Building coalitions: the 'Together' campaign

Another, very different area where The Climate Group has been able to move things forward quickly and practically has been in the UK consumer market with the creation of the 'Together' campaign. This campaign gives UK businesses a credible joint platform to develop and offer new products and services to customers that make it easy for them to reduce their personal carbon footprints.

As outlined above, Together was triggered by evidence — from our own research and that of others — that, although people frequently express a personal desire to do something about climate change, this has not so far been matched by significant levels of individual action. People often say that they find the challenge of tackling global climate change too daunting; they doubt that their efforts could be of any value and they fear that reducing their personal greenhouse gas emissions will be not only expensive but also inconvenient. Above all, consumers do not want to feel that they are acting alone.

The message behind Together is: 'doing something is easier than you think'. The aim is to help every UK household to reduce its greenhouse gas emissions by one tonne; a total of around 24 million tonnes over the next three years — more than the combined household emissions of Scotland and Wales.

As the name suggests, Together is built on partnerships between consumers and businesses, and between the businesses themselves. B&Q, Barclaycard, British Gas, Marks & Spencer, O2, Royal & Sun Alliance, BSkyB and Tesco have all united behind the campaign and are providing effective ways for people to take action. In January 2008, for example, six million energy-saving bulbs went into homes through a number of partner initiatives. The first was the light bulb amnesty run by the Mayor of London with B&Q and British Gas, followed by *The Sun* and Cool NRG giving away 4.5 million bulbs on the 19 January. Finally, Tesco ran a buy-one-get-one-free promotion throughout the month.

Since Together was launched, over 445,000 tonnes of CO_2 have been saved through 16 million individual actions — that's more than a third of households in London saving one tonne. This total is updated regularly on the 'Togetheriser', posted on the campaign website.[21] The Togetheriser adds up the emissions savings associated with every action taken by individuals through the brands in the campaign. It means people can see that

21 www.together.com

their contribution, however small, is helping towards a significant total saving when it joins the efforts of others.

Our future

The Climate Group's experience to date demonstrates how supporting leadership builds confidence and accelerates momentum towards change. Based on lessons from this experience, and on findings from the latest analysis and research, our focus going forward is to continue to support and facilitate progress among our network by developing our leadership programme and seeking out game-changing initiatives which can help to deliver step change across a critical mass of organisations. Across our remit, we will look for the highest-impact opportunities, applying this logic to the partners we work with, the regions we work in and the initiatives we work on.

One of our key challenges is to tread the fine line between working with influencers and leaders in a way that aligns interests — demonstrating that low-greenhouse-gas-emitting choices are good for business and that brands can benefit from communicating positively on this issue — while also ensuring that these organisations are making the appropriate steps forward to back this up. As well as our membership principles, the development of our monitoring and improvement programme is integral to this.

The scientific evidence shows clearly that action is urgently required if we are to avoid the worst impacts of climate change, which are likely to include both ecological and humanitarian disasters. The policy, technology and capacity lag mean that we have a very small and closing window of opportunity to tackle this problem.

We believe that our work at The Climate Group over the last four years has put in place strong foundations for taking things to the next level and helping to catalyse change on the scale required. Although 20% cuts in greenhouse gas emissions are achievable relatively easily through good housekeeping and implementation of best practice, the larger-scale cuts needed to deliver the low-carbon economy of the future will require new coalitions. The regional and international networks we have put in place, and which we seek to develop, will help to seed these partnerships across businesses, sectors and nations.

The purpose of these networks is to support existing public policy processes: for example, the G8 and Kyoto Protocol dialogues. There is no doubt that strong public policy — in particular a global international deal — is essential if we are to succeed in tackling climate change. But one of the most significant policy barriers is a fear of harming business. Therefore, helping key constituents in the process to view the low-carbon economy as synonymous with opportunity and prosperity, and demonstrating how this works practically and in different contexts, is of vital importance.

Over the next months and years, The Climate Group will strive to make its networks and their communications as effective and influential as possible, to positively drive forward the step change required to achieve a low-carbon future.

11
Climate protection partnerships: activities and achievements*

Oliver Salzmann, Ulrich Steger and Aileen Ionescu-Somers
IMD, Switzerland

Setting the scene

A series of high-profile climate protection partnerships — including initiatives such as The Climate Group, the Carbon Disclosure Project and the Corporate Leaders Group on Climate Change — have been established over the past five or six years. The supporters of these partnerships have argued that these partnerships have made a substantial contribution to the mitigation of greenhouse gas emissions as well as providing a range of other benefits such as stimulating innovation, developing corporate knowledge on climate change-related issues and, for the participants, the strengthening of corporate brand and reputation.

However, the overall contribution of these partnerships to climate change mitigation has not been studied in detail. Much of the literature has focused on individual partnerships in isolation, seeking to attribute changes in corporate performance to the actions of that partnership alone. In fact, the reality is much more complex.

From an analytical perspective, we are wary of any efforts to derive simple causal relationships between the activities of a partnership, changes in corporate performance and actual climate protection (e.g. a reduction in greenhouse gas emissions). Causality is difficult to assess because:

---

* This chapter is part of a wider study on the role and significance of sustainability partnerships in which we also looked at partnerships in the areas of human rights, public health and sustainable food. For further information see Steger *et al.* 2008.

- Even in the absence of the partnerships, the reality is that companies would be taking action anyway since they are also driven by regulatory pressures, media and non-governmental organisation (NGO) attention, and increasing investor interest in corporate climate change performance and risk management

- Companies are frequently members of a number of partnerships; hence it is difficult to say which, if any, partnership is the dominant influence on corporate performance

- Some partnership effects are impossible to isolate or to attribute directly to the partnership, such as effects on the policy debate

Despite these limitations, we believe that there is significant value in analysing the activities of the key partnerships in this area and the outcomes attributed to them. But, rather than trying to analyse the partnerships in isolation, this chapter seeks to analyse climate change partnerships in the round, comparing and contrasting their key activities and assessing their overall contribution to climate change protection.

In the research on which this chapter is based, we focused on nine of the major climate protection partnerships (see below), interviewing a total of 34 people across these nine partnerships. We complemented these interviews with 17 interviews with experts and stakeholders (such as representatives of the International Energy Agency [IEA], WWF, the US Energy Information Administration and the UK Department of Trade and Industry[1]).[2]

This chapter is divided into three parts as follows:

- An overview of the types of partnerships being seen in the climate change debate

- An overview of the main activities of the major climate protection partnerships

- An evaluation of the outcomes achieved to date from these climate protection partnerships

Climate protection partnerships

Some general comments on partnerships

For the purposes of our research we have taken a broad definition of partnership. We defined the term 'climate protection partnership' as a non-philanthropic form of collaboration between one or several corporate actors (excluding industry associations) and any other combination of non-profit (NGO) and public actors (local, regional or national governments or other public authorities) that aims to mitigate climate change: for example, by developing and commercialising renewable energy technologies, reducing gas flaring, or developing and implementing emissions inventories and ambitious emission reduction plans.

---

1 Now the Department for Business, Enterprise and Regulatory Reform (BERR).
2 For a more comprehensive description of the methodology see Steger *et al.* 2008.

Partnerships can be classified in a number of different ways — by their composition, mission or objectives, geographic scope, size, etc. For the purposes of our research, we classed partnerships according to their objectives — primarily because objectives represent the starting point for setting a partnership's activities and for measuring its performance and achievements. The typology (see Table 11.1) is the result of an extensive pre-study, in which we reviewed some 80 partnerships.

TABLE 11.1 **Partnership approaches**

Source: Based on Steger *et al.* 2008

Partnership approach to the issue	Mission/activities
Quasi-regulation	Partnership develops certifiable standard (e.g. Gold Standard, Voluntary Carbon Standard). Complementary activities may include promotion of the standard and certification
Advocacy	Partnership advocates the introduction of legislation (to reduce strategic uncertainty and achieve a level playing field), e.g. US Climate Action Partnership (USCAP), the Corporate Leaders Group on Climate Change
New business	Partnership develops, tests or uses new business — in the form of new technologies, new products or services, new business models and new markets (geographically, vertically, horizontally), e.g. the Green Power Market Development Group
Best practice	Partnership establishes a platform to develop and test best practices (e.g. Climate Savers, Climate Leaders)

Climate protection partnerships

To identify the key climate protection partnerships we employed a combination of desk research and expert interviews. We started off with an extensive internet search including examining the partnership websites and media coverage.

The result was a pre-selection of key initiatives in the field, which we then discussed with three experts. We compiled the final sample (see Table 11.2) on the basis of partnerships' innovativeness and significance.[3] We also tried to balance the sample across a variety of sampling dimensions such as partnership composition, level (pan-industry, industry, company level) and regional focus (see Table 11.3).

Classification of climate protection partnerships

Table 11.3 gives an overview how the nine partnerships studied were categorised across the four sampling dimensions (partnership approach, partnership composition, partnership level and geographic focus) used in our research.

3 The field of climate protection partnerships is extremely dynamic and, during our research, a number of interesting new initiatives were established (e.g. Climate Savers Computing Initiatives). However, these initiatives are too new — as yet — to allow a meaningful assessment of their effects.

TABLE 11.2 Partnership sample

Final sample	Partnerships additionally targeted
● Carbon Disclosure Project (CDP)	● Asia-Pacific Partnership on Clean Development and Climate
● The Climate Group	
● Climate Leaders	● Carbon Mitigation Initiative
● Climate Savers	● Centre of Excellence for Low Carbon and Fuel Cell Technology
● Corporate Leaders Group on Climate Change	
● Gold Standard	● CO_2 Capture Project
● Green Power Market Development Group	● Global Climate and Energy Project
● Global Gas Flaring Reduction Partnership	● International Climate Change Partnership
● US Climate Action Partnership (USCAP)	● London Climate Change Partnership
	● Low Carbon Vehicle Partnership
	● Methane to Markets Partnership
	● New Iceland Energy
	● Partnership for Climate Action
	● Pew Center
	● Renewable Energy and Energy Efficiency Partnership
	● UK Emissions Trading Group
	● UNEP Finance Initiative (Climate Change Working Group)

We recognise that this categorisation is not clear cut. For example, the Green Power Market Development consists of a group of businesses (energy buyers) that is convened by the World Resources Institute (WRI; a non-profit think-tank). The WRI works in collaboration with The Climate Group, which is a business–public partnership. The resulting network could be considered a tri-sector partnership. Furthermore it has, for example, a business leadership programme. However, we categorised The Climate Group as an advocacy partnership as we consider its activities in that area to be the core part of its mission – and the most high-profile of its activities.

Classifying the partnerships by approach

Business development partnerships (see Table 11.4) focus primarily on managerial and technical capacity-building. Compared with the other three approaches, these partnerships have had less of a focus on external communication and advertising. This seems to primarily reflect the current development of these partnerships with most of their projects (at the time of conducting this research) being in a pilot or business development mode. However, in some instances, we have seen external facing or advocacy activities such as participating in the public and political debate through leveraging the partnership's pool of knowledge to call for more supportive legislation.

Advocacy partnerships focus on lobbying key stakeholders (primarily governments and business) to either support the introduction of climate change-related legislation and/or to incorporate specific measures into this legislation. These activities are generally supported by research and capacity-building. This is particularly the case in policy–design partnerships, which contribute to policy discussions at a much more

TABLE 11.3 Partnership categorisation

Approach	
Advocacy	• The Climate Group • Corporate Leaders Group on Climate Change • US Climate Action Partnership (USCAP)
Best practice	• Climate Leaders • Climate Savers • Global Gas Flaring Reduction Partnership
Business development	• Green Power Market Development Group (EU and US group)
Quasi-regulation	• Carbon Disclosure Project (CDP) • Gold Standard
Composition	
Business only	• Carbon Disclosure Project (CDP) • Corporate Leaders Group on Climate Change
Business–NGO	• Climate Savers • US Climate Action Partnership
Business–public actor	• The Climate Group • Climate Leaders • Global Gas Flaring Reduction Partnership
Tri-sector	• Green Power Market Development Group (EU and US group) • Gold Standard
Level	
Pan-industry	• The Climate Group • Climate Leaders • Climate Savers • Corporate Leaders Group on Climate Change • Gold Standard • Green Power Market Development Group (EU and US group) • US Climate Action Partnership
Industry	• Carbon Disclosure Project (CDP) • Global Gas Flaring Reduction • Green Power Market Development Group (EU and US group)
Company	None
Region	
Industrialised countries	• Climate Leaders • Climate Savers • Corporate Leaders Group on Climate Change • Green Power Market Development Group (EU and US group) • US Climate Action Partnership
Developing and emerging countries	None
Global	• Carbon Disclosure Project (CDP) • The Climate Group • Global Gas Flaring Reduction Partnership • Gold Standard

TABLE 11.4 Partnership activities by approach

Approach	Activities
Advocacy	• *Policy demand:* demanding the introduction of a regulatory framework for climate protection (The Climate Group, Corporate Leaders Group on Climate Change)
	• *Policy design:* contributing to design of specific policy instruments (USCAP)
Best practice	• *Emissions inventories:* developing and promoting emission inventories (Climate Savers, Climate Leaders)
	• *Emission reduction targets:* setting and meeting emission reduction targets through changes in processes and products, e.g. increasing the uptake of green power, energy efficiency, offsetting (Climate Savers, Climate Leaders)
	• *Gas flaring reduction:* developing and promoting best practice for gas flaring reduction (Global Gas Flaring Reduction Partnership)
	• *Verification:* promoting third-party verification of inventories and emission reduction activities (Climate Savers)
Quasi-regulation	• *Standard development and promotion:* developing and promoting a quality standard for the social and environmental integrity of emission reduction projects
	• *Transparency:* facilitating dialogue between companies and stakeholders by improving the available information on business opportunities and risks associated with climate change (Carbon Disclosure Project)
Business development	• *New business models:* developing testing and promoting new business processes, new business models and innovative financing models (Green Power Market Development Group)

detailed level than policy–demand partnerships. Such partnerships may be time-limited as the introduction of the regulation that meets the partnership's objectives may make the partnership obsolete. Our interviewees were frequently uncertain as to the post-regulatory role of such partnerships; some expected that the introduction of legislation would signal the end of the partnership, whereas others saw a potential ongoing role for the partnership in monitoring policies and their implementation.

Quasi-regulatory partnerships start off with a clear focus on research and development to define a solid and credible standard or benchmark. Depending on the partnership composition (i.e. how inclusive it is) and the level of shared interest among partners, this development requires a significant amount of internal dialogue and engagement with external key stakeholders. Once the standard or label has been introduced, the focus generally switches to marketing and 'brand building' to ensure that the standard or label is swiftly taken up by the market.

Finally, best-practice partnerships concentrate on managerial and technical capacity-building, with the ultimate aim of defining and implementing best practice (emission inventories, ambitious emission reductions, etc.). But, compared with business development partnerships, external communication and advocacy (in particular) take a more prominent role. This probably reflects the fact that the best practices that have been developed through these partnerships are closer to the mainstream than the pilot projects emerging from the business development partnerships.

Partnership activities

General comments

Broadly speaking, all nine partnerships in our study have two characteristics (capacity-building and engagement and dialogue) in common. Capacity-building includes the generation and application of technical and managerial knowledge to:

- Formulate and employ best practice
- Develop and test new business models
- Define new standards or labels

Engagement and dialogue includes activities such as simple sharing of information among partners, communicating with key stakeholders and direct or media-based lobbying for climate legislation — which we refer to as advocacy.

We found that the key activities of a partnership will shift over its life-cycle. This is particularly obvious in the case of quasi-regulation-type partnerships where, once the partnership has developed its standard or label, its focus naturally shifts to other activities such as marketing, fundraising, brand building and maintenance.

Across the nine partnerships, we identified a number of key activities (discussed in more detail below). These activities were common to virtually all of the partnerships, although their relative importance varied between the partnerships.

Generating and sharing technical and managerial knowledge

The climate protection partnerships we studied have generated managerial and technical knowledge through:

- Conducting feasibility studies (for pilot projects) and market research (e.g. to identify key stakeholders and technologies)
- Developing new business models, processes and standards/labels

In addition to generating knowledge, the partnerships have developed partner capacity through activities such as:

- Establishing emission inventories
- Formulating and benchmarking emission reduction targets
- Developing and sharing standards, processes, technologies and business models (including new ways of financing) to reduce emissions
- Developing informed advocacy positions

This generated knowledge then feeds into other partnership activities. For example, in best-practice and business development partnerships, successful implementation on the ground (which, in turn, provides the impetus for partners and others to replicate the management processes and business models) is the ultimate litmus test. A similar approach can be seen in advocacy partnerships, where the generated knowledge (e.g. the six principles developed and made operational by USCAP) guides the partnership's advocacy approach.

If the required knowledge and expertise does not exist in the partnership, it is often brought in through brokering (i.e. bringing in a third party with the knowledge required). Brokering can be very efficient since it allows the partnership participants to tap expert knowledge rather than having to spend the time to develop this knowledge themselves. We also found that the credentials of a trustworthy, independent and knowledgeable institution (an 'honest broker'), as well as skilful facilitation, create an atmosphere in which the partners are willing to share existing knowledge. This aspect of knowledge sharing should not be underestimated, even if the partnership is convened and facilitated by an organisation that is able to devote specific resources to knowledge generation.

Securing and transferring knowledge

Knowledge generated from partnerships should be documented and used to its full potential. This means it should be passed on where it is needed and, ideally, put to use within the individual partner organisations and, where applicable, beyond. The stronger the emphasis on capacity-building, the more crucial this kind of knowledge management becomes, i.e. it naturally holds greater importance for best-practice and business development partnerships, and is less crucial to advocacy partnerships.

Our interviews suggest that securing and transferring learning is a major concern for partnerships in general. There are two reasons for this. First, it is largely based on soft factors, i.e. it does not so much rely on systems and processes that could be standardised but on experienced and well-connected individuals. Second, the partner organisations frequently lack the necessary time to capitalise on and exploit new knowledge. This latter point reflects the practical challenge faced by many companies where daily business priorities are frequently seen as more important than the wider issue of climate change.

There are no 'silver bullet' solutions to these challenges. Our interviewees pointed to a range of options that may help including:

- The involvement of third parties (e.g. consultancies) that could be specifically contracted for this form of knowledge management

- A conscious decision (time- and staff-wise) by company management to systematically document key learning in-house

- Regular reviews and contact with staff (essentially an attempt to standardise the process of securing and transferring knowledge)

Engagement and dialogue

In the partnerships we studied, engagement and dialogue occur in many ways such as event management, press releases, working with partners on internal and external communication, and updating and expanding websites.

The prominence of this engagement and communication is contingent on the partnership approach: it is the *raison d'être* for advocacy partnerships and quasi-regulatory partnerships (obviously after the standard and label has been defined), but is less important in best-practice and business development partnerships other than in so far as it is seen as a means to recruit new members, obtain positive media coverage, etc.

Interestingly, in the course of our research, we found that advocacy is being discussed as future strategy in the other partnership approaches (business development and best practice). This is because the adoption of certain business practices and models can be accelerated through new legislation, which would create first-mover advantages for partners (relative to their 'outsider' peers).

Fundraising

Fundraising is an important support activity for the partnerships we studied with, in the climate protection partnerships, funds mainly being raised through membership fees and donations (private and/or public, members and/or non-members).

The importance of fundraising as a specific activity depends on the size and activities of the partnership, and its sources of funding. However, several interviewees acknowledged that they had initially underestimated the significance of financial bottlenecks: in particular, at the point where the partnership was seeking to significantly scale up its activities and, in retrospect, would have put a stronger and earlier emphasis on fundraising. A number of interviewees noted the importance of understanding the financial potential of the different players and of having the flexibility of differentiated membership contributions (e.g. 'global heavyweights' versus companies with more modest financial returns).

Measurement

Alongside fundraising, measurement is one of the key support activities in partnerships. Its importance in the partnerships we examined was primarily driven by two factors:

- Corporate membership (i.e. where companies are striving for efficiency and transparency in their partnerships)
- NGO competition for funding and members

Interestingly, we found that public actors were less concerned about measuring effects. Our interviewees attributed this to their more reactive attitude and relative lack of expertise in this area.

Our analysis of the climate protection partnerships and sustainability partnerships more generally suggests that the level and rigour of measurement processes are critically dependent on the partnership's objectives. Many of the climate protection partnerships have the explicit objective of achieving reductions in corporate greenhouse gas emissions. This generally requires stringent measurement processes and a significant upfront investment of time and effort in setting up indicators and systems, and training staff to monitor and review performance.

In contrast, advocacy partnerships tend to rely on 'soft' (or non-quantitative) performance measures. The reason is that it is generally difficult to measure the specific effects (e.g. a change of opinion, voting behaviour, etc.) of these partnerships. Furthermore, causality is particularly difficult to demonstrate in the case of these partnerships.

Our interviewees pointed to a number of significant challenges in measuring partnership effects. Although the challenges are obviously partnership-specific, there were a number of recurring themes:

- Managerial and technical capacities are frequently insufficient. For example, setting up emission inventories is resource-intensive. However, it is also fair to note that after this upfront investment, the time and cost of performance measurement decreases

- Some partnerships need to deal with multiple co-existing systems (e.g. variations in reporting requirements on greenhouse gas emissions in different schemes and registries). The resulting problem of compatibility and redundancy also occurs across different countries and facilities. These variations have the effect of increasing the partnership transaction costs

- It is generally difficult to measure indirect and soft effects such as changes in mind-sets and organisational cultures. Some partnerships are currently experimenting with survey-based approaches (e.g. asking their members how they see or value the partnership's contribution) as a means of gathering this information

- Data availability is a significant bottleneck due to unwillingness (rather than inability) to disclose information. For example, data on gas flaring is a sensitive issue and other emissions-related information may be commercially sensitive

Finally, we found that approaches to measurement change at different stages in partnerships. At the point of formation, simple measures such as the number and diversity of members, and the number and kind of activities carried out may suffice. In addition, proxy measures such as media coverage are also useful in assessing the contribution of the partnership. As the partnerships mature and thus effects materialise (acknowledging that there is always a lead time between actions and outcomes), the focus can turn to measuring the actual partnership effects.

Rollout

Successful rollout (i.e. leveraging and expanding partnership activities and thus influence) is the ultimate challenge. Our interviewees acknowledged that, in order to successfully roll out their partnerships, they need to gain knowledge about new members, industries, regulatory frameworks and markets, while often lacking the resources to cover the full scope of these issues. This issue of lack of resources appeared particularly prevalent in those partnerships that rely heavily on hard data to assess key issues and set priorities.

In the following paragraphs, we discuss several key dimensions of partnership rollout, namely:

- Communication and marketing
- Internal versus external rollout
- Quantitative or qualitative growth
- Variations in rollout as a function of partnership approach

Communication and marketing

Communications and marketing are generally important since they ensure that partnership activities become recognised within the partner organisations (most importantly among those individuals not dealing with climate change and partnership activities) and beyond (e.g. among key stakeholders and potential new partners).

However, we found that communication and marketing have different levels of importance across the different partnership approaches. Although external communication is an integral part of advocacy (e.g. getting the message out to key decision-makers to change policies), it appears to be of secondary importance for other partnership approaches (e.g. business development and best practice).

In many cases, the partnership hub organises activities such as PR campaigns and lobbying activities, with the partner organisations also conducting their own communication activities. We found that not all corporate partners are equally keen to take a strong public approach on the issue of climate change. From our research, it appears that this can be primarily attributed to differences in organisational cultures (outspoken versus reserved) and the company's position on the learning curve in the area of climate protection, with partners that are new to the topic and to the partnership tending to be less confident and less willing to 'stick their neck out'. This latter point also seems to explain why some of these companies are happy for the partnership hub to lead on the lobbying process.

Internal versus external rollout

We identified several kinds of rollout activities during our research. Internal rollout takes place within the existing settings of the partners and includes activities to promote and advertise partnership activities such as through roadshows supported by the partnership hub or through the corporate owner/standard bearer of the partnership. Partnership influence can also be strengthened through external expansion by, for example, recruiting new members (also in new geographical locations) or taking on board new activities. Our respondents had different views about the need to expand their partnership, though the specific views depended on factors such as the partnership approach and corresponding external barriers. For example, one of the non-advocacy partnerships surveyed had considered the possibility of establishing an Asian office. However, it eventually decided against it since the national settings failed to provide a reliable framework and thus a solid business case for partnership activities in this region.

We identified three different external rollout strategies: namely, vertical integration, internationalisation and diversification. Vertical integration and internationalisation can go hand in hand: for example, if a partnership considers including upstream emissions (those embedded, for example, in products purchased or in supplier activities) or downstream emissions (those associated with product use, logistics and disposal). This breadth of scope often means that new geographical areas may have to be covered. This can be a substantial challenge, particularly if the value chain is complex and suppliers/customers in developing countries lack the necessary technical and managerial capacity.

Internationalisation may also be a strategy on its own; this applies in particular to advocacy and business development partnerships. By establishing satellites in other locations, such partnerships hope to be more effective through an improved local or regional presence tailored to the legislation, culture and capacity. This strategy also sets the partnership onto a broader base; this is a factor that may become of significance when it comes to developing a more stringent international legislative framework on climate change in the future. An internationalisation of advocacy partnerships can also occur without the establishment of actual satellites when the approach to advocacy includes both national and international dimensions. For example, USCAP is providing input to the *US* political debate on how to navigate towards *international* climate change legislation.

Diversification is an option our interviewees outlined in the context of widening the partnership focus. For example, a satellite may be formed by a more proactive subset of partner organisations to embrace new and more ambitious partnership activities. Diversification also occurs when partnerships venture into areas initially untouched: for example, when approaching a major tipping point in the partnership's development such as changes in legislation, reaching a critical membership mass, emerging consensus, a flattening learning curve. Such tipping points may call for changes in both missions and activities. Finally, diversification can be an essential means to secure partnership comprehensiveness, and thus legitimacy and power. In the case of USCAP, after a period of fast (and less controlled) partnership growth, the partners subsequently focused their attention on getting key industries on board.

Quantitative versus qualitative growth?

Rollout activities are pursued to achieve different kinds of growth. Some of the partnerships surveyed put their emphasis on growing quantitatively by, for example, boosting the number of partner organisations — and going about it in a somewhat opportunistic way.

Others have adopted a clearly qualitative approach to partnership growth. For example, the Climate Savers programme strives to establish *one* model company for greenhouse gas emissions reduction in every industry, i.e. extending the partnership beyond the leaders is not an aspiration. The same applies to several other partnerships in which the recruiting processes were similarly selective. Examples include:

- USCAP's efforts to achieve a broad representation from key industries and avoid 'naysayers'

- The Climate Group's prioritisation of expansion in key geographical areas — specifically areas that are either characterised by a high (current or future) carbon footprint or by a vast availability of energy resources

Overall, across the nine climate protection partnerships, we detected a strong preference for strategic rather than qualitative growth. The primary reason was the downside of having to manage greater diversity, which was widely seen as outweighing the potential upside of achieving greater leverage through sheer partnership size.

Rollout by partnership approach

We identified different definitions of rollout across the four partnership approaches:

- Business development partnerships aim to mainstream products, technology or business models

- Advocacy partnerships seek to strengthen their advocacy position and effectiveness

- Quasi-regulatory partnerships promote their respective standard or label

- Best-practice partnerships promote the best practice identified

Overall, we found similar rollout activities across the nine partnerships, with vertical integration and internationalisation being the most common. Most of our interviewees saw significant rollout potential for their particular partnership approach. They also reported significant external barriers such as a lack of political endorsement and inadequate regulatory frameworks (e.g. a failure to provide necessary incentives for green power). They also acknowledged significant internal barriers — most importantly a lack of financial resources.

Across most of the partnerships examined, we found a gradual shift from capacity-building and pilot projects to more engagement and lobbying (i.e. advocacy). This tends to reflect the partners' increased confidence in the partnerships (e.g. in the technical and managerial capacity built over time), more comprehensive knowledge about key stakeholders and the more intensive public and political debate on climate change since the beginning of 2007. Simply put, the partnership participants now have a better understanding of what constitutes an effective and realisable regulatory framework, and hope to influence policy-makers accordingly.

Partnership achievements

In this section, we report on the outcomes of the partnerships examined. Several key findings emerged, although they did not apply necessarily across all partnerships. We emphasise that we report on partnership achievements as *perceived* by our interviewees, with the risk that these data may be subject to biases such as social desirability and self-representation. We have sought to reduce this risk by, as discussed above, triangulating our data through:

- Interviews with stakeholders and experts

- Complementary desk research (databases and media)

Effects on policy-making, intangibles and management practices

Evidently, partnership outcomes are determined by the partnership mission, activities and resources. Hence it is possible, broadly speaking, to differentiate between advocacy partnerships and other partnership approaches.

Advocacy partnerships primarily seek to influence the public and political debate. This influence can be direct (through immediate contact with policy-makers and opinion leaders) and indirect through, for example, media advocacy. Direct advocacy also occurs behind the scenes and thus is virtually impossible to assess.

In contrast, the three other partnership approaches primarily affect intangibles (knowledge, mind-set, corporate culture, etc.), with the resulting new managerial and technical capacity then leveraged to improve management practices. In the partnerships studied, for example, companies established emissions inventories, formulated reduction targets that were benchmarked and adjusted where necessary and, to deliver on these targets, implemented measures to increase energy efficiency and purchase more renewable energy. We also came across changes in organisational structure such as strengthened links between health, safety and environment (HSE) teams and facility or operational managers, and in technology and products. In relation to this latter point, an example is Nike which cushions its shoes with pressurised gases. As a Climate Saver organisation, it substituted sulphur hexafluoride (SF_6) – a potent greenhouse gas – with nitrogen and other gases that have significantly lower climate impacts.

There also appear to be knock-on effects on public policy from the three partnership approaches. For example, quasi-regulatory partnerships are able to stimulate the debate in their individual field, with specific examples including the social and environmental integrity of emission reduction projects (Gold Standard) and carbon disclosure practices and their relevance (Carbon Disclosure Project).

Moreover, all four partnership approaches contributed positively to an improved licence to operate, brand and reputation. Although none of the partnerships surveyed was designed to provide regulatory relief, several interviewees argued that their association with a climate protection partnership had improved their reputation, helped build trust with regulators and helped improve their relationships with other stakeholders.

Low-hanging fruits — flattening learning curves

From the nine partnerships studied, it seems that best-practice partnerships in particular tend to be initially about picking low-hanging fruits. This can be seen from the relative ease with which several corporate partners were able to meet and even exceed their emission reduction targets. It also suggests that companies tend to be inappropriately equipped to deal with new and complex issues such as climate protection. It can often come down to a question of priorities and thus managerial and technical knowledge; as one of our interviewees commented: 'They often simply have the wrong person in charge – a PR guy rather than a more technical person.'

The learning curve of partnerships tends to flatten. This has two implications. First, if partnerships want to persist, they need renewal. Second, partnerships are attractive for late entrants since they benefit from experienced peers and capacity already built. However, it can also be risky to bet on becoming a 'smart follower' later in the process given partnerships' preference for qualitative growth, which may mean that the partnership is closed to new entrants if the leading companies are already on board.

Minor effects on the bottom line, strategy and business systems

Despite the profile of climate change, we found only very small effects on the corporate bottom line. The strongest effects are visible on the cost side, with several companies reporting that they have saved millions of dollars every year on their energy bills — though these are marginal cost savings relative to other financial metrics (turnover, net income, etc.).

From our research, it also appeared that the partnerships had limited direct influence on companies' business strategies or day-to-day operations. However, we recognise that this may understate the influence of some of the partnerships as the broader effects on intangibles such as organisational culture and knowledge may also have influenced companies' overall sustainability strategies and internal attitudes towards advocacy.

Advocacy partnerships have made significant contributions to legislative processes (most prominently in the UK). They removed a stalemate in which governments were struggling to develop and implement legislation without a clear supportive signal from the corporate sector. Furthermore, business development and best-practice partnerships have managed to strengthen networks along the energy value chain (e.g. energy producers and users) and among key stakeholders. These changes may have the effect, over the longer term, of changing the manner in which:

- Companies manage their greenhouse gas emissions at the operational level

- Climate change is considered in discussions around business strategy

Unexpected effects

We also asked our respondents about unexpected partnership effects they had experienced. The vast majority reported that the partnership had delivered positive effects that had gone beyond their initial expectations. Among other things, they mentioned:

- Strong support by the partnership hub

- Increases in membership

- Quick access to key stakeholders and resulting endorsement (e.g. advocacy by non-advocacy partnerships, which increased political traction)

- High willingness to share information

- The visibility of partnership projects (e.g. a large solar installation which became a local tourist attraction)

Overall, this suggests that:

- The expectations in climate protection partnerships are not overwhelmingly high

- Organisations and individuals are still at the lower end of the learning curve

- Previous experiences with partnerships cannot have been overwhelmingly positive

Overall effectiveness

It is difficult to make a meaningful assessment of the overall effectiveness of the partnerships studied because each has quite different objectives and contributes to climate protection in different ways. For example, according to our interviewees, the Climate Savers programme provides for credible targets and emission reduction (through third-party verification, robust emission inventories), whereas the Green Power Market Development delivers more concrete information and networking opportunities (with a focus on building new business models for green power).

It appears that the advocacy approach has more far-reaching implications than the other three approaches. We base this conclusion on two factors. First, advocacy requires a minimum level of expertise and confidence and thus tends to be a step that logically follows from capacity-building, which is one of the key elements of the three other approaches (business development, best practice and quasi-regulation). Second, there is only so much that can be achieved at the grassroots level since such activities require relatively resource-intensive (local or regional) interaction within the partnership and beyond and, inevitably, companies require clearer incentives for climate protection from their stakeholders (through regulation in particular). In contrast, an advocacy partnership interacts with fewer institutions (primarily governments) and, if successful, can affect a wider circle of actors since it changes the rules of the game for all.

Conclusion

Given that we examined some of the *leading* partnerships in the field of climate protection in this study, it is interesting that we did not come across any *dramatic* effects; participants' business models do not exhibit any ground-breaking developments in the context of climate change other than the growth of niche markets (e.g. renewable energy technologies, emissions trading). This is not entirely unexpected given that climate change is yet to develop significant business relevance such as a clear effect on the bottom line (be it direct on costs or revenues or indirect on brand and reputation) in most industries. We also conclude that measuring the specific outcomes from the nine partnerships studied is difficult due to various measurement challenges; most importantly, internal capacity deficits and the high complexity of the climate change debate (e.g. multiple actors and systems, intangible and indirect effects).

Although the partnerships we looked at did not have any dramatic effects, we came across several meaningful and complementary contributions to driving corporate and policy responses to climate change. First, advocacy partnerships have helped to resolve a significant Catch 22 situation in climate change policy: namely, that, while companies need regulatory certainty for their investments, governments are hesitant since they are uncertain as to how positively companies (and thus their national economies) would react to climate policies.

It is interesting to note that we may be witnessing the emergence of a new business perspective on climate change-related regulation with, for example, the Bali Communiqué, which was signed by 150 CEOs from global corporations, calling for a legally binding UN framework. Second, the other three partnership approaches provide platforms for innovating and testing products, business models and processes under existing regulatory regimes. Hence they supply essential input to the policy arena by generating essential data on issues such as possible emission reductions and the profitability of specific technologies by:

● Building trust and consensus across different sectors and industries

● Boosting the confidence of individuals and organisations to take action on climate change

Finally, from a business perspective, if there is no alternative to stricter legislation in the mid to long term, partnerships involving a critical mass of companies can influence policy in a positive way through, for example, contributing to the creation of favourable test markets for low-carbon business models (e.g. Europe) or to a more level playing field globally (if a tougher post-Kyoto framework were to emerge). Partnerships may also help to build networks across industries and value chains to drive technological advances and innovative business models.

If public and regulatory pressure is expected to increase, climate protection partnerships have a meaningful and important role to play, and should not be underestimated as a critical part of the public policy toolkit.

Reference

Steger, U., A. Ionescu-Somers, O. Salzmann and S. Mansourian (2008) *Sustainability Partnerships: The Manager's Handbook* (Basingstoke, UK: Palgrave Macmillan).

12

The evolution of UK institutional investor interest in climate change*

Rory Sullivan
Insight Investment, UK

Stephanie Pfeifer
Institutional Investors Group on Climate Change, UK

Setting the scene

Climate change is probably the most significant environmental issue faced by the world today. Yet it is probably fair to say that the majority of institutional investors have been slow to play an active role in the climate change debate. This may seem surprising given the potentially major financial implications of climate change for companies as a consequence of the physical impacts of climate change (e.g. the effects of extreme weather events on water utilities and on the insurance and reinsurance industries) and from government action to encourage businesses to reduce greenhouse gas emissions (e.g. the effects of emissions trading on electricity utilities). Recent years have, however, seen institutional investors start to play a more proactive role, with some using their influence to encourage companies to adopt more proactive approaches to managing the risks and opportunities presented by climate change and others starting to explicitly factor climate change risks and opportunities into their investment analysis. Investors have also started to play a greater and more constructive role in the public policy process.

* This chapter is based on a research project that sought to analyse the factors that influenced the attitudes of UK institutional investors to climate change over the period 1990–2005 (though the material is updated to account for developments over the period 2005–2007) (Pfeifer 2005; Pfeifer and Sullivan 2008) and on the authors' experience in a major UK asset manager (Sullivan) and with the Institutional Investors Group on Climate Change (Sullivan, Pfeifer).

This chapter reviews the evolution of UK institutional investor interest in climate change from 1990 to the present, focusing in particular on the relative contributions of 'soft' policy measures such as information disclosure and awareness raising, and 'hard' policy measures such as regulation and market-based instruments. The chapter canvasses the broader implications of these findings for government efforts in the UK and elsewhere to encourage investors to play a more proactive role in the climate change debate. The chapter also discusses the barriers to the greater integration of climate change concerns into investment decision-making and the potential for institutional investors to contribute to public policy efforts directed at reducing greenhouse gas emissions.

Social and environmental issues in institutional investment

Institutional investment in the UK: an overview

The focus of this chapter is on the institutional market. This is a general term for:

- Investments managed or controlled by insurance companies, pension funds and investment managers
- Investment managers in pension funds, mutual funds and other pooled investment vehicles

The key actors

Pension funds invest on behalf of large numbers of individuals (the beneficiaries) who have their pensions and savings invested in these funds. Most pension funds in the UK operate under trust law, which places a fiduciary duty on their trustees to act in the best interest of their beneficiaries. The obligation is usually interpreted exclusively in financial terms as the optimisation of investment returns (Allianz Group and WWF 2005; Sullivan and Mackenzie 2006a; WEF 2005),[1] with the result that environmental issues such as climate change were historically ignored by most institutional investors when making investment decisions.

Investment consultants provide a range of services to pension fund trustees including advice on the selection and monitoring of investment managers, the selection of benchmarks against which fund managers' performance can be judged, investment time horizons and asset allocation analysis (Dlugolecki and Mansley 2005; Mercer 2005; WEF 2005; Whitaker 2006). Investment consultants are, therefore, a critical influence on pension fund and fund manager decisions.

---

1 For a challenge to this view, see Freshfields Bruckhaus Deringer 2005.

While some pension funds manage their funds in-house, the vast majority use external investment managers (or fund managers) and specify the financial targets (expected returns, risk, volatility) to be met by the fund managers. Investment performance is usually measured against a benchmark index and usually over relatively short periods, typically quarterly or annually (WEF 2005; Sullivan and Mackenzie 2006a). An important consequence is that longer-term (i.e. beyond 2–3 years) impacts and risks tend to be significantly discounted or even excluded from the scope of analysis by fund managers.

Finally, brokers sell research and investment recommendations to active fund managers, who invest on the basis of financial analysis. Brokers are referred to as being on the 'sell side' as opposed to fund managers who are referred to as being on the 'buy side'.

Addressing social and environmental issues in investment processes

In broad terms, there are four main strategies for addressing social and environmental issues in investment processes (Table 12.1).

TABLE 12.1 **Strategies for addressing social and environmental issues in investment processes**

Type	Description	References
Screening	Involves either excluding companies (negative screening) or including companies (positive screening) in investment portfolios based on a range of social and/or environmental criteria	Michelson *et al.* 2004; Solomon *et al.* 2004
Best-in-class	Investment is allowed in all sectors, but investments are preferentially made in those companies with the best performance on specific social or environmental issues	Michelson *et al.* 2004
Engagement or activism	Investors — individually or collectively — use their formal rights as shareholders (e.g. voting rights) and informal influence to encourage improved corporate performance on these issues	Just Pensions 2005; Sparkes and Cowton 2004; Sullivan and Mackenzie 2006b
Investment integration	Social or environmental risks or opportunities are taken into account in conventional financial analysis	For examples see Lim 2006 and Sullivan *et al.* 2006

Why should climate change be of concern to institutional investors?

The *prima facie* for investors to be concerned about climate change is reasonably clear. Companies may be affected by the weather-related impacts of climate change (e.g. droughts, floods, storms, rising sea levels), public policy measures directed at reducing greenhouse gas emissions, and consumer or other pressures to take action on climate change. There may also be opportunities for companies to improve their brand or reputation through taking a proactive approach to responding to climate change or through developing new products or technologies in areas such as renewable energy. The manner in which companies respond to these risks and opportunities may have significant financial implications and may therefore affect the performance of investment portfolios.

There is a wider dimension to the argument that investors should be concerned about climate change, which is that institutional investors generally hold shares in many different companies and thus they are exposed to the economy-wide consequences of climate change. That is, if the economy-wide impacts of climate change are negative, this in turn is likely to have negative consequences for the value of investors' investments. Furthermore, given that investors are invested across a range of economic activities, they will therefore be affected by the impacts of the actions of individual companies on other companies held in their portfolios.[2] An example, albeit quite simplified, could be an investor who holds shares in an electricity utility that emits greenhouse gases but also holds shares in a water utility that has to spend money to address the flood risks resulting from changed patterns of rainfall. In this type of situation, the investor could encourage the electricity utility to reduce its greenhouse gas emissions to the benefit of its portfolio as a whole.

The evolution of UK investor interest in climate change

Late 1980s to mid-1990s

Scientific evidence of climate change increased from the mid-1980s. The establishment of the Intergovernmental Panel on Climate Change (IPCC) in 1988 and the subsequent publication of its first Assessment Report in 1990 (IPCC 1990) helped to push climate change onto the political agenda (Leggett 2000). The press coverage of climate change showed a peak around this time (SustainAbility *et al.* 2002) and some companies began to set themselves emission reduction targets and to search for new business opportunities focused on energy efficiency.

However, institutional investors showed limited interest in climate change during this period. 'Socially responsible investment' or SRI (also referred to as 'ethical investment') fund managers were the first institutional investors to address climate change issues in their investments. For example, the Jupiter Ecology Fund and the NPI Global Care Growth Fund (set up in the late 1980s and early 1990s) used screened approaches to avoid investments in large oil and coal companies, and sought to invest in companies providing solutions to the problem of global warming (Sparkes 2002). These screened funds won some institutional mandates, primarily from local authorities (Sparkes 2002; Sparkes and Cowton 2004). However, the vast majority of pension funds and fund managers, and their brokers and investment consultants, were paying no attention to climate change at this time (Pfeifer 2005).

---

2 This is a very abridged version of the 'universal owner' argument proposed by Hawley and Williams (2000).

It was these SRI fund managers and analysts who, in the early 1990s, began the process of engaging with companies on social and environmental issues, including climate change. Reviewing this period, Mark Campanale, Head of SRI Business Development at Henderson Global Investors, argued:

> Engagement was generated by asset managers with active fund management SRI teams, Jupiter and NPI, and describes the lobbying impact they had with all companies held in the broader parent company investment portfolios. It was done out of curiosity, but also out of the deeply held belief that companies are ultimately accountable to shareholders — and to get companies to be more responsible, we needed to convince them that shareholders were genuinely interested in climate change.

NGOs made some attempts at influencing institutional investors on climate change during this period. For example, a 1994 Greenpeace report on the financial implications of climate change on fuel companies (Mansley 1994) helped to raise investors' awareness of the issue.

Mid to late 1990s

The IPCC released its second Assessment Report in 1995 (IPCC 1995). The subsequent adoption of the Kyoto Protocol in 1997 marked a major milestone in the political debate about climate change and resulted in another peak in press coverage of the issue (SustainAbility *et al.* 2002). In addition, more companies started to talk publicly about climate change mitigation with BP, for example, setting itself an emissions reduction target during this period (BP 2005).

Despite the growing profile of climate change, the issue still had limited visibility for institutional investors during this time. In an interview, David Russell, SRI adviser at the University Superannuation Scheme (USS) — one of the largest pension funds in the UK — suggested that:

> The policy implications of the Kyoto Protocol were unclear at this point so that from a mainstream investment perspective there was no reason for it to be an issue at the time . . . Most pension funds would have been completely unaware of the issue.

Some pension funds (e.g. Nottinghamshire County Council pension fund) published reports on the financial implications of climate change but these received little attention in the wider pension fund community (Pfeifer 2005).

The mainstream investment community continued to ignore the issue of climate change during this period, except for a small minority of brokers who analysed what are referred to as 'small caps' (i.e. companies with a relatively small market capitalisation) in areas such as solar energy.

2000-2004: government pressures on pension funds

UK pension funds were encouraged to be more active in relation to social and environmental issues generally by the amendment to the Pension Act 1995, effective from July 2000, which required trustees to disclose in their Statement of Investment Principles

(SIPs) 'the extent (if at all) to which they have taken into account social, environmental or ethical considerations in their investment process'. Most pension funds responded by changing their SIPs to state that they (or their fund managers) would engage with companies to encourage improved performance on these issues and/or would explicitly analyse these issues in their investment decision-making. While the amendment to the Pensions Act is widely recognised as having catalysed increased interest in these issues in investment, the principles articulated in SIPs were generally not incorporated into the specific mandates awarded to investment managers (Just Pensions 2004).

The 2001 Treasury Review of institutional investment — commonly referred to as the Myners Review (Myners 2001) — encouraged institutional investors to be more active in relation to their ownership of companies and in particular to take account of governance, social and environmental matters within companies. The investment industry responded by producing the Institutional Shareholders Committee (ISC)'s code, which provides guidance on how investors should monitor the performance of and establish regular dialogue with the companies in which they are invested. The ISC is widely supported by the institutional investment industry.

Although the amendment to the Pensions Act and the Myners Review were important starting points in encouraging investors to take account of social and environmental issues in their investment processes, these were not the most important drivers for investors to focus specifically on climate change. Of much greater significance was the pressure from the beneficiaries of one of the UK's largest pension funds (USS) which resulted in the publication of a report which was an important step in alerting pension funds to the implications of climate change. The report, *Climate Change: A Risk Management Challenge for Institutional Investors*, was published in August 2001 (Mansley and Dlugolecki 2001).

2000–2004: the emergence of collaboration

Over the period 2000–2004, fund managers and pension funds set up several collaborative initiatives to address investment risks and opportunities associated with climate change. The most significant of these initiatives were:

- The Institutional Investor Group on Climate Change[3] (IIGCC) — established in 2001 to engage with companies on their climate strategies and to emphasise to policy-makers the need for long-term policy targets in order to provide an environment facilitating long-term investment decision-making (Sullivan *et al.* 2005)

- The Carbon Disclosure Project[4] (CDP) — launched in 2000, was a disclosure request from fund managers and pension funds to the largest companies in the world on their greenhouse gas emissions

3 www.iigcc.org
4 www.cdproject.net

● The Enhanced Analytics Initiative[5] (EAI) — established in 2004, committed its members (a number of large UK and European asset managers and pension funds) to allocate 5% of their respective broker commission budgets to brokers who are effective at analysing material governance, social and environmental issues

2000–2004: changes in brokers and asset managers

Several large asset management companies recruited or expanded teams in the late 1990s/early 2000s to analyse social and environmental issues and to engage with companies on these issues (Sparkes 2002, 2006). Around this time, Friends Provident and NPI — followed by a small number of other large asset managers — also announced that they were extending this engagement to all their funds under management and not just their SRI/ethical funds (Sparkes 2002).

In the late 1990s and early years of the 21st century, funds that focused on renewable energy/clean technology were able to attract pension fund investment because of the perceived financial returns resulting from government regulation in these areas. But when these funds fell in value during the bear market of 2001/2002, pension fund interest faded and only revived again from 2003 onwards when the market recovered (Sparkes 2006).

A significant change in the attitude of mainstream investors to climate change came in 2003 when brokers on the sell-side began to investigate policy risk from climate change — in particular the implications of the impending EU Emission Trading Scheme (EU ETS)

TABLE 12.2 Broker reports considering climate change in 2003

Source: Pfeifer 2005

Month	Broker	Report title
March	DrKW	*Emissions Trading: Carbon Derby Part I*
May	UBS	*Focus on UK Utilities*
July	West LB	*EU Emission Trading Scheme: Bonanza or Bust?*
September	UBS	*European Emissions Trading Scheme*
September	HSBC	*European Utilities Pathfinder II*
October	Citigroup Smith Barney	*Utilities: The Impact of Carbon Trading on the European Sector*
October	CSFB	*EU Carbon Trading: Utilities to Get a Carbo-boost*
October	DrKW	*Emissions Trading: Carbon Derby Part II*
November	DrKW	*Aviation Emissions: Another Cost to Bear*
November	ABN Amro	*Research Process, Climate Change and Analysis*
December	HSBC	*Aviation and Climate: Prepare to Trade*

5 www.enhancedanalytics.com

for the electricity utilities sector. The first broker report on the issue was published on 12 March 2003 by Dresdner Kleinwort Wasserstein (DrKW) and was followed by others, some of which are presented in Table 12.2.

DrKW then formally incorporated the price of carbon into its financial models for electricity utilities from October 2003, and a number of other brokers followed suit shortly afterwards (Pfeifer 2005).

On the basis of the investment recommendations from their brokers, fund managers increasingly began to integrate a carbon dioxide (CO_2) price into their investment decisions on electricity utility companies. At this point, climate change became a legitimate factor in buy-side investment decision-making. This occurred despite the absence of explicit requests from pension funds for their fund managers to take climate change concerns into account and without advice on the issue from investment consultants (Pfeifer 2005).

2000–2004: other actors

Over this period, a number of other actors initiated activities to encourage institutional investors to pay greater attention to climate change. For example, the Carbon Trust[6] published a number of reports on the issue (Carbon Trust 2004a, 2004b, 2005a, 2005b) and hosted a series of investor lunches in 2004. NGOs also tried to encourage mainstream investors to incorporate climate change and climate change policy into their engagement and investment strategies.

Perhaps more importantly, investment consultants started paying some attention to the issue of climate change. For example, Mercer Investment Consulting (one of the largest investment consulting companies) expanded its research of mainstream investment managers in 2004 to assess their approach to integrating environmental, social and corporate governance issues within investment processes, with a particular focus on climate change (Pfeifer 2005). In 2005, Mercer was commissioned by the Institutional Investors Group on Climate Change (IIGCC) and the Carbon Trust to produce a training guide for pension fund trustees on the subject of climate change, with the aim of '[raising] awareness amongst pension fund trustees about the relevance of climate change as a fiduciary issue' and 'to communicate the opportunities that exist for them to address it' (Mercer 2005: 2). This report probably represented the first time that investment consultants had seriously engaged with the issue of climate change (Pfeifer 2005).

2005 to end of 2007: continued progress

The period from the end of 2004 to the end of 2007 saw further evolution of the initiatives above and greater attention being paid by investors to climate change, driven by a range of factors, including:

● The implementation of EU ETS, which began operation in early 2005

---

6 The Carbon Trust (www.carbontrust.co.uk) was set up by the UK government in 2001 as an independent company to accelerate the transition to a low-carbon economy by working with organisations to reduce their greenhouse gas emissions and to develop commercial low-carbon technologies.

- The high profile of the *Stern Review on the Economics of Climate Change* (Stern 2006)

- The Fourth Assessment Report of the IPCC (IPCC 2007)

- Growing interest — triggered by strengthening regulatory support — in the investment opportunities presented by renewable energy and climate change solutions (see, generally, Impax 2006)

There are a number of specific points to highlight.

First, the volume of broker research on climate change grew significantly. Much of the research continued to focus on electricity utilities and the EU ETS, but research was also published on:

- The investment implications of other European policy initiatives (e.g. aviation and emissions trading, proposals to regulate CO_2 emissions from automobiles)

- Country-specific policies on wind and renewable energy

- Proposed emissions trading or other regulatory measures in countries outside the EU

Most of the research continued to focus on regulation or potential regulation directed at reducing greenhouse gas emissions. Far less attention — with the notable exception of the insurance and reinsurance sectors — was being paid to the physical impacts of climate change. While a number of broker reports examined issues such as drought in Australia and specific types of extreme weather events, these reports were few in number and paid relatively little attention to the specific causes of these events or the likelihood of such events being repeated in the future.

This picture is reflected in the manner in which investment managers looked at climate change. Although investment managers were increasingly paying attention to climate change and building their capacity to analyse the financial implications of new regulations, the main areas of focus remained those where there were investment opportunities (e.g. clean energy or low-carbon technologies) and those sectors (electricity utilities, cement and other heavy industries) directly exposed to climate change regulation (IIGCC 2008). Even though some investment managers reported that they had developed a system of environmental rankings or analysed climate implications for their portfolios as a whole, this information only affected investment decision-making in situations where there were significant costs associated with greenhouse gas emissions (IIGCC 2008).

The second point is that the number of investors participating in collaborative initiatives continued to grow. The number of investors supporting the Carbon Disclosure Project grew from 35 investors representing $4.5 trillion in assets under management in 2003 to 385 investors representing $57 trillion in assets under management in 2007. The membership of the Institutional Investors Group on Climate Change also grew, from around 25 investors at the end of 2005 to 45 at the start of 2008. Perhaps of greater significance has been the expansion in the geographic membership of IIGCC from a predominantly UK base to one that now encompasses asset managers and asset owners from the Netherlands, Germany, France and Switzerland.

The third point is that investors started to show renewed interest in investing in the areas of renewable energy and clean technology and/or establishing funds for clients to invest in these areas. For example, it has been estimated that, in 2007 alone, global investment in renewable energy exceeded $100 billion. Similarly dramatic growth has been seen in global carbon markets (i.e. markets for the trading of greenhouse gas emission permits) with the value of these markets estimated at $30 billion in 2006 – a threefold increase over 2005 (World Bank 2007). There were a number of drivers for this growth in investor interest in these areas. The most obvious were the twin influences of an increasingly supportive policy environment in many countries and the rising costs of conventional energy sources (with oil rising to almost $100 per barrel over this period). In addition, many of these renewable energy technologies had significantly improved, with wind, hydro, solar and some biomass conversion technologies regarded as reasonably mature technologies with significantly reduced technical risks than had been the case in the past. This interest was reinforced by the trend in the institutional investment market towards diversification (specifically, the search for asset classes whose risks and returns were not correlated with other major asset classes such as equities or corporate bonds). Renewable energies were seen as one strand of this diversification (as a consequence of the underlying regulatory support for companies in this area and the potential growth profile of many of these companies) and so attracted an increasing amount of investment – albeit still small in absolute terms.

The fourth point is that investors appeared increasingly willing to engage with companies on climate change-related issues, in particular in relation to improving reporting and disclosure, and on integrating climate change into business strategies (see, for example, the data presented in IIGCC 2008). However, the scope of engagement was relatively limited with many issues – including adaptation to unavoidable climate change and climate-friendly product design – receiving relatively limited attention to date. In relation to adaptation, four major UK institutional investors – two asset managers (Insight Investment and Henderson) and two pension funds (USS and Railpen) – initiated a collaborative research project on adaptation in mid-2007. The project, which is expected to be completed by the end of 2008, seeks to identify the major adaptation-related risks for a number of sectors and to use this information as the starting point for investment research on this issue, as well as providing a basis for investor dialogue with companies and policy-makers (see, further, Sullivan *et al.* 2008).

The fifth point is that investors were increasingly willing to contribute to the public policy debate around climate change, primarily (for European investors) under the auspices of the IIGCC. IIGCC's public policy work is aimed at encouraging public policy-makers to take into account the long-term interests of institutional investors. IIGCC has made submissions to government inquiries and consultations relating to issues such as Phase II of the EU ETS, the inclusion of aviation in the EU ETS and EU Action on Climate Change post-2012.[7] In these submissions, IIGCC has emphasised:

- The importance of policy clarity
- The need for long-term policy targets directed at significant reductions in greenhouse gas emissions

7 See the IIGCC website for a comprehensive listing: www.iigcc.org/activities/activity4.aspx (accessed 20 August 2008).

- The desirability of extending the use of economic instruments to internalise the costs of greenhouse gas emissions

- The need for better corporate disclosures on climate change-related risks and opportunities

IIGCC has also met with key climate change policy-makers to ensure that investors' views on these issues are clearly heard.

This focus on public policy and on the wider climate change debate was reinforced by the launch of the IIGCC Investor Statement on Climate Change in October 2006, which is now supported by 22 institutional investors (pension funds and asset managers) managing assets worth over €2 trillion. The Statement represents perhaps the most significant call by UK and European investors for action on the threat posed by greenhouse gas emissions. It signals investors' willingness to strengthen their focus on climate change in their investment processes and to actively engage with companies and governments to develop appropriate policy solutions to climate change. The Statement encourages companies to (see Box):

- Set management responsibilities for climate change

- Make commitments to greenhouse gas emission reductions

- Integrate climate change considerations into their business strategies

---

Investors' expectations of companies (from the IIGCC Investor Statement on Climate Change)

Companies are encouraged to:

- Define board and senior management responsibilities for climate change

- Integrate climate change risks and opportunities into business strategy

- Set high-level policy commitments in support of action on climate change including commitments to greenhouse gas emission reductions

- Provide appropriate disclosures on climate change risks and opportunities that allow investors to assess the financial implications of these risks and opportunities for the company

- Prepare and report comprehensive inventories of greenhouse gas emissions (both directly from operations and activities and indirectly from, for example, the use of the company's products). These inventories should allow historic performance to be assessed and should include projections of likely changes in future emissions

- Integrate climate change into product design and operational management. This should include setting targets and timelines for reducing greenhouse gas emissions and impacts along the value chain

- Proactively engage with public policy-makers and other stakeholders in support of policy measures to reduce greenhouse gas emissions, and not lobby to obstruct legitimate attempts to reduce greenhouse gas emissions or mitigate the effects of climate change

---

The Statement also encourages governments to:

- Set short-, medium- and long-term targets for greenhouse gas emissions reductions that will enable atmospheric concentrations of greenhouse gases to be stabilised at a level that averts the most significant risks of climate change

- Provide the necessary mechanisms and institutions for the delivery of these targets

Finally, the Statement outlines the actions that investors themselves should take. The asset owners (i.e. pension funds) that sign the Statement agree to:

- Encourage their asset managers to consider climate change risks and opportunities in their investment research, analysis and decision-making and shareholder ownership activities

- Consider climate change in their processes for the appointment and evaluation of their asset managers

More generally, the signatories agree to work together to share information and to incentivise and/or support research on the risks and opportunities of climate change and climate policy.

Despite significant progress, the reality at the end of 2007 was that investment decision-making and share ownership activities still did not fully account for the risks and opportunities presented by climate change. There were a number of specific areas where progress remained limited.

First, most of the dialogue between investors and companies on climate change issues continued to be conducted by SRI fund managers or analysts, with interviewees from companies reporting that what they regard as 'mainstream investment analysts' were not focused on these issues (Sullivan 2006). The subject of this engagement was relatively modest, focusing on improved disclosures and risk assessment but with limited evidence of investor appetite for encouraging companies to significantly reduce their greenhouse gas emissions.

Second, as discussed above, most investment research focused on regulation directed at reducing greenhouse gas emissions with relatively little attention being paid to the physical impacts of climate change. Furthermore, this research was concentrated on the implications of climate change for equity investments – rather than across the range of asset classes that institutional investors are invested in.

Third, pension funds appeared to be paying limited attention to the manner in which their investment managers take account of climate change in investment decision-making; the incentives for investment managers to take a proactive approach remained limited. For example, in a survey of some of the leading UK pension funds, while ten of the 12 surveyed said that they encouraged their internal or external asset managers to take climate change into account in exercising their voting rights and seven encouraged their managers to integrate climate change considerations in investment analysis and decision-making, only three of the 12 had formally integrated climate change (or wider social and environmental issues) into their processes for appointing and evaluating fund managers (IIGCC 2008). Moreover, although investment consultants have increased the resources allocated to this area (or to the area of 'responsible investment' more generally), pension funds are not proactively seeking this information from their investment consultants (Sullivan 2006; IIGCC 2008).

Discussion

Drivers for action

The primary drivers for the growth in investor interest engagement with companies over the period 2000–2004 were media attention and the leadership role of USS (in particular USS's role in establishing the IIGCC) (Pfeifer 2005). While other factors such as government awareness-raising programmes, NGO campaigning activity and policy measures directed at encouraging investors to pay more attention to corporate governance also played some role, the influence of these factors appears to be have been modest. Some care is required with this conclusion as it may do a disservice to the contributions of the other parties, in particular NGOs, which have been very effective at stimulating press interest in climate change.

However, in analysing the integration of climate change into investment analysis, a very different picture emerges. Virtually none of the interviewees for the research on which this chapter is based (Pfeifer 2005) mentioned any of the drivers for activism listed above as influencing their decision to integrate climate change into their investment analysis. Virtually without exception, the EU ETS was cited as the critical — and, in many cases, the only — reason that they decided to explicitly consider climate change in their investment analysis. In an interview, David Russell of USS articulated the views of many of those interviewed for this research when he suggested that:

> The ETS has been a milestone by giving carbon a value. It moved climate change from being a long-term issue to being a short-term issue and this is affecting the value of companies.

Barriers

There are a number of reasons why climate change risks and opportunities are, even now, not being fully factored into mainstream investment analysis. The two most important are the practical difficulties in assessing the financial implications of climate change and the short-term nature of investment mandates (Pfeifer 2005; Sullivan 2006).

In relation to investment analysis, it is difficult for investors to make robust assessments of the financial impact of climate change. Outside those sectors that are included in the EU ETS or supported by government policy (e.g. renewables), the significant uncertainties in the future direction of climate change policy make it very difficult to quantify the implications of climate change for asset values over the longer term. In relation to the physical impacts of climate change, there are still significant uncertainties in relation to the specific impacts on companies and the timing of these impacts — though the adaptation research project mentioned above (Sullivan et al. 2008) may help to address this. The assessment of risks and opportunities is compounded by the general inadequacy of corporate disclosures in this area (Sullivan and Kozak 2006; Sullivan 2008).

Addressing these weaknesses in corporate disclosures has been a particular focus for the IIGCC. The IIGCC published the IIGCC Disclosure Framework for Electricity Utilities in 2007 and subsequently collaborated with US-based Ceres and the Australia/New Zealand Investors Group on Climate Change on a disclosure framework appropriate for electricity utilities globally (IIGCC 2007; IIGCC *et al.* 2008). These disclosure guidelines provide a standard format for disclosing climate change-related risks and opportunities for electricity utilities and, if widely adopted, will make it easier for investors to make comparisons between individual companies in the sector. The IIGCC has also engaged with the CDP to adopt these guidelines within the CDP reporting framework.

Even if the difficulties in investment analysis could be addressed, it is not necessarily the case that this will affect investment decisions. There are a number of dimensions to this.

The first is that certain of the impacts from climate change are likely to be too far into the future (and hence heavily discounted) for investors to take account of them in decision-making. The short-term nature of investment mandates means that longer-term risks and opportunities tend to be excluded from analysis.

The second is the lack of pension fund demand. With the exception of a few public funds such as the Environment Agency Pension Fund and the London Pension Fund Authority (LPFA), there has been limited demand from UK pension funds for their fund managers to explicitly consider climate change in their investment processes and/or to engage on their behalf (Pfeifer 2005). The need to deliver investment performance over, at maximum, one year means that investment managers inevitably have an incentive to focus much of their attention on short-term drivers of investment rather than on longer-term value drivers (Sullivan and Mackenzie 2006c). Furthermore, the general absence of rewards for investment managers to engage with companies means that there is limited incentive for investment managers to devote significant resources to this type of activity (Sullivan and Mackenzie 2006c).

The third is that there are broader changes in the investment market that may undermine efforts to encourage investment managers to take greater account of climate change in their investment processes. The most significant of these are the growth of hedge funds and private equity funds – which tend to be much less transparent than traditional equity funds – and the move away from defined benefit to defined contribution plans. This latter change may lead to share ownership becoming even more widely dispersed, with the proportion of investments owned or managed by large pension funds and investment managers falling, thereby reducing the potential influence that these 'traditional investors' may be able to exert over corporate strategy.

Institutional investors and the public policy process

The most important reason for investors to factor climate change into their investment analysis is either government support for new climate change-related industries or the existence of an explicit financial value for climate change risks and opportunities. The implication for policy-makers is that stronger policy measures will be necessary to encourage investors to integrate climate change considerations into their investment decisions; conversely, the absence of clear market signals militates against investors looking at these issues.

Public policy is also critical to ensuring that investor engagement with companies is effective. There is strong evidence that shareholder engagement (also commonly referred to as investor activism) can contribute to improvements in corporate responsibility performance. By and large, this engagement has been conducted on the basis that there is a financial case for companies to improve their social and environmental performance, i.e. where the interests of companies and society are aligned.

So far discussions on the subject of climate change between companies and their investors have followed this pattern, with this engagement primarily focused on reporting and, to a lesser extent, on encouraging companies to ensure that climate change is integrated properly into corporate strategies and risk management processes. The primary assumptions underpinning this engagement are that:

● Better reporting will encourage investment managers to better manage the financial risks associated with climate change

● The request for disclosure will signal to companies that investors are concerned about climate change, thereby providing an incentive for companies to reduce their greenhouse gas emissions

As yet, however, investors have not asked companies to reduce greenhouse gas emissions beyond those that would be justified in financial terms. Such action would – unless there is a compelling business case otherwise – run counter to at least the short-term interests of companies. From a public policy perspective, it is therefore necessary to ensure that the incentives faced (or likely to be faced) by companies are structured in a way that reducing greenhouse gas emissions is the profit-maximising course of action to take. This will enable investors concerned about the long-term prospects of companies to send a clear signal that they support actions directed at reducing greenhouse gas emissions as this is likely to be critical to the longer-term success of the business.

The question is whether investors have (or could have) a role to play in the development of appropriate public policy responses to climate change? The *prima facie* answer is yes. As discussed above, there is a growing body of evidence that investors will in fact engage with public policy-makers to encourage the development of an appropriate long-term policy framework including public policy measures directed at achieving significant reductions in greenhouse gas emissions and at allowing society to adapt to the physical impacts of climate change. For example, the IIGCC has encouraged the EU to set credible long-term targets for greenhouse gas emission reductions and to increase the role of emissions trading as a key instrument for the delivery of these targets, arguing that this would have the effect of encouraging investors to properly integrate climate change issues into investment analysis and investment decision-making.

This type of public policy engagement does not necessarily mean that investors are not acting in a rational manner. There are a number of strands to this. First, on a macro-economic basis, the short-term costs should be outweighed by the longer-term economic benefits. For example, the IEA (2006) argues that countries could implement a range of measures directed at reducing energy demand growth and greenhouse gas emissions, and increasing energy security, where the benefits of using and producing energy more efficiently significantly outweigh the costs incurred. Second, reflecting the universal owner argument (see further Hawley and Williams 2000), the actions that may be in the interest of an individual company – in this particular case, allowing increasing greenhouse gas emissions — may not be in the long-term interests of the economy as a whole as such emissions may expose other companies to the physical impacts of climate change. Third, well-designed public policy should provide significant opportunities for companies. Benefits should be realised through, for example, identifying new technologies, capturing new markets or reducing the need for defensive expenditures (e.g. to respond to increased risks of floods or extreme weather events).

Conclusions

There has been significant progress in the level of attention being paid by investors to the issue of climate change, with an increasing number of fund managers actively engaging with companies on climate change issues, the emergence of collaborative initiatives, and investors (fund managers and brokers) explicitly integrating the financial implications of the EU ETS into their investment analysis. However, significant gaps remain. The most important of these are:

- The relative lack of demand from pension funds for their investment managers to take account of climate change

- The lack of reward from pension funds for fund managers who do significant work on this issue

- The short-term nature of many investment mandates

- The lack of appropriate price signals to encourage investors to explicitly include these issues in their investment analysis

There are positive signs that investors will play a more constructive role in the climate change debate going forwards, in particular in the public policy process. For example, the IIGCC has stated:

> We see that investor collaboration (on research, corporate engagement and public policy) will be most effective — and most likely to be seen as credible by analysts, policy makers and other stakeholders such as NGOs — if it is geared towards achieving a smooth transition to specific policy goals that lead to a low carbon economy.[8]

This support for public policy action should help encourage policy-makers to take action which in turn will increase the incentives for institutional investors to:

- Fully integrate climate change into their investment analysis

- Send strong and unequivocal signals to companies about the critical importance of significantly reducing greenhouse gas emissions

---

8 See the IIGCC Public Policy Position Statement at www.iigcc.org/activities/activity4.aspx (accessed 20 August 2008).

References

Allianz Group and WWF (2005) *Climate Change and the Financial Sector: An Agenda for Action* (London: Allianz Group).

BP (2005) 'Future Growth and Sustainability: BP and Sustainability'; www.bp.com/genericarticle.do?categoryId=98&contentId=2000431 (accessed 6 June 2008).

Carbon Trust (2004a) *Core Ratings Investor Engagement Scoping Study* (London: Carbon Trust).

— (2004b) *The European Emissions Trading Scheme: Implications for Industrial Competitiveness* (London: Carbon Trust).

— (2005a) *Investor Guide to Climate Change* (London: Carbon Trust).

— (2005b) *Brand Value at Risk from Climate Change* (London: Carbon Trust).

Dlugolecki, A., and M. Mansley (2005) *Climate Change and Asset Management* (Tyndall Centre Technical Report; Manchester: Tyndall Centre).

Freshfields Bruckhaus Deringer (2005) *A Legal Framework for the Integration of Environmental, Social and Governance Issues into Institutional Investment* (Geneva: United Nations Environment Programme Finance Initiative).

Hawley, J., and A. Williams (2000) *The Rise of Fiduciary Capitalism* (Philadelphia: University of Pennsylvania Press).

IEA (International Energy Authority) (2006) *World Energy Outlook* (Geneva: IEA).

IIGCC (Institutional Investor Group on Climate Change) (2007) *IIGCC Disclosure Framework for Electricity Utilities* (London: IIGCC).

— (2008) *Investor Statement Report 2007* (London: IIGCC).

—, Ceres and Investor Group on Climate Change Australia/New Zealand (IGCC) (2008) *Electric Utilities: Global Climate Disclosure Framework* (London: IIGCC; www.iigcc.org/docs/PDF/Public/Globalelectricutilitiesdisclosureframework.pdf [accessed 6 June 2008]).

Impax (2006) *Investment Opportunities in a Changing Climate: The Alternative Energy Sector* (London: Institutional Investor Group on Climate Change).

IPCC (Intergovernmental Panel on Climate Change) (1990) *Scientific Assessment of Climate Change: Report of Working Group I* (Geneva: United Nations Environment Programme/World Meteorological Organisation).

— (1995) *IPCC Second Assessment: Climate Change 1995* (Geneva: United Nations Environment Programme/World Meteorological Organisation).

— (2007) *Climate Change 2007: The Physical Science Basis – Summary for Policy-makers* (Geneva: IPCC).

Just Pensions (2004) *Will UK Pension Funds Become More Responsible? A Survey of Trustees* (London: Just Pensions).

— (2005) *Responsible Investment Trustee Toolkit* (London: Just Pensions).

Leggett, J. (2000) *The Carbon War: Global Warming and the End of the Oil Era* (London: Penguin Books).

Lim, R. (2006) 'Morley Fund Management's Approach to Investment Integration', in R. Sullivan and C. Mackenzie (eds.), *Responsible Investment* (Sheffield, UK: Greenleaf Publishing): 81-91.

Mansley, M. (1994) *Long-Term Financial Risks to the Carbon Fuel Industry from Climate Change* (London: Delphi International).

— and A. Dlugolecki (2001) *Climate Change: A Risk Management Challenge for Institutional Investors* (London: Universities Superannuation Scheme).

Mercer (Mercer Investment Consulting) (2005) *A Climate for Change: A Trustee's Guide to Understanding and Addressing Climate Risk* (London: Carbon Trust).

Michelson, G., N. Wiales, S. van der Laan and G. Frost (2004) 'Ethical Investment Processes and Outcomes', *Journal of Business Ethics* 52.1: 1-10.

Myners, P. (2001) *Institutional Investment in the UK: A Review* (London: HM Treasury).

Pfeifer, S. (2005) *Institutional Investors and Climate Change: An Analysis of the Integration of Climate Change Risks and Opportunities into Investment Decision-making* (dissertation submitted as part of an MSc in Environmental Change and Management; Oxford, UK: University of Oxford).

—— and R. Sullivan (2008) 'Public Policy, Institutional Investors and Climate Change: A UK Case Study', *Climatic Change* (online), 22 January 2008; www.springerlink.com/content/313p531185855p30 (accessed 20 August 2008).

Solomon, A., J. Solomon and M. Suto (2004) 'Can the UK Experience Provide Lessons for the Evolution of SRI in Japan?', *Corporate Governance* 12.4: 552-66.

Sparkes, R. (2002) *Socially Responsible Investment: A Global Revolution* (Chichester, UK: John Wiley).

—— (2006) 'A Historical Perspective on the Growth of Socially Responsible Investment', in R. Sullivan and C. Mackenzie (eds.), *Responsible Investment* (Sheffield, UK: Greenleaf Publishing): 39-54.

—— and C. Cowton (2004) 'The Maturing of Socially Responsible Investment: A Review of the Developing Link with Corporate Social Responsibility', *Journal of Business Ethics* 52.1: 45-57.

Stern, N. (2006) *Stern Review: The Economics of Climate Change* (Cambridge, UK: Cambridge University Press).

Sullivan, R. (2006) *Managing Investments in a Changing Climate* (London: IIGCC).

—— (2008) *Taking the Temperature: Assessing the Performance of Large UK and European Companies in Responding to Climate Change* (London: Insight Investment).

—— and J. Kozak (2006) *The Climate Change Disclosures of European Electricity Utilities* (London: Insight Investment).

—— and C. Mackenzie (2006a) 'Introduction', in R. Sullivan and C. Mackenzie (eds.), *Responsible Investment* (Sheffield, UK: Greenleaf Publishing): 12-19.

—— and C. Mackenzie (2006b) 'Shareholder Activism on Social, Ethical and Environmental Issues: An Introduction', in R. Sullivan and C. Mackenzie (eds.), *Responsible Investment* (Sheffield, UK: Greenleaf Publishing): 150-57.

—— and C. Mackenzie (2006c) 'The Practice of Responsible Investment', in R. Sullivan and C. Mackenzie (eds.), *Responsible Investment* (Sheffield, UK: Greenleaf Publishing): 332-46.

——, N. Robins, D. Russell and H. Barnes (2005), 'Investor Collaboration on Climate Change: The Work of the IIGCC', in K. Tang (ed.), *The Finance of Climate Change* (London: Risk Books): 197-210.

——, C. Mackenzie and S. Waygood (2006) 'Does a Focus on Social, Ethical and Environmental Issues Enhance Investment Performance?', in R. Sullivan and C. Mackenzie (eds.), *Responsible Investment* (Sheffield, UK: Greenleaf Publishing): 56-61.

——, D. Russell and N. Robins (2008) *Managing the Unavoidable: Understanding the Investment Implications of Adapting to Climate Change* (London: Insight Investment, Henderson Global Investors, University Superannuation Scheme and Railpen).

SustainAbility, Ketchum and the United Nations Environment Programme (2002) *Good News and Bad: The Media, Corporate Social Responsibility and Sustainable Development* (London: SustainAbility, 1st edn).

WEF (World Economic Forum) (2005) *Mainstreaming Responsible Investment* (Geneva: WEF).

Whitaker, E. (2006) 'Evaluation and Research of SRI Managers', in R. Sullivan and C. Mackenzie (eds.), *Responsible Investment* (Sheffield, UK: Greenleaf Publishing): 284-89.

World Bank (2007) *State and Trends of the Carbon Market* (Washington, DC: World Bank).

13
Reporting on climate change: the case of Lloyds TSB

Andrea B. Coulson
University of Strathclyde, UK

Climate change is increasingly recognised as a business risk (as a consequence of, for example, the physical impacts of changing weather patterns and government regulations directed at reducing greenhouse gas emissions) and also as presenting business opportunities (e.g. the development of new investment and insurance products). As a result, there is increased interest in corporate climate change disclosures with a range of actors — investors, non-governmental organisations (NGOs), the media, customers — requesting information on how companies are responding to the risks and opportunities presented by climate change. Yet, while there is a growing consensus that companies should report more, it is less clear what specific information is required and how this information will be used.

This chapter focuses on one dimension of this debate: namely, the role that investor demand for information plays as a driver for reporting. Building on the information gathered from a detailed research study currently being conducted within Lloyds TSB, this chapter will examine how Lloyds TSB reports on climate change and why it has chosen to report in this way. The study provides a reflection on both strategic and practical responses to climate change reporting drawn from interviews, document review and observations on the climate discourse within the Lloyds TSB Group.

The chapter is divided into four main sections. The first discusses investors' expectations of corporate reporting on climate change and some of the potential technical difficulties facing companies in meeting these expectations. The second discusses the response of Lloyds TSB to climate change including a brief description of its key activities, climate change impact and drivers for action. The third is a detailed discussion of how and why Lloyds TSB reports on climate change. The chapter concludes with an assessment of the value of reporting to the business and some reflections on the broader challenges faced by companies when reporting on climate change.

Investor expectations of climate change reporting

Investors have been at the forefront of demands for increased corporate disclosure and transparency. For example, the Europe-wide Institutional Investors Group on Climate Change (IIGCC),[1] which currently represents 45 leading asset owners and asset managers with over €4 trillion assets under management, has called for the 'assessment and active management of the investment risk and opportunities associated with climate change' (IIGCC 2007: 3). There are similar investor initiatives across the globe most notably:

● Investor Network on Climate Risk (INCR)[2] in North America

● Investor Group on Climate Change (IGCC)[3] in Australia/New Zealand

● The international Carbon Disclosure Project (CDP)[4]

● International climate working groups hosted by the United Nations Environment Programme Finance Initiative (UNEP FI)

Even though these, and other, investor initiatives have stressed the need for and importance of climate change-related disclosures, there is no real consistency in the information being demanded, how this information should be reported and what use will be made of such information. While documents such as the IIGCC *Global Framework for Climate Risk Disclosure* (IIGCC 2006) provide a multi-dimensional reporting framework covering regulatory risk, physical risks, competitiveness, litigation and reputation, and encourage reporting on the implications of these risks for future cash flows and the financial position of the company, these documents do not provide specific guidance on how these are to be calculated and assessed.

Even in relation to emissions, there is still a lack of clarity on the data that is to be reported and how this data is to be presented. Although the World Business Council for Sustainable Development (WBCSD)/World Resources Institute (WRI) *Greenhouse Gas Protocol* (WBCSD/WRI 2004) is widely cited as the standard reporting framework, this document does not prescribe which emissions sources are to be included or excluded from reporting, or the form in which emissions data should be reported.

Despite its support from over 315 institutional investors representing $41 trillion in assets under management, similar concerns have been expressed about the Carbon Disclosure Project (see, for example, the critique in Sullivan 2006). There are some signs of progress with the establishment of the international Carbon Standard Disclosure Board which is charged with clarifying some of these areas of inconsistency. Recent investor

1 See Chapter 12.
2 www.incr.com
3 www.igcc.org.au
4 www.cdproject.net

initiatives have begun focusing their attention more closely on providing sector-specific guidance on greenhouse gas emissions reporting; see, for example, the *Electric Utilities Global Carbon Disclosure Framework* (IIGCC *et al.* 2008).

A related issue is that, even if the specific data required by investors was agreed, it is not clear whether or how this data aligns with financial accounting and reporting rules. In January 2008, the International Accounting Standards Board (IASB) began project work on establishing a framework for the appropriate accounting treatment of carbon trading.[5] Although this topic had previously been worked on by IASB and the International Financial Reporting Interpretations Committee (IFRIC), there had been limited progress because of difficulties in achieving consensus about how data (emissions allocations, emissions surpluses and shortfalls) could be incorporated into conventional profit and loss accounts.

The headline objectives of the current IASB project provide a good illustration of the difficulties in incorporating the financial implications of emissions trading into conventional corporate financial reporting frameworks. The project centres on the following questions:[6]

- Are emissions allowances assets?

- What is the corresponding entry for an entity that receives allowances from government free of charge?

- How should allowances subsequently be accounted for?

- When should an entity recognise its obligations in emissions trading schemes and how should they be measured?

- What are the overall financial reporting effects of the above decisions?

Of course, accounting for greenhouse gas emissions is just one part of a wider discussion around how climate change (and social and environmental issues more generally) are to be reported. For example in relation to the EU Accounts Modernisation Directive, which requires companies to prepare a business review that contains information on the financial and non-financial performance indicators necessary to properly understand a company, there have been a whole series of documents that offer advice on how to report this information (see, for example: CICA 2005; Hill and McAulay 2006; ACCA and FTSE 2007; Deloitte 2007; PwC 2007).

This lack of clarity regarding investors' requirements for climate change-related information and the lack of a common framework for corporate reporting on climate change presents a real challenge for companies that need to report this information in a manner that is useful to their investors. It raises questions around the scope of reporting, and the specific data and performance measures (or indicators) that are to be provided. For companies, it also opens up the risk that they may not report in a manner that is useful to investors and may find themselves penalised by the market as a result.

---

5 www.iasb.org/Current+Projects/IASB+Projects/Emission+Trading+Schemes/Emission+Trading+Schemes.htm (accessed 20 August 2008).
6 *Ibid.*

The response of Lloyds TSB to climate change

Lloyds TSB Group offers financial services in international, wholesale, retail and personal banking, insurance and fund management. In 2006, it had over 16 million customers and made £4.2 billion profits from its operations across 25 different countries. Climate change action and reporting fall under the umbrella of the Group's corporate responsibility policy on the environment. The ultimate responsibility for this policy is a steering group chaired by the Deputy Group Chief Executive and made up of senior executives from all business divisions. Day-to-day responsibility for climate change rests centrally with the Head of Corporate Responsibility, and the Public Policy and Regulation team.

Given the service nature of its operations, Lloyds TSB is generally considered to have a relatively low direct climate change impact in terms of emissions, with 2006 emissions relating to group property and travel totalling 189,373 tonnes of CO_2 equivalent.[7] The Group's direct emissions are not currently regulated, but the Group is likely to face mandatory emissions targets under the forthcoming UK Carbon Reduction Commitment.

In addition, the Group and its customers are exposed to the physical risks of climate change including extreme weather events, energy and water shortages, and sea level rises in low-level coastal sites. From a business perspective, these risks represent the company's biggest financial exposures, although they are mitigated through its risk assessment and management processes.

The Group's climate change strategy is focused primarily on its direct impact and engaging employees to reduce the Group's carbon footprint and achieving carbon neutrality. This commitment involves reducing the Group's total direct greenhouse emissions by 30% by 2012 and achieving carbon neutrality through offsetting its remaining emissions. The primary drivers for the Group's action on emissions are government policy and demand from employees.

Secondary drivers for action include engagement with customers 'to do the same' in reducing their greenhouse gas emissions and managing increasing customer demand for climate change-related products. Competitor reactions to the climate change agenda are also influencing the Group's climate action. For example, HSBC's commitment to carbon neutrality secured the first-mover advantage for itself in the market but also raised expectations that other banks would follow suit. Although Lloyds TSB does have a carbon neutrality commitment, it is seeking to differentiate itself from its competitors by prioritising greenhouse gas emissions reductions over offsetting.

The drivers for the Group's actions on the physical risks associated with climate change relate primarily to managing its exposure and the need to minimise these risks to the business. The Group recognises that managing these risks effectively may also provide it with broader business benefits such as allowing it to differentiate itself through:

---

7 2006 travel data will be restated in the 2007 report to reflect the fact that data for Cheltenham & Gloucester (C&G) is now included. To provide a more accurate comparison between 2006 and 2007, the comparative total for 2006 will be 191,847 tonnes.

● Working with its customers to develop climate change-related products based on emissions reductions

● The assessment and management of physical risks such as flooding within its environmental credit risk assessment system (thereby enabling it to make better-informed decisions about the climate change risks faced by the business)

For example, while the 2007 floods in the UK resulted in large financial costs for the Group's general insurance function, the Group also benefited from positive public relations as Lloyds TSB was seen as one of the first to respond to the floods and to provide appropriate service at a local level. In terms of climate adaptation, the Group recognise the biggest climate change risks and opportunities facing businesses come from exposure to extreme weather and the associated opportunities to expand the current business and develop new products.

Reporting on climate change

Drivers for reporting

There are various drivers for action on climate change and for reporting on climate change. However, these drivers are not necessarily the same. At present, the Group faces most demand for climate change-related information from its own employees. Employees are primarily interested in the direct greenhouse gas emissions of the Group and, to a lesser extent, their role in pursuing indirect climate opportunities with customers. Hence, Lloyds TSB sees its employees as the primary audience for climate change reporting, followed by investors and then the general public.

While customers are an important driver for the Group's action on climate change, the Head of Corporate Responsibility notes: 'Our research suggests that consumers aren't looking for their bank to engage them on climate.' However, Lloyds TSB recognises increasing demand from corporate customers to demonstrate its corporate responsibility credentials as a basis for doing business. For example, a section on corporate responsibility is included in all of the Group's corporate pitch documents and, in some cases, this information forms a major part of the Group presentation. In an interview, the Head of Corporate Responsibility commented that: 'While such credentials are not the sole determinant of doing business they could be used to screen us out of business.' He also anticipated increasing demand for reporting from customers over time on climate change-specific issues.

Mainstream investors are seen to have, at best, minimal interest in the climate change performance of the Group. For example, in an interview, the Director of Investor Relations noted: 'In the past two years in over 400 meetings with mainstream investors the issue of climate change has never been raised.' This lack of interest is attributed to the belief that climate change is simply not material to investors' views of financial performance in the sector. This is not necessarily surprising given the financial sector has a relatively low direct climate change impact (i.e. in terms of emissions per unit of turnover or profit) and is in the relatively early stages of developing financial services based on climate change performance (i.e. the sector has yet to make climate change and related matters a material driver of financial performance). However, in an interview, the Head of Corporate Responsibility argued that investors should not focus only on financial performance but also on quality of management, reflecting:

> It is important to highlight climate change as part of our long term business strategy to demonstrate that management has a grasp of this issue and takes a long-term view, managing the company accordingly.

Most of the investor demand for climate change-related information is seen as coming from the 'SRI [socially responsible investment] community' (i.e. investors running funds or with an explicit ethical or environmental mandate). Reflecting on SRI feedback on reporting, the Public Policy and Regulation team believes that Lloyds TSB is now at the point where SRI analysts are aware of the Group's climate change policy and strategy, and are now either demanding updates on performance or tackling the company on particular issues arising in the media or through third-party rating agencies. This latter point suggests that investors' demand for information may not be fully met through the Group's formal reporting channels (detailed below) but is being supplemented through engagement activities and indirectly through third-party disclosures. There are various possible reasons for this disconnect. It may be because Lloyds TSB does not properly understand investors' needs or because the company does not prioritise requests for information from SRI analysts. Or it may simply reflect the reality that corporate reporting cannot cover every item of data or information that may be required by investors (and so engagement is critical to these investors to gather all of the information that they require). In relation to the question of the influence of SRI analysts, the Head of Corporate Responsibility noted:

> A key question I have is whether SRI analysts are actually reporting into their mainstream colleagues? SRI analysts represent only a tiny proportion of investment houses. It would be interesting to know what use the mainstream make of them. If SRI research is only being factored into investment decisions for SRI funds, it has limited worth.[8]

Reporting framework

The Group's strategy, policy, responsibilities and achievements are reported through four different channels, designed to meet the demands of a number of different audiences. These are:

● The mandatory Annual Report and Accounts prepared for shareholders

● An annual Corporate Responsibility Report aimed primarily at shareholders

8 One objective of the research project on which this chapter is based is for Lloyds TSB to more actively lead engagement with investors on climate. In particular, Lloyds TSB perceives that greater work needs to be done to integrate the views of SRI analysts with those of buy- and sell-side (or 'mainstream') analysts who frequently lack interest in corporate responsibility in general and climate change in particular.

● An annual Corporate Responsibility Review aimed at other stakeholders such as customers, business partners and community representatives

● A monthly magazine to employees entitled *Upfront*

All these 'reports' are seen as complementary to, rather than replacements for, other forms of engagement. Anyone seeking to examine Lloyds TSB's public position on climate change, for example, has these reports available to them plus further information published on the web, including the Group's corporate responsibility policies.

Given the Group's perception that a lack of symmetry exists between information that is demanded by external stakeholders (in particular investors) and that produced by companies, Lloyds TSB sees direct engagement with these stakeholders as an important element of its overall reporting processes.

Finally, in addition to the reports listed above, the Group also reports to initiatives such as the Carbon Disclosure Project; this reporting may be seen as a fifth strand of its climate change reporting.

Annual Report and Accounts

The Annual Report and Accounts includes climate change in a number of places. Taking the example of the Group's 2006 Annual Report and Accounts:[9]

● The Chairman's Statement presents the Chairman's view on the business challenges presented by climate change and the importance of greenhouse gas emissions reduction as a corporate responsibility

● The Business Review section on corporate responsibility includes a sub-heading on environment and, within it, an explicit discussion of climate change

The overarching comment on environment stresses that the Group:

> first introduced a formal environmental policy in 1996 and was also one of the first UK banks to develop an environmental risk assessment system for all of our business lending.

The Business Review extends the Chairman's Statement as it highlights not only the cost saving potential of greenhouse gas emission reductions but also:

> the need to fully understand the potential financial impact of climate change on others that we may lend to or invest in, so that we can manage the risks and opportunities.

This reflects the Group's position that future climate change could have a material financial impact on business performance. The Review goes on to explain that the challenge of understanding the financial impact of climate change is a primary objective behind:

---

9 www.investorrelations.lloydstsb.com/media/pdf_irmc/ir/2006/2006_LTSB_Group_R&A.pdf (accessed 20 August 2008).

establishing a group wide Climate Forum, led by the Deputy Group CE [Chief Executive] to develop a holistic approach to managing climate related risks and opportunities.

To emphasise the link between the environment, climate change and risk, this sub-section in the Business Review is followed directly by 12 pages on risk management. But, despite the Group's recognition that risk management is a key interest of investors and that the Annual Report and Accounts is the primary vehicle for conveying information on business risks to investors, no specific reference is made to the company's exposure to physical risks of climate change and how these risks are being, or may be, managed. In an interview, the Head of Corporate Responsibility confirmed that this is because the impact of climate change is not currently considered material to the Group's financial position (i.e. physical risks are not material) but, as noted previously, this is an issue currently under consideration by the Group's Climate Forum.

The Business Review links to a table on Corporate Responsibility Key Performance Indicators that provides data on greenhouse gas emissions from property and travel for the period 2003–2006.

TABLE 13.1 Lloyds TSB greenhouse gas emissions as presented in its Annual Report and Accounts 2006

	Tonnes CO_2			
	2003	**2004**	**2005**	**2006**
Property	195,175	188,624	177,047	181,086
Property renewable	(730)	(4,438)	(14,606)	(18,944)
Travel	26,998	29,499	29,540	27,231
Total	**221,443**	**213,685**	**191,981**	**189,373**

Note: We have reported our greenhouse gas emissions arising from our operations since 1999 using the Defra *Guidelines for Reporting Greenhouse Gas Emissions*. The key activities that contribute to our Global Warming Impact are energy used in managing our buildings (lighting, building controls and IT) and in business travel (road, rail and air).

The Head of Corporate Responsibility commented: 'These [the provision of climate change-related data] provide a tick box for quality of management.' He further noted that:

Our report will focus on the fact that we are a AAA rated bank ploughing a safe route and ensuring we are not exposed to the big hits recently witnessed in the market. From a mainstream investor point of view, they may spot our comments on climate change in the Annual Report but I am not sure that they look any further. I anticipate that it is only SRIs [SRI analysts] who will look as far as the Corporate Responsibility Report.

Corporate Responsibility Report

In contrast to the Annual Report and Accounts, the voluntary nature of the Corporate Responsibility Report allows the company much more freedom in terms of content and design. As noted by the Head of Corporate Responsibility:

> In the Corporate Responsibility Report we have more flexibility with respect to the coverage that can be devoted to individual issues whereas in the Annual Report we are asked to keep our comments down to relatively short statements. We have to acknowledge that the Report gives us the opportunity to set out the material issues but it should not be an endless list of things that might appeal to someone. That is why analyst meetings are valuable so we can put more detail to them.

The 2006 Report[10] provided a description of the Group's leadership, policy and strategy, people, partnership and resources, and processes together with a more detailed section on results and an assurance statement. The report contained 19 pages in total with charts and tables demonstrating clearly the management framework and detailed results. In terms of overall corporate responsibility, the Head of Corporate Responsibility noted: 'Climate change reporting for a bank is never going to be as important as other areas, again it needs to be considered in perspective.'

Climate change-related information is provided in three of the seven main sections in the report:

- The management framework at Lloyds TSB is described under 'Policy and Strategy — Corporate Responsibility Management'. A detailed table outlines the company's self-assessment of its areas for improvement. This includes 'develop climate change strategy' as an area for improvement with the specific objectives being to 'review carbon management programme, environmental performance indicators and develop targets'

- A half-page narrative is provided under 'Partnerships & Resources — Environment — Climate Change' on the historical development of the Group's environmental policy and environmental risk assessment system for all of its business lending. Building on this foundation, drivers for action and targets are detailed and matched to management initiatives such as a five-year carbon management programme

- 'Society Results — Objectives for 2007' include further reference to the Group's greenhouse gas emission reduction targets

The report reflects much of what is advocated by the FORGE[11] Group as best practice for climate change reporting in this sector (FORGE 2007). This is not surprising given

---

10 www.lloydstsb.com/media/lloydstsb2004/pdfs/corporate_responsibility_report_06.pdf (accessed 20 August 2008).

11 FORGE is an informal group of UK-based banks and insurance companies that, since 1998, have collaborated periodically to debate common interests of corporate responsibility. Further details on FORGE are available through the websites of the Association of British Insurers (ABI) and British Bankers' Association (BBA). Their recent collaboration offering guidance on climate change reporting can be found at www.bba.org.uk/content/1/c6/01/16/92/FORGE%20V%20Guidance%20 Framework%20FINAL.pdf (accessed 20 August 2008).

that Head of Corporate Responsibility at Lloyds TSB chaired the FORGE climate change initiative.

Explicit reference to climate change in the Corporate Responsibility Report is almost exclusively devoted to direct greenhouse gas emissions, although there are a number of references in the document to indirect environment impact and environmental credit risk assessment processes.

Corporate Responsibility Review

The Corporate Responsibility Review is designed for all stakeholders other than investors. It is intended to provide the 'general public and others' with management and performance-related information across the Group's four central corporate responsibility policy themes (marketplace, workplace, community and environment) and is presented in a 'reader friendly design'.

While of equal length to the Corporate Responsibility Report, the visual style of the Corporate Responsibility Review 2006[12] is very different. It is designed around Lloyds TSB's public marketing campaign 'for the journey' with animated illustrations on the cover and throughout the document. Research on reporting recognises animated pictures and reference to people and animals as common and effective ways of appealing to the emotions of readers and legitimising the content for a broader audience.

In contrast to the Corporate Responsibility Report, the Review concentrates on the Group's activities and seeks to personalise the Group by identifying — with the help of quotes, photographs and boxed text — the senior management, their roles and responsibilities. It begins with 'A message from Sir Victor Blank', the Chairman. The other sections in the report are:

● Who we are and what we do

● The future of banking

● Investing pensions responsibly

● Environment and banking

● Working in the community

● Reflecting the diversity of our customers

● Balancing priorities

● Career paths

● Race for opportunity

The Review ends by providing a list of contacts for further information.

Climate change-related information is contained in the Chairman's message and in the section on environment and banking. The Chairman's message focuses on the Group's

---

12 www.lloydstsb.com/media/lloydstsb2004/pdfs/corporate_responsibility_review_06.pdf (accessed 20 August 2008).

commitments to reducing greenhouse gas emissions and carbon neutrality. The section on environment and banking is personalised and was prepared on behalf of Ian Thompson, Director of Group Operational Services and a member of the Corporate Responsibility Steering Group. It includes the headline 'Can we really save the polar bears by switching off the lights?' and emphasises that 'every action, no matter how small, is badly needed if we are to continue to make a better environment'. This section also includes examples of actions taken by the Group on emissions reductions over the last five years and presents its plans and targets for the next five years. It mirrors the information included in the table on greenhouse gas emissions reduction contained in the Annual Report and Accounts (see above). Reference is made to the Group's work with outside parties such as the Carbon Trust to achieve its objectives and its climate change commitment is described as 'a positive contribution to the UK government's target to reduce CO_2 emissions by 60% by 2050'.

The Review is the only one of the Group's formal reports that discusses the physical risks posed by climate change or the Group's indirect climate change impact. Reference to the physical risks is limited to social issues of climate safety and acknowledgement of indirect impact is expressed in terms of the Group's '16 million customers' — appearing more as a promotional statement than an issue of product development or risk management.

Upfront (employee magazine)

Commenting on the content of Upfront, the Head of Corporate Responsibility noted:

> Our climate strategy is to set stretching targets and use these to engage employees because we believe they are interested and will respond positively. If we engage employees with the company in terms of what they think and do we believe this is better for recruitment and retention, employees respond if they are proud to work for a company with a good reputation.

Taking a typical issue of Upfront from June 2007 (Issue 25), two out of the 16 pages were devoted to outlining the reasons for the Group's commitment to reduce its CO_2 emissions by 30% by 2012 and support for this commitment throughout the Group. The issue included a quote by the Deputy Group Chief Executive who stated:

> Thirty per cent is a challenging target; it will need commitment across the business and the support of all our staff if we are to meet that challenge.

He went on to highlight the example of the Group's property management, procurement and IT divisions which have committed to consider the environmental impact of all major projects and to emphasise the Group's achievements to date:

> We have already begun progress, reducing our carbon footprint by almost 36,000 tonnes of CO_2 in the last five years.

Upfront is the only one of the Group's formal reports to make direct reference to its competitors and its strategy for market leadership. In Issue 25, the Deputy Group Chief Executive notes:

> We've seen a lot of companies including some of our competitors becoming carbon neutral. Many of them have not actually reduced their CO_2 emissions

> at all but have bought carbon credits or offsets or planted trees. If we are going to make a real contribution to fighting climate change, we believe that it is very important to achieve actual reduction in our emissions — this is our priority.

In terms of the time-frame for action by employees, reference is made to the UK government's medium-term emissions targets for 2020. Targets are set out in terms of workplace practice such as energy use and travel.

As with the other reports produced by the Group, *Upfront* is seen as complementary to other forms of engagement. For example, the Group operates a comprehensive programme of engagement with its employees including providing information on carbon footprints and suggestions on emissions reductions as part of the Group's intranet communications.

Third-party disclosure

In addition to its own reports and publications, Lloyds TSB provides additional disclosures through its support for initiatives such as the Business in the Community (BiTC) Environment Index and FTSE4Good, and through its participation in climate change-specific exercises such as the Carbon Disclosure Project (CDP).

To examine the nature of this third-party disclosure and benchmarking in more detail, the example of the CDP disclosure by Lloyds TSB is considered in more detail. Lloyds TSB has responded to CDP each year since its introduction.[13] Reading through its CDP5 response (for 2006), it is evident that the company is providing climate change-related disclosures in its CDP response that are not being provided elsewhere in its multi-tier reporting framework. These additional disclosures include, for example, a breakdown on climate risks according to the Group's risk strategy framework and detailed examples of the Group's indirect impact and climate change-related opportunities.

The first section of the CDP questionnaire examines climate change risks, opportunities and strategy. The answers that Lloyds TSB provides to the question on regulatory risks broadly reflect the information in its Annual Report and Accounts, Business Review disclosure. However, its response to the question on physical risks contains the following information which is not disclosed within its multi-tier reports:

> Physical risks to our business operations may arise from extreme weather events, energy and water shortages and also from sea level rise in low level coastal areas. For example, where offices are situated in flood plains we need to consider vulnerability, possible workforce disruption and overall risk of business interruption. Over 95% of our business is UK based and we have detailed knowledge of areas at risk of flooding and comprehensive business continuity plans in place. Such plans are regularly tested in scenario planning exercises and recently have been tested in a live situation (not climate related) where staff were successfully relocated to a 'warm site' with minimal disruption to business. Environmental considerations feature in all new property decisions, including vulnerability, location and energy efficiency.

Lloyds TSB also provides additional information on other climate change-related risks (see Table 13.2) and on the business opportunities presented by climate change (see Table 13.3). In Table 13.2, examples of other risks are categorised under the headings

13 Lloyds TSB's responses to CDP2 to CDP5 are in the public domain at www.cdproject.net.

TABLE 13.2 Lloyds TSB's response to the CDP5 question on other climate change-related risks

Type of risk	Response
Credit risk	• Exposure to high-risk sectors • Customer defaults due to climate impacts • Impact on security values
Financial soundness	• Need for increased liquidity to deal with potential shocks
Insurance risk	• Morbidity risk, mortality risk and heightened property insurance risks
Market risk	• Exposure to sudden interest rate movements and foreign exchange risk brought about by climate impacts • Bond, commodity and equity risks
Operational risk	• Regulatory — as above • People risks, e.g. recruitment and retention, health & safety, etc. • Physical risks — as above
Strategy risk	• Risk appetite • Risk of failure to ensure alignment of climate strategy across all Group businesses • Product development and pricing — failure to develop products to meet evolving customer demand

TABLE 13.3 Lloyds TSB's response to the CDP5 question on climate change-related opportunities

Area	Response
Products and services	• We already offer environmental and ethical investment products through Scottish Widows Investment Partnership • We are examining opportunities to develop green products and services although, to date, we have not witnessed significant demand for green banking products • We do not consider it appropriate to simply brand a product or service as green if it has no different features to standard products and services. We have always been willing to provide loans to assist customers make energy efficiency improvements to their homes • C&G issues advice on energy efficiency with all new mortgage packs • General Insurance provides cover for renewable energy appliances such as wind turbines, solar panels and ground source heating pumps at no extra cost on Lloyds TSB buildings insurance policies
Corporate markets	• Opportunities exist to develop specific products and services, such as climate hedging, carbon trading, venture capital for the 'green economy' and project finance for renewable energy projects • Project Finance and Lloyds TSB Corporate have financed biofuel and renewables projects
Support for small business	• Our long-established environmental credit risk assessment process has helped hundreds of small businesses to identify and address environmental risks to their business operations

used in Lloyds TSB's risk management framework. These examples are not intended to be exhaustive.

The Group's responses to the CDP5 questions on strategy and emission reduction targets mirror the information provided in its other corporate reports. But, while the company's response to the questions on its greenhouse gas emissions accounting methodology also mirrors the information detailed within its other reports, it provides additional commentary noting:

> i. As a financial services company we consider the CO_2 emissions associated with the use and disposal of the company's products and services to be negligible. Through our environmental credit risk assessment process we help our business customers identify and address environmental risks to their businesses. ii. We do not measure the emissions of our supply chain and we believe that this would be impractical and would also lead to double counting. We do, however, assess our suppliers' environmental and corporate responsibility performance as part of the tender process and engage with those suppliers that do not meet our criteria to agree a practical solution. In our Group Code of Business Conduct we state that we expect our suppliers to have broadly similar principles to our own. Our Scope 3 emissions therefore relate only to employee business travel.

The disclosures in the BiTC and the FTSE4Good Index series repeat much of the environment and climate change information relating to the Group's strategic management framework and environmental policy outlined above, though the specific information varies reflecting the different objectives and requirements of each data gatherer.

Discussion and conclusions

Lloyds TSB recognises that it needs to manage both direct greenhouse gas emissions as well as the direct and indirect (through customers) impacts of climate change on the business. From a purely financial perspective, the physical impacts represent by far the greater business exposure (though this exposure is significantly reduced by the company's risk assessment and management processes).

However, the focus of Lloyds TSB's reporting is its own carbon footprint and actions its employees can take; little is presently being explicitly reported on the physical risks of climate change despite its potentially greater financial significance. This emphasis reflects the fact that reporting is demand-driven and that most demand is being seen to come from employees rather than other stakeholders. Employee feedback on the integration of climate criteria in business development opportunities suggests they believe there is increasing potential to reduce their own emissions and develop brand value from reporting on climate. Furthermore, Lloyds TSB sees employees as a key stakeholder and, hence, reporting on climate change is an important dimension of its reporting on its social and environmental performance.

In contrast, Lloyds TSB sees limited demand for climate change reporting from stakeholders such as investors and customers. While the Group responds to investor-supported initiatives such as the CDP, it does not see that there is investor or customer demand for the provision of more information on climate change-related risks and opportunities in its corporate communications.

Based on its investor engagement experience, Lloyds TSB believes demand for climate change information from investors is at best minimal. The Group classes investors in two broad categories: mainstream investors and SRI analysts.

Mainstream investors are perceived to have little or no interest in climate change information, while SRI analysts have a degree more interest but in largely different things. Mainstream investors are perceived to be interested in financial impact and, in particular, what is financially material to business performance. The Group's engagement experience has led it to believe that mainstream investors do not see climate change as a material risk for the business nor do they seem to recognise that quality of management on climate change is/could be proxy for quality of management overall. However, this perception is probably reinforced by Lloyds TSB concentrating its reporting on greenhouse gas emissions (e.g. in its Annual Report and Accounts); the reality is that, even if the Group was to offset all its 180,000 tonnes of emissions, this would probably cost significantly less than €5 million (and so is not a significant exposure for the business). In fact, there is a risk that not only will mainstream investors disregard emissions-related information but they may even see this reporting as an attempt by the Group to distract attention from the real risks faced by the business. That is, the Group may not deliver on its objective of demonstrating that it has a clear understanding of the climate change issue and is managing it accordingly.

The information being sought by SRI analysts is, almost inevitably, quite different to that sought by mainstream investors, reflecting the specific funds that they are managing and, in many cases, their interests as social investors. Although SRI analysts are an important constituency (in terms of ensuring that Lloyds TSB is included in ethical funds and in indices such as FTSE4Good), they are just a small part of the wider investment community.

Investor response to CDP reporting is interesting as it is here that Lloyds TSB discusses − albeit in very general terms − the most significant business risks and opportunities presented by climate change. Thus this reporting would be expected to stimulate much more attention in the performance of Lloyds TSB in this area. However, it is surprising to find that its CDP5 disclosure raised only one follow-up query from an SRI analyst. The absence of feedback on its CDP disclosures has led the Group to believe that mainstream investors simply have no knowledge of CDP or, if they do, that CDP data are simply not being used by investors. Indeed, the impression is that CDP is primarily a means for investors to show that they are 'doing something' on climate change.[14]

In conclusion, this chapter illustrates the challenges faced by companies in trying to report to multiple stakeholders, even when there are multiple communication routes/ vehicles. Overall, the Lloyds TSB case suggests that a lack of symmetry exists between information provided by the Group and that demanded by investors. This is particularly relevant to the manner in which mainstream investors view climate change reporting by Lloyds TSB. Mainstream investors have requested information from companies on the risks, opportunities and financial implications of climate change through initiatives such

---

14 The next phase of the research programme reported in this chapter will be aimed at engaging Lloyds TSB investors to evaluate their response to what has been reported by the Group and to discuss investor demand for climate change information in more detail.

as CDP, but it is not clear what (if anything) they then do with this information. Determining the value of reporting for investors is difficult due to the limited amount of feedback or engagement taking place with investors on climate change issues. In the case of Lloyds TSB, it is not clear whether this reflects a lack of investor demand for this information, a lack of investor satisfaction with information supplied by the Group or simply that investors are gaining information on issues such as indirect climate change risks through third-party sources such as CDP. Given that investors are so important to Lloyds TSB, better understanding of what investors are looking for in this area is a priority for the organisation.

Reflecting more generally on the climate change debate, investors are recognised as a critical influence on the manner in which companies operate both through their role as 'allocators of capital' and through their ability to influence corporate strategy and decision-making. Yet the experience of Lloyds TSB suggests that, despite the high profile of initiatives such as IIGCC and CDP, investor engagement with the issue of climate change remains largely symbolic and confined to SRI analysts or specialists rather than being a mainstream investor interest.

References

ACCA (Association of Chartered Certified Accountants) and FTSE (2007) *Improving Climate Change Reporting: An ACCA and FTSE Group Discussion Paper* (London: ACCA).

CICA (Canadian Institute of Chartered Accountants) (2005) *MD&A Disclosure about the Financial Impact of Climate Change and other Environmental Issues* (Discussion Brief: Canadian Performance Reporting Board; Ontario: CICA).

Deloitte (2007) *Accounting for Emission Rights: Discussion Paper* (London: Deloitte).

FORGE (2007) *Managing Climate Change in Financial Services: A Guidance Framework* (London: FORGE).

Hill, M., and L. McAulay (2006) *Emissions Trading and the Management Accountant* (Chartered Institute of Management Accountants Research Report; London: CIMA).

IIGCC (Institutional Investor Group on Climate Change) (2006) *Global Framework for Climate Risk Disclosure: A Statement of Investor Expectations for Comprehensive Corporate Disclosure* (London: IIGCC).

—— (2007) 'IIGCC and IGCC on Bali Climate Change Negotiations. Letter to the UNFCCC at Bali, 30 November 2007' (London: IIGCC).

——, Ceres and IGCC (2008) *Electric Utilities: Global Carbon Disclosure Framework* (London: IIGCC, January 2008).

PwC (PricewaterhouseCoopers) (2007) *Building Trust in Emissions Reporting: Global Trends in Emissions Trading Schemes* (London: PwC; https://www.pwc.com/extweb/pwcpublications.nsf/docid/8df4237f6b2f7fcf8525728300503b70).

Sullivan, R. (2006) *Climate Change Disclosure Standards and Initiatives: Have they Added Value for Investors?* (London: Insight Investment).

WBCSD (World Business Council for Sustainable Development) and WRI (World Resources Institute) (2004) *Greenhouse Gas Protocol: A Corporate Accounting and Reporting Standard* (Washington, DC: WRI).

Part IV
Corporate responses and case studies

14
Curbing greenhouse gas emissions on a sectoral basis: the Cement Sustainability Initiative

Timo Busch
ETH Zurich, Switzerland

Howard Klee
World Business Council for Sustainable Development, Switzerland

Volker H. Hoffmann
ETH Zurich, Switzerland

Origins of sectoral considerations

To date, discussions on climate policy have focused mainly on government as the central actor responsible for greenhouse gas emission reduction approaches necessary to avert the most serious consequences of global warming. International, regional (e.g. EU) and national policy and regulatory frameworks are seen as the critical means for the delivery of these emission reductions. However, policy efforts that do not require action on a worldwide scale (e.g. a global cap-and-trade system) must address two problems.

First, from an economic perspective, different regulatory standards may result in different competitive environments for companies, with companies in countries with strict emission regulations potentially facing higher costs which might, in turn, result in a loss of competitiveness. Second, from an environmental perspective, 'carbon leakage' (Felder and Rutherford 1993) might occur; companies may decide to relocate carbon-intensive processes to countries where emission regulation is less stringent. However, such a movement of emission-intensive industrial activities is counterproductive because manufacturing emissions still occur and are potentially compounded by the emissions associated with the subsequent shipment of products back to the original country. Furthermore, the emissions relative to product output might, in fact, increase if production standards are less technologically developed in those countries.

In response to these problems, it has been suggested that policy frameworks should focus on global industrial sectors instead of on single national or regional economies: in particular, sectors exposed to international competition. Such sectoral approaches have even been considered as 'needed to complement an economically sound cap and trade system to create additional incentives to invest in low greenhouse gas approaches in key sectors' (USCAP 2007). This chapter reviews the current debate regarding the potential role of sectoral approaches in international climate policy using the case of the Cement Sustainability Initiative (CSI), an industry-led sectoral approach, to illustrate the practical application of this idea and some of the challenges that need to be addressed for sectoral approaches to be seen as a credible part of the overall international climate policy.

Proposals for sectoral approaches

There is no common agreement about the exact content, scope and scale of pursuing a sectoral approach: the term is used and interpreted differently by different parties and in different contexts (Baron *et al.* 2007). Table 14.1 illustrates a few prominent examples within the debate.

The common idea behind sectoral approaches is that new policies on climate change mitigation can be based on sectoral considerations which take into account specific circumstances and conditions for greenhouse gas reductions of one individual or several industries (Baron 2006). The recent discussions about sectoral approaches in climate change policy have centred on the question of how a sector could come to an agreement to take on a greenhouse gas reduction goal.

For the purposes of this chapter, we define sectoral approaches broadly, encompassing all of the actions initiated and taken by an industry group directed at greenhouse gas emission reductions. Based on this definition, we focus specifically on how collective actions can go beyond voluntary agreements and be embedded in post-2012 climate legislation. Within this debate, four main reasons have been advanced in support of sectoral approaches:

- **Participation of developing countries.** A central issue for global climate change mitigation is the participation of both developing and developed countries. If the right incentives for a less greenhouse gas-intensive development are provided, developing countries might overcome the risk of being locked into a carbon-intensive path and adopt a sector-wide emission target (Bosi and Ellis 2005; Baron 2006)

- **Incorporation of competitiveness concerns.** Companies in developed countries that have adopted binding emission targets are facing a constraint (and associated costs) that their international competitors are not facing. A sectoral

TABLE 14.1 Examples of sectoral approaches

Type	Objective	Concept	References
Triptych sectoral approach	Distributing the burden of greenhouse gas emission reductions	Differentiation between the power-producing sector, internationally operating energy-intensive industries and the remaining domestically oriented sectors	Phylipsen et al. 1998; Groenenberg et al. 2004
Sectoral crediting mechanisms	Extending flexible project mechanisms to a sector-wide approach	If emission reduction targets are exceeded, Emission Reduction Credits are granted. These can be traded on national or international markets and offer flexibility to reduce emissions where most cost-effective	Philibert and Pershing 2001; Bosi and Ellis 2005
Center for Clean Air Policy (CCAP) sector-based proposal	Developing a global framework for effective emission reductions targeting sectors with a high reduction potential	Differentiation between developing countries (voluntary non-binding targets) and developed countries (binding targets) Provide incentives and crediting mechanisms for developing countries to meet targets	Schmidt et al. 2006
Asia-Pacific Partnership (APP) sector task forces	Accelerating the development and deployment of clean energy technologies	Development, diffusion, deployment and transfer of cost-effective, cleaner, more efficient technologies and practices Sectoral assessments, capacity-building and best-practice identification	APP 2007
Industry-driven initiatives such as the steel, aluminium and cement industry (see Box opposite)	Establishing an industrial sector-specific approach for greenhouse gas emission reductions	Development of common reporting standards and protocols, targets, implementation of joint projects, R&D pooling	For example: IISI 2007; WBCSD 2007

There has been a lot of work done on the concepts of and options for sectoral approaches. This table provides an overview of important contributions. But, due to the sheer number of studies and policy suggestions, this table is not exhaustive. For a general review of approaches for international climate change efforts, see Bodansky 2004.

Sectoral approaches in the aluminium and steel industry

Participants in the sectoral approach of the International Aluminium Institute (IAI), a group representing 70% of worldwide aluminium production, have set themselves 13 voluntary objectives of which some pertain to climate change. These include an 80% reduction in emissions of perfluorocarbons (PFCs) per tonne of aluminium produced by 2010 versus 1990, and a 10% reduction in average smelting energy use per tonne of aluminium produced by 2010 versus 1990. Furthermore, the industry will monitor its transport activities with the objective of identifying emission reduction opportunities, and will seek to reduce emissions from the production of alumina per tonne produced.

For further information see www.world-aluminium.org.

The International Iron and Steel Institute (IISI), a group of the major global steel companies that produce around 75% of the world's steel (excluding China) has proposed a sectoral approach for the steel industry focusing on developing new and breakthrough technologies. The current focus is on the collection and reporting of carbon dioxide emissions data by steel plants in all the major steel-producing countries and on the transfer of the best available steelmaking technologies to developing countries. The aim is to set commitments on a national or regional basis for emissions measured in CO_2 per tonne of steel.

For more information see www.worldsteel.org.

approach could reduce these competitiveness concerns and lessen the risk of carbon leakage. This may involve setting different types of targets for developing and developed countries (see, for example, Schmidt *et al.* 2006) and/or the use of flexible mechanisms such as sectoral crediting mechanisms (see, for example, Bosi and Ellis 2005)

● **Incentives for accelerated technology diffusion.** Substantial reductions of greenhouse gas emissions could be achieved by the employment of currently best available technologies. Sectoral approaches focus attention on where the key breakthroughs have to be made (Watson *et al.* 2005). Presuming the right incentives are set, sectoral approaches may facilitate technology diffusion through increased cooperation of companies in a given sector (Bosi and Ellis 2005)

● **Construction of sector-tailored policies.** Sectoral approaches offer the opportunity to construct policies tailored to the individual characteristics of industrial sectors (e.g. the number and market share of the major players, the number and location of countries where companies are based, the general status of technology development in the sector)

Designing sectoral approaches: some key features

The development of any sectoral approach has to take into account the specific characteristics of the sector. In this regard, the target setting process, the scope of coverage and the mode of governance are of particular importance (Baron *et al.* 2007). In the following section, we briefly discuss each of these key elements before focusing specifically on the case of the Cement Sustainability Initiative.

Target setting

Emission targets within a sectoral approach can be defined either by addressing greenhouse gas emissions directly (Baron and Ellis 2006) or by indirect means such as technological standards (Watson *et al.* 2005). There are four main methods:

- Emission-based absolute targets are typically based on historical emissions (e.g. baseline year or average over a defined period of time) or emission scenarios (e.g. business as usual emission projections)

- Emission-based dynamic targets are similar to absolute targets, but contain a further measure that allows for target adjustments (e.g. the development of a country's GDP)

- Emission-based relative targets require the definition of benchmarks for different facilities and/or production processes. For this purpose, the target is formulated as a ratio between the emissions and an input or output variable (e.g. the unit of production)

- Technology-based approaches focus on:
  - Common standards such as the use of best available technologies
  - Process or production standards
  - Technology targets such as the obligation for all companies in a sector to use a certain type of technology

Targets can be binding or non-binding. Binding targets are a legal obligation to meet a certain target for all involved parties, whereas non-binding targets do not impose any consequences in the case of non-compliance (Bosi and Ellis 2005). Furthermore, a mixture of target-setting approaches can be adopted: for example, if companies from developed countries have to meet binding targets whereas companies from developing countries agree to voluntary, non-binding targets (so-called 'no-lose pledges'). Companies in developed countries could be given incentives to adopt such targets if, for example, emission reductions below an agreed baseline could be sold in an international carbon market (Schmidt *et al.* 2006). In addition, financial and research agreements may be used to indirectly address the sources of the emissions through, for example, supporting the diffusion of efficient technologies, fostering the use of alternative fuels, or raising the level of research and development projects in that area (Watson *et al.* 2005).

Scope of coverage

Sectoral approaches could be established on the international (more than one country), national (one country) or regional level (e.g. concentrated on one economic area). Among these possibilities, a sectoral approach on the international level is the most effective way of curbing global greenhouse gas emissions.

If an approach on this level was adopted, the participants would need to decide whether targets should be the same for all participating countries or whether the targets should reflect country-specific circumstances (Bosi and Ellis 2005).

Furthermore, when establishing the scope of such an approach, a decision would be needed on whether all installations within a sector or just a limited number (perhaps installations above a certain threshold) should be involved.

Mode of governance

Institutionalising a sectoral approach requires a mechanism — whether through regulatory or other means — to create obligations (binding targets) and adequate incentives (in case of non-binding targets) for the participants. Furthermore, for the development of a sectoral approach, close industry involvement is required due to the inevitable information asymmetries between governments and industry (Baron *et al.* 2007).

Industry participation offers the advantage that companies themselves are most familiar with technical details of their industry and, thus, are able to suggest appropriate steps on how to best curb greenhouse gas emissions from the industry. On the other hand, policy-makers and other stakeholders would need to be reassured that the steps or actions proposed are sufficient to meet the overall emission target.

Three general options for the negotiation and governance of sectoral approaches can be considered (Baron *et al.* 2007):

- National policies with intergovernmental coordination — where the national government would agree in close collaboration with an industry sector's representatives on how to set the targets and define the mechanisms and governance structures. Competitiveness concerns could be addressed through intergovernmental coordination

- Global agreements — where an industry sector works within the UNFCCC framework to determine the targets and mechanisms. This would require formal recognition of the industry's effort by governments and governments to confirm that the actions proposed as a result of such a process align with domestic policy goals. This option may also include the development of monitoring or enforcement mechanisms

- Global action — where an industry sector takes unilateral action to define emission targets and sectoral approach mechanisms. This option provides an umbrella for the coordination of the efforts by individual companies and could either offer a monitoring, verifying and reviewing process or rely on domestic policy frameworks. However, with respect to governance, it would not be subject to any official government oversight or compliance measures

These three options differ significantly regarding the involvement of national governments and the UNFCCC. We now discuss the Cement Sustainability Initiative which, in its current form, is an example of 'global action' by an industry sector, though it does include elements of the other two options (global agreements and national policies with intergovernmental coordination) in terms of its further development.

The Cement Sustainability Initiative

In 1999, ten leading cement companies — representing at that time one-third of the world's cement production — voluntarily embarked on a common project. The initiating companies had realised the need for the industry to address the global challenge of sustainability and that addressing this challenge could be achieved most effectively through a collective effort of the industry (WBCSD 2005). Therefore, the companies decided to search for new ways for the industry to reduce its ecological footprint, understand its social impact and increase stakeholder engagement.

The companies chose to conduct their work under the auspices of the World Business Council for Sustainable Development (WBCSD). Over time (see Table 14.2), this project evolved into the CSI. Currently, 18 major cement producers, with operations in more than 100 countries, participate in the initiative. These companies account for 60% of global cement production outside of China, which equals approximately 640 million tonnes of cement and 450 million tonnes of CO_2 emissions.

TABLE 14.2 **Overview of important events of the CSI**

Date	Event
1999	**CSI formation.** Ten of the world's biggest cement producers joined forces under the auspices of the WBCSD to investigate ways for a more sustainable cement industry
2000–2002	**Evaluation.** On behalf of the group, a US-based not-for-profit consulting company conducted an independent review of the cement industry and suggested ways for the industry to meet its sustainability challenges
2002	**Results.** The report, *Toward a Sustainable Cement Industry*, discussed the major sustainability issues facing the industry over the next 20 years. The companies responded with their 'Agenda for Action' to address issues raised by the research and stakeholder consultations
2005	**Progress.** Three years into the process, the CSI released a progress report documenting its first action steps. During the same period, good-practice guidelines were completed and a number of new members joined the group
2007 and beyond	**Future.** Updated report and future priorities, most importantly in the area of an international climate change-focused sectoral approach for the cement industry

In 2005, the cement industry produced about 2,000 million tonnes of cement and 4.6% of global anthropogenic greenhouse gas emissions (Baron *et al.* 2007): half of these emissions from the chemical process of clinker production, 40% from burning fuel and the remaining 10% from electricity use and transportation (WBCSD 2005).

It typically requires the equivalent of 60–130 kg of fuel oil and 110 kWh of electricity to produce one tonne of cement. This equals an average amount between 0.65 and 0.93 tonnes CO_2 per tonne of cement for energy consumption and calcination (IEA 2007). As Figure 14.1 illustrates, global cement production is forecast to grow by more than 150% by 2050. This figure emphasises the need to actively manage the sources of CO_2 within the cement sector.

FIGURE 14.1 Forecast for global cement production

Source: IEA as cited in WBCSD 2007

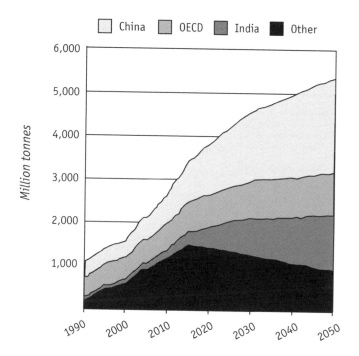

In 2002 the cement industry's Agenda for Action was agreed as a 20-year programme. As part of the Agenda, the CSI Charter sets out the minimum requirements for membership. The key actions with respect to climate change are that:

- Companies report their greenhouse gas emissions in a standardised format using the WBCSD's *Greenhouse Gas Protocol* (WBCSD/WRI 2004)

- Commit to a proactive approach to carbon management

- Set targets for reducing their CO_2 emissions (see Table 14.3)

TABLE 14.3 CSI companies as of 2007 and their reduction targets

Company	CO$_2$ reduction target (per tonne cementitious product)	Baseline	Target year
CEMEX (Mexico)	25%	1990	2015
Cimpor (Portugal)	15%	1990	2015
CRH plc (Ireland)	15%	1990	2015
Votorantim (Brazil)	10%	1990	2012
Holcim (Switzerland)	20%	1990	2010
Lafarge (France)	20%	1990	2010
HeidelbergCement (Germany)	15%	1990	2010
Titan Cement SA (Greece)	15%	1990	2010
Taiheiyo Cement (Japan)	3%	2000	2010
Italcementi (Italy)	711 kg	–	2008–2012
Siam Cement (Thailand)	670 kg	–	2010

Under the CSI Charter, member companies have four years from joining to meet their charter obligations. In 2007, the following companies were also members of the CSI but not yet required to voluntarily set and publicly report reduction targets: Ash Grove Cement (USA), Cementos Molins (Spain), Cementos Portland Valderrivas (Spain), Cimentos Liz (Brazil), Grasim Industries Ltd (India), Secil Cement Company (Portugal) and Shree Cement (India).

Among other concerns, the CSI has chosen not to set a common target because of both antitrust and governance concerns. CO$_2$ emissions are directly related to each company's production and antitrust authorities may be concerned that a common emissions target could translate into some production targets. In addition, the WBCSD and CSI are not enforcement organisations and so have taken the view that the responsibility for setting and meeting targets is better left to those with direct responsibility for these actions. However, the CSI does require that each company publishes details of its target and reports annually on progress towards meeting the target. The companies have further agreed to independent third-party verification of their CO$_2$ emissions, beginning with data collected in 2006.

Beyond the activities undertaken so far, the CSI has started to consider the potential for a sectoral approach in the cement industry for the post-Kyoto climate agreement. In its current thinking, the following key elements have been formulated:

● Targets and incentives to be set by governmental authorities based on output-based efficiency benchmarks

● Simple measurement metric of CO$_2$/cementitious product (tonne/tonne) for proposing consistent but possibly regionally differentiated targets

- Sectoral crediting mechanisms to reward improved efficiency and to promote use of waste fuel and blended cement production

- Support for research and development projects to develop technology and build capacity by public–private partnerships

CSI members would adopt these actions on the understanding that new regulations on greenhouse gas emissions will emerge and it is important the cement industry to be regulated on a global scale. From an economic point of view, every policy that is not anchored on the global level will bring up competitiveness concerns. The proposed sectoral approach offers the opportunity to deal with the whole sector in a way that does not commit single economies to broader actions. From an ecological point of view, about 80% of the emissions in 2050 from cement production will come from developing countries, which currently do not have emission limitations in place. Consequently, actions in countries that already control emissions will have only limited impact.

Evaluation of the CSI

The CSI is an important initiative for two reasons:

- Its potential to influence greenhouse gas emissions from one of the most significant sectors

- Its potential as a model for other sectors to take action to reduce their emissions

Following the framework developed by Sullivan (2005) for the evaluation of voluntary approaches in environmental policy, we assess the outcomes from CSI to date and the potential for CSI to make a substantive contribution to emission mitigation in the future.

Environmental effectiveness

The environmental effectiveness of a sectoral approach requires the participation of major emitters and depends on the nature of the targets (binding or non-binding). Thus far, 11 of the 18 CSI companies have set individually binding targets on a voluntary basis (see Table 14.3). Other companies are obliged to do so within four years of joining the CSI.

Although the magnitude of these targets has been set individually by the companies, they have already contributed to emission reductions and thus demonstrated the environmental effectiveness of current measures. For instance, Holcim achieved about a 15% reduction in its specific CO_2 emissions per tonne cementitious product by 2005 compared with 1990 emissions.

In the longer term, the companies recognise that a voluntary approach will need to become mandatory, with appropriate enforcement mechanisms to ensure compliance. Many key performance indicators of the CSI members are verified by independent third-party assurance, beginning with CO_2 data reported from 2006 onwards. If a company does not meet its obligations under the CSI Charter, such as setting targets and reporting publicly on progress towards those targets, the CSI issues first a reminder and then a caution. If the company fails to meet its Charter obligations for a third time, the company is asked to resign. These mechanisms will help to maintain the effectiveness of the emission reduction activities under the CSI umbrella.

For the intended post-2012 sectoral approach, the CSI will deliver a framework methodology and necessary emission data, which are prerequisites for the development of effective sector-wide emission benchmarks and baselines (Bosi and Ellis 2005). To ensure the credibility of the achieved environmental improvements, participating companies will provide the appropriate data to an independent third-party service provider, which will collate the data and develop and maintain the database.

Beyond that, the CSI does not intend to formulate targets itself, nor does it intend monitoring performance against these targets. Rather, the CSI targets will need to be set by governments or the international community. This governmental involvement constitutes the core pillar for the overall environmental effectives of the post-2012 approach. But, even with a goal of emission reductions that go beyond 'business as usual', total emissions from the sector are likely to increase as driven by increasing demand for cement (as indicated in Fig. 14.1).

Moreover, the establishment of an independent facility for verification and monitoring will be important to ensure the effectiveness of the approach. Such a facility does not necessarily have to be a governmental body, but might also be an international institute supervised and advised by the UNFCCC or independent auditors with experience in emissions verification. Furthermore, governments endorsed by the UNFCCC should also be entitled to enforce the objectives of the approach through the use of sanctions or penalties if individual companies are in breach of their obligations.

Economic efficiency

An economically efficient climate regime would allow (if not actively encourage) reductions to take place where they can be achieved with the lowest cost (Baron et al. 2007). At the current stage of the CSI — without a crediting system — it is up to the companies to determine their reduction strategy. While this should mean that the initiative is economically efficient at the company level, it may not be efficient across the sector as a whole; for example, companies may have different criteria for defining what is economically efficient and there is no structured mechanism for differentiation (e.g. through the trading of emissions reduction credits) between companies based on which has the lowest marginal abatement costs.

The post-2012 sectoral approach will need to cover cement companies on a global level. If the regime is designed properly, it should allow emission reductions to be realised where this is the most efficient in the sector. A focus on economically efficient solutions will be critical to maximise industry participation in a sectoral approach.

Transaction costs

Typically, transaction costs are determined by search, negotiation, approval, monitoring, enforcement and insurance costs (Woerdman 2001). Since 1999, about US$6 million has been invested in the CSI mainly due to search and monitoring costs. Beyond that, there have not been any additional transaction costs for the CSI as the individual companies have mainly organised their carbon management activities internally.

The transaction costs of the post-2012 sectoral approach will depend largely on the design of the approach. Key factors influencing these costs will include the effort needed to monitor, review and enforce the approach.

For sectoral crediting mechanisms, the transaction costs of establishing and implementing will vary according to the mechanism type, its characteristics (e.g. type of targets) and number of participants (Bosi *et al.* 2005). Although the transaction costs are likely to be significant, these costs are dwarfed by the scale of the challenge faced by the industry and the likely costs (and benefits) of achieving significant reductions in greenhouse gas emissions. The industry sees the costs incurred to date as necessary to allow it to develop its strategies for action. The costs of monitoring emissions are, in the main, costs that probably would have been incurred anyway (i.e. there are limited additional transaction costs associated with implementing CSI-related commitments).

Competitiveness

Within the current CSI activities, the main competitiveness effects come about because member companies are better prepared for future legislation compared with companies that have not yet started to manage their greenhouse gas emissions. As such, the CSI paves the way for its member companies to generate a competitive advantage.

More importantly, in particular given the discussion around carbon leakage, the CSI post-2012 sectoral approach seeks to provide a more level playing field through:

- The participation of major cement-producing companies
- The use of a benchmarking system (including incentives)
- Allowing developing countries to participate in the CSI — albeit on the basis of 'no-lose' pledges

In general, the inclusion of such pledges appears to be problematic as participating companies in developing countries face no binding obligations. Hence, the question whether the intended approach will indeed provide a level playing field will depend significantly on the incentive structures: companies in developing countries need adequate incentives to emit less than the target in order to generate, for example, certificates for a sectoral crediting system.

Innovation and soft effects

One important effect of current CSI activities is that member companies have, through their participation to date, developed management competence to respond to climate change. The CSI has put special emphasis on establishing a common set of CO_2 management tools and practices (WBCSD 2007). The CSI has also developed industry-wide knowledge about the climate change performance of the sector. As part of this, the CSI members have built a database consisting of all CO_2 emissions, fuel use and other data from more than 700 cement plants.

Although there is no well-known or proven technology to make the process of cement production carbon neutral, CSI companies actively pursue emission reductions. This has stimulated innovation and beyond business as usual efforts, in particular in terms of process optimisation.

Some of the measures taken by the member companies to reduce their CO_2 emissions include (Watson *et al.* 2005):

- Reducing the clinker-to-cement ratio through the use of additives such as slag or fly ash

- Increasing operating efficiency by optimising production technology

- Reducing the reliance on fossil fuel sources (e.g. by using alternative fuels and biomass instead of coal)

- Utilising drier limestone as lower moisture content reduces energy use

The increasing uptake of these options in combination with accelerated technology development, diffusion and transfer will further reduce emissions from the cement sector under a post-2012 sector framework. For example, in cooperation with the Asia-Pacific Partnership (APP), the CSI has started to build up the foundation for this by offering training and workshops for companies in China and India.

Acceptability and public participation

The development of a sectoral approach requires the consideration of its acceptability by stakeholders. First, it is important for the legitimacy of the approach that national governments, which might have mitigating strategies and policy objectives of their own, are involved in the development phase. Second, the participation of non-governmental organisations (NGOs) and other concerned stakeholders is important to assure the public about the integrity and effectiveness of the sectoral approach. It is, therefore, important to include the perceptions and suggestions of these groups when defining the targets and mechanisms (Buysse and Verbeke 2003; Hart and Sharma 2004). Consultation with stakeholders has been an important feature of the CSI's activities.[1]

Furthermore, the CSI has worked closely with a number of international governmental and NGOs (e.g. IEA, WRI) to increase the acceptability of the approach by addressing various aspects of the work programme. Finally, the work and progress of the CSI is evaluated by a senior advisory board, which consists of former senior members from WWF International, the International Institute for Sustainable Development in Canada and the United Nations Environment Programme (UNEP).

To maintain this level of acceptability and public participation within the post-2012 process, the CSI will need to continue to work in close cooperation with stakeholders and the UNFCCC, which is seen as the primary global organisation for responding to climate change. For example, the CSI recently discussed issues of a future sectoral approach

---

1 Dialogues have been conducted in the USA, Europe, Asia, Middle East and South America with representatives from government, residents' groups, employees, consumer organisations, suppliers and NGOs. Details of the CSI's stakeholder dialogues can be found at www.wbcsdcement.org/stakeholder_dialogues.asp.

in a multi-stakeholder task force convened under the auspice of the Centre for European Policy Studies. These activities will become increasingly important for ensuring the effective establishment of the sectoral approach within an international policy framework.

Law and public policy issues

One of the key challenges for any sectoral approach is how it interacts with other climate change-related initiatives. The CSI post-2012 sectoral initiative will seek to complement other mechanisms and approaches through building closer links with other initiatives such as the Asia-Pacific Partnership.

With respect to the EU Emission Trading Scheme (EU ETS), one particular challenge is how to include companies that are already covered by the scheme. A related issue is the role of the Kyoto Protocol's flexible mechanisms in the post-2012 CSI initiative. One option could be for reductions below the established targets to be eligible for sale as Emission Reduction Credits in a given trading system (Bosi *et al.* 2005). However, the manner in which such credits would be integrated into an existing trading scheme across industries and how they would relate to established flexibility mechanisms (e.g. the Clean Development Mechanism) is complex, and would require further negotiation and agreement between the involved parties (Bosi *et al.* 2005; Baron 2006; Baron *et al.* 2007). These are ongoing questions for the CSI.

Conclusion

Climate change is a major challenge of the 21st century that requires immediate action. It requires thinking about different ways to reduce emissions effectively on a global scale. Sectoral approaches represent one way for curbing greenhouse gas emissions beyond national climate policies. With current technological knowledge, the cement industry has a limited emission reduction potential, given that the chemical reaction of the calcination process emits CO_2 and is energy-intensive. However, or especially due to this fact, the CSI plays an important role in tackling this issue for an important and growing sector.

The efforts of the CSI thus far demonstrate that sectors emitting significant amounts of greenhouse gases can take voluntary action to increase their 'carbon efficiency' by reducing emissions per tonne of output — albeit acknowledging that it is most unlikely that the cement sector will ever be entirely 'carbon independent'. We see four central challenges for the CSI to be successfully embedded in the post-2012 climate policy framework:

- The CSI seeks to provide a level playing field through ensuring the participation of major cement producing companies, the use of a benchmarking system and some form of sectoral 'no-lose' crediting. In order to reach this goal, major companies internationally — notably those in China — have to participate and support a system with stringent targets and sanction mechanisms

- The 'no-lose' pledges in particular raise the risk that developing countries do not contribute fully to the global target. Hence, companies in developing countries have to be encouraged to meet their targets through the adoption of adequate mechanisms (e.g. accelerated technology diffusion) and incentives (e.g. tradable permits)

- To be in line with future international climate policy, the approach has to be complementary to other schemes such as the existing EU ETS or even a possible global cap-and-trade system

- While an alternative technology has not yet been developed, the projected increase in cement demand calls for the cement industry to accelerate its efforts to develop and commercialise low-carbon production processes such as oxyfuel combustion in kilns or blended cements. Furthermore, carbon capture and storage could constitute a possible option for reducing the industry's climate impact

The idea of sectoral approaches is currently being discussed in other sectors. A successful CSI sectoral approach can help guide the way ahead for other industries by demonstrating the following potential benefits:

- Given that major parts of the sector worldwide could be tied to similar obligations, competitiveness impacts of nationally linked emissions caps could be reduced

- Assuming the rejection of national emission caps by developing countries, an approach based on 'no-lose' pledges could serve to break a potential deadlock in international negotiations

- As there is a lack of high-quality data at both sectoral and national levels in most developing countries, the approach can improve data and information on emissions performance in developing countries – a critical factor for determining technology needs in those countries

- The approach can deliver technology transfer benefits through raising awareness of benchmark performance and broaden the application of best available technologies

Finally, the CSI recognises that voluntary actions alone will not be sufficient to meet the climate change challenge. In this regard, CSI overcomes a major criticism of most voluntary approaches by explicitly locating itself as part of the move towards mandatory requirements. Should a sectoral agreement be reached, governments will be required to set the overall targets to be achieved by the industry rather than relying on targets set by the industry itself. However, policy has to take into account the fundamental barrier that, with currently available technologies, the emission reduction potential is limited. Therefore, it is also important to look more closely at the consumption of cement, as the projected drastic increase in global cement demand is almost certain to swamp any efficiency gains that will be achieved.

References

APP (Asia-Pacific Partnership) (2007) 'Asia-Pacific Partnership on Clean Development and Climate: Fact Sheet'; www.asiapacificpartnership.org/FactSheet-Oc-07version.pdf (accessed 6 June 2008).

Baron, R. (2006) *Sectoral Approaches to GHG Mitigation: Scenarios for Integration* (Paris: OECD/IEA).

—— and J. Ellis (2006) *Sectoral Crediting Mechanisms for Greenhouse Gas Mitigation: Institutional and Operational Issues* (Paris: OECD/IEA).

——, J. Reinaud, M. Genasci and C. Philibert (2007) *Sectoral Approaches to Greenhouse Gas Mitigation: Exploring Issues for Heavy Industry* (IEA Information Paper; Paris: OECD/IEA).

Bodansky, D. (2004) *International Climate Efforts Beyond 2012: A Survey of Approaches* (Arlington, VA: Pew Center on Global Climate Change).

Bosi, M., and J. Ellis (2005) *Exploring Options for 'Sectoral Crediting Mechanisms* (Paris: OECD/IEA).

Buysse, K., and A. Verbeke (2003) 'Proactive Environmental Strategies: A Stakeholder Management Perspective', *Strategic Management Journal* 24.5: 453-70.

Felder, S., and T. Rutherford (1993) 'Unilateral CO_2 Reductions and Carbon Leakage: The Consequences of International Trade in Oil and Basic Materials', *Journal of Environmental Economics and Management* 25.2: 162-76.

Groenenberg, H., K. Blok and J. van der Sluijs (2004) 'Global Triptych: A Bottom-up Approach for the Differentiation of Commitments under the Climate Convention', *Climate Policy* 4.2: 153-75.

Hart, S., and S. Sharma (2004) 'Engaging Fringe Stakeholders for Competitive Imagination', *Academy of Management Executive* 18.1: 7-18.

IEA (International Energy Agency) (2007) *Tracking Industrial Energy Efficiency and CO_2 Emissions* (Paris: IEA/OECD).

IISI (International Iron and Steel Institute) (2007) *A Global Sector Approach to CO_2 Emissions Reduction for the Steel Industry* (Brussels: IISI).

Philibert, C., and J. Pershing (2001) 'Considering the Options: Climate Targets for all Countries', *Climate Policy* 1.2: 211-27.

Phylipsen, G., J. Bode, K. Blok, H. Merkus and B. Metz (1998) 'A Triptych Sectoral Approach to Burden Differentiation: GHG Emissions in the European Bubble', *Energy Policy* 26.12: 929-43.

Schmidt, J., N. Helme, J. Lee and M. Houdashelt (2006) *Sector-Based Approach to the Post-2012 Climate Change Policy Architecture* (Washington, DC: Centre for Clean Air Policy).

Sullivan, R. (2005) *Rethinking Voluntary Approaches in Environmental Policy* (Cheltenham, UK: Edward Elgar).

USCAP (US Climate Action Partnership) (2007) *A Call for Action – Consensus Principles and Recommendations from the US Climate Action Partnership: A Business and NGO Partnership* (Washington, DC: USCAP).

Watson, C., J. Newman, R. Upton and P. Hackmann (2005) 'Can Transnational Sectoral Agreements Help Reduce Greenhouse Gas Emissions? Background paper for the meeting of the Round Table on Sustainable Development, 1–2 June 2005' (Paris: OECD).

WBCSD (World Business Council for Sustainable Development) (2005) *The Cement Sustainability Initiative: Interim Progress Report* (Geneva: WBCSD; www.wbcsdcement.org [accessed January 2008]).

—— (2007) 'The Cement Sustainability Initiative Progress Report 2007'; www.csiprogress2007.org (accessed 5 June 2008).

—— and WRI (World Resources Institute) (2004) *The Greenhouse Gas Protocol: A Corporate Accounting and Reporting Standard* (Geneva: WBCSD, rev. edn).

Woerdman, E. (2001) 'Emissions Trading and Transaction Costs: Analysing the Flaws in the Discussion', *Ecological Economics* 38.2: 293-304.

15

Novartis: demonstrating leadership through emissions reductions

Helen Mathews
University of Basel, Switzerland

Claus-Heinrich Daub
University of Applied Sciences Northwestern, Switzerland

Novartis is one of the world's largest pharmaceutical companies with operations in over 140 countries and around 100,000 employees. It is the 30th largest corporation by market capitalisation worldwide and has annual revenues nearly one-fifth as large as those of the Australian government. In June 2005, Novartis committed to reducing its average absolute annual greenhouse emissions in the years between 2008 and 2012 to a level that was 5% lower than its emissions in 1990. In this chapter, we discuss the company's motivations for setting this goal, its strategy for delivering on the goal and the implications of the goal for the Novartis business.

Motivations for action

The decision by Novartis to set and publish a greenhouse gas emissions reduction target was motivated by two factors:

- The effects of climate change on its business
- The company's sense of its moral responsibilities

We discuss these motivations briefly below as they are critical to understanding the actions taken and the business benefits expected to result from taking action.

Effects of climate change on Novartis

Climate change is expected to affect both production processes and sales markets. With respect to its production processes, Novartis is likely to experience input shortages as a consequence of the negative impacts of climate change on biodiversity and water. In relation to biodiversity, the Intergovernmental Panel on Climate Change (IPCC) has stated that temperature increases of 1.5–2.5°C above pre-industrial levels, which are expected by 2050, will lead to the extinction of 20–30% of known plant and animal species (IPCC 2007). Biodiversity is an important source of new drugs with over 60% of all (global) new anti-cancer and anti-infective agents in the period 1984 to 1995 coming from natural products or their derivatives (Cragg *et al.* 1997). Novartis, in particular, will suffer from a reduction in biodiversity as it is still actively involved in screening natural substances for medicinal properties and maintains a large library of natural product samples. Recent Novartis medications based on natural compounds include the top-selling brands Neoral, Sandostatin and Miacalcic, which together brought in nearly US$700 million in net sales in 2006.

The other important input to the company's production processes likely to be affected by climate change is water. Novartis uses substantial amounts of water in its operations, in particular in central Europe, where its ten biggest sites currently use 80% of its total water input. In this region, a 20% reduction in precipitation is expected in the long term (2071–2100), but with more rain falling in winter and less in summer (Christensen *et al.* 2007). Water availability in summer will also be lower due to less water from snow melt (Jasper *et al.* 2004). The potential shortage could be compounded by competing demand from farmers and hydroelectric and nuclear energy companies (Lehner *et al.* 2005).

Climate change will also affect future consumer demand for pharmaceuticals by changing the spatial distribution and frequency of diseases. The IPCC (IPCC 2007) and the World Health Organisation (WHO 2007) both state that the effects of climate change on human health are likely to include an increased frequency of heart disorders due to higher levels of ground-level ozone and increases in infectious diseases such as malaria and dengue fever. Novartis is actively engaged in the fight against infectious diseases: for example, through its Institute for Tropical Diseases in Singapore and through providing the malarial treatment drug Coartem without profit to those in need. However, as many of the increases in infectious diseases are expected to occur in countries least able to pay for treatment, the potential for Novartis to turn the changed patterns of demand into increased profits is low.

Moral imperative

Although climate change could impact Novartis in a variety of ways, the most significant of the predicted impacts occur at points in time outside its planning horizon and are, therefore, not yet considered relevant for the company. In addition, the reality is that climate change cannot be prevented by the actions of a single entity acting alone — particularly for relatively low-energy industries such as pharmaceuticals. Thus, any efforts by Novartis to reduce its greenhouse gas emissions are not driven by the hope of reducing the impacts of climate change on its business. Rather, the motivation for Novartis to reduce its greenhouse gas emissions lies in its view of the moral imperative to take action against the risk of climate change.

So why does Novartis feel a moral imperative to help combat climate change? There are a number of different factors at play. The first is the broadening of the public's expectations of companies. It is often no longer considered sufficient to produce good financial returns. Companies are also expected to earn their 'licence to operate' by providing positive social outcomes and by minimising harm to the natural environment. This change in expectations has increased the pressure on politicians to enact legislation to protect the environment, in turn providing further incentive for companies to act proactively to reduce the impact of new laws on their business. Advances in scientific research on climate change and increasingly tangible effects such as more frequent and intense extreme weather events (Rosenzweig et al. 2007) have also strengthened the public pressure for action to be taken on climate change.

These external factors affect all companies, though those companies with particularly high emissions (e.g. oil and electricity companies) have received the most public attention. Therefore the question is why Novartis, as a company with low energy intensity (in terms of energy use per unit of turnover), felt compelled to set a target? The answer involves looking at some of the internal factors within Novartis that created a receptive environment for the company to take action on climate change.

When Ciba-Geigy and Sandoz merged to form Novartis in 1996, both brought experience in environmental mishaps including contaminated sites and a poisoned river. These experiences, along with the general trend towards corporate responsibility, resulted in both companies introducing company-wide environmental targets in the 1990s and being forerunners in public environmental reporting.[1] This commitment to the environment continued in the merged entity and has contributed to the strong focus on environmental issues in today's Novartis.

The environmental responsiveness of the company is also shaped by the personal beliefs of its senior management. The decision to set a greenhouse gas emissions reduction target was made by Daniel Vasella, the Chairman and CEO, who immediately made it a strategic priority for the company. One reason for his commitment to action on climate change may be fact that the head office is located in Switzerland, one of the most environmentally conscious countries worldwide.[2] The decade-long planning time-frame in the pharmaceutical industry (from drug synthesis to government approval) is also conducive to implementing environmental improvements with long payback periods (Berry and Rondinelli 2000).

There are two final comments to be made about the ethical motivations of Novartis. The first is that, as the cost of energy generated on-site and purchased electricity represent less than 1% of its operating costs, the potential for cost savings through reduced

---

1 Ciba started publishing official company environmental reports in 1991 and Sandoz in 1992. The novelty of such reporting is evidenced by the fact that SustainAbility identified only 70 such reports worldwide in 1993 (SustainAbility et al. 2004). Ciba was recognised externally as a leader in environmental management and was awarded the 1995 World Environment Council Gold Medal for its commitment to sustainability.

2 The International Social Survey Program rated Switzerland as the third most environmentally conscious country of the 26 countries surveyed in 2000 (Franzen 2003).

energy use was not a significant motivational factor. The second is that ethical values provide little guidance when it comes to weighing up competing interests such as investments in research and development or in saving energy. This created a need for Novartis to translate its ethical principles into concrete guidelines for action. Novartis therefore chose the Kyoto Protocol's overall reduction target by 2012 for industrialised countries as a proxy for the international moral consensus about the reductions that it as a business should be seeking to achieve.

Turning good intentions into concrete targets

In June 2005, Novartis committed itself to an emission reduction target based on the Kyoto Protocol, stating its intention to reduce its average absolute greenhouse gas (GHG) emissions by 5% from the 1990 baseline in the time period 2008–2012. The boundaries of the target were drawn at GHG emissions from sources owned or controlled[3] by the company, excluding emissions from company-owned vehicles. In the language of the *Greenhouse Gas Protocol* (WBCSD/WRI 2004) this target applies to 'Scope 1' emissions excluding cars. A separate target for vehicle emissions – to reduce absolute CO_2 emissions from vehicles to 10% below the 2005 figure by 2010 – was also set.

Novartis expects to meet its target through a mix of internal and external emission reduction projects, with up to 75% of the reductions until 2012 coming from external projects. If the option of offsetting emissions by undertaking external emission reduction projects had not existed at the time of setting the goal, it is likely that the goal would have been expressed in terms of efficiency (e.g. in terms of greenhouse gas emissions per US$1,000 sales) rather than in absolute terms. This would probably have resulted in fewer reductions overall, though potentially more internal reductions. Beyond 2012, Novartis aims to achieve a greater proportion of its emissions reductions through internal emission reduction projects.

'Scope 2' emissions (i.e. emissions from the generation of purchased electricity, steam, heating and cooling) are not included in the target, but are currently being measured and reported upon in the company's Annual Report. Novartis has not yet set an emission reduction target for these sources, but such an action has not been ruled out for the future. Setting a target for Scope 2 emissions would be a logical extension of its emissions programme, as it would ensure that all emissions caused by Novartis on- and off-site are considered.

Are the targets challenging?

When Novartis set its 'Kyoto target' in 2005, its total relevant emissions were over one-fifth higher than the target. Using projections of expected sales growth and an assumption of a 'business as usual' linkage between sales and emissions, an estimate was made

---

3 Control is defined as operational control – meaning that the target included all greenhouse gas emissions from subsidiaries over which Novartis had full authority to introduce and implement its operating policies.

of the company's expected emissions in 2008–2012 if no emission reduction programme was implemented.

This calculation found that, over the five-year target period, the average difference between business as usual and target emissions would be 152,000 tonnes of CO_2 equivalents [CO_2(eq)] per year or 760,000 tonnes over the five years. This compares with the company's Scope 1 emissions in 2006 (excluding cars) of 400,000 tonnes, indicating that its 'Kyoto target' is indeed significantly more difficult than business as usual. The target's difficulty arises from the high growth rate in the pharmaceutical industry since the 1990s, which has led to large increases in production volumes and greenhouse gas emissions.[4]

Similarly, the target for the car fleet is also quite challenging, as it demands an absolute reduction in emissions in five years, despite an expected 10% increase in the number of company cars over this time.

Impact of regulatory framework

As a global company operating in over 150 countries Novartis is exposed to a number of different greenhouse gas emission-related regulatory schemes, with the key ones being the EU Emission Trading Scheme (EU ETS), the Swiss CO_2 Law and, until 2008, the UK Emissions Trading Scheme. The company's sites in the USA are not at present affected by any of the state schemes there.

Six of the company's European sites have been included in Phase 1 of the EU ETS and eight will be included in Phase 2. The CO_2 emissions from these eight sites in 2006 made up 25% of the company's total Scope 1 emissions, including vehicle emissions. As a significant proportion of its total emissions falls under the EU ETS regulatory framework, the impact of the EU ETS must be considered to see how 'voluntary' the Novartis target really was.

An analysis was conducted to compare the estimated emissions of the eight EU ETS sites in 1990 with the emissions allocations for these sites for Phase 2. The analysis showed that the Phase 2 allocations allowed these sites to increase their emissions to around 50% above their emissions in 1990. Thus, for the quarter of the company's emissions that are subject to the EU ETS, the regulatory target is much more modest than the overall 5% reduction goal the company has set itself.

The path beyond 2012

Novartis has not yet set a greenhouse gas emission target for the period after 2012. Following the tradition it began with its first target, management has stated that the company is likely to base its target on the emission reduction goals of the leading industrialised countries. At the time of writing, the governments of these countries had not yet committed themselves to a reduction percentage, but expectations are for a reduction target of 20–40% by 2020 versus a 1990 baseline.

---

4 The pharmaceutical industry was the second-fastest-growing industry in Europe over the period 1988–1997 (Böheim *et al.* 2000).

Internal actions to reduce greenhouse gas emissions

Awareness and motivation

To raise staff awareness and to motivate them to participate actively in saving energy and reducing emissions, Novartis established a number of initiatives including an annual Energy Excellence Award, annual energy workshops and energy-specific internal communications.

The first Energy Excellence Award ceremony was held in 2004 to reward projects that reduced energy use and emissions, and to promote these projects across the company. Eighty-nine projects were submitted in the first three years of the Awards, with the 32 projects from 2006 expected to save about 4% of the company's total energy costs.

Two-day energy workshops were first held in 2004 and were institutionalised in 2006 as annual workshops in three regions (Europe, North America, Asia). All interested employees — including those in utilities, facilities management and health, safety and environment (HSE) functions can attend these workshops. They function not only as a knowledge-sharing forum, but also as a networking and motivational tool. The topics discussed include the progress made on energy efficiency and renewables use, and how to improve the internal tools for their implementation. External speakers from other companies with emission reduction programmes and from energy consulting agencies are often invited make presentations at the workshops.

To share successes and lessons learned in environment and energy projects across the company, an intranet-based Environment and Energy Case Study Database, as well as a quarterly *Energy Newsletter*, were rolled out in late 2007.

Embedding energy efficiency in the daily business

Due to the decentralised nature of the Novartis business, the company's decision-making framework had to be changed to encourage investments by its operating divisions in emission reduction projects. This has been achieved by including an emissions target in the annual goals of each division and by changing some rules regarding capital asset acquisitions to favour energy-efficient assets.

Each division's emissions target is an energy efficiency target, requiring improvements in energy inputs per selected unit of output (e.g. tonnes of product produced, sales dollars) of 2.5% per year. This form of target was chosen instead of a long-term absolute reduction target as it makes it possible to measure progress on an annual basis and, because it is expressed in 'efficiency' terms, is not seen as hindering the growth of individual divisions. Progress on the divisional targets is assessed annually by the head of corporate affairs, who is a member of the Novartis Group Executive Committee.

To meet their energy efficiency target, each division has appointed energy experts and created annual 'action plans' of energy efficiency projects identified at sites. Around half of the manufacturing sites have site energy experts, mostly site engineers within the Facilities Management team, who work either full-time or part-time on energy issues. Their role is to collect energy data, identify improvements, launch awareness campaigns and conduct energy audits. But, as discussed below, their influence depends on their individual persuasive abilities and their motivation, as their role has little direct authority. Although they serve to reinforce the importance of energy efficiency at a site level, they cannot single-handedly undertake improvement projects.

In addition to site energy advisors, the Sandoz Division and the Ciba Vision Business Unit have appointed global energy advisors to oversee their emission reduction programmes. Sandoz's focus on reducing energy use can be partly explained by the desire to reduce costs, as it is a generic medicine producer facing intense price competition. It also has the largest energy costs as a percentage of total costs within the Novartis business.

In recognition of the fact that 'end-of-pipe' improvements in the energy efficiency of existing assets are more expensive than integrating energy efficiency into the purchasing and installation process, Novartis has altered its capital asset acquisition process. It now allows payback periods up to the lifetime of the asset for assets/projects that save energy. This has had the effect of significantly increasing the chances of an energy efficiency asset acquisition being approved. Previously, the company's practice had been to demand payback periods of 2–3 years, thereby effectively excluding investments in energy-saving devices with long lifetimes such as a new building heating system with a useful life of 10–20 years. In addition, Novartis has made the sign-off from an 'energy expert' on Capital Appropriation Request (CAR) forms mandatory. Use of a CAR form is necessary for all asset acquisitions larger than 20,000 Swiss francs, meaning that all purchases above this value should be reviewed for their energy efficiency.

Both the payback rule and the CAR form change have experienced teething problems. Some renewable energy projects have been approved only once corporate HSE became involved in the lobbying process to persuade site managers of the benefits and not all asset acquisitions have used the latest CAR form. These problems are being tackled by corporate HSE through information campaigns and targeted audits.

Car fleet

The 180,000 tonnes of greenhouse gas emissions from owned/leased company cars form a significant portion of the company's total emissions. Therefore reducing these emissions represents an important part of the company's overall greenhouse gas emission reduction strategy.

The emission reduction target will be achieved by progressively switching to less polluting cars as car lease periods end. In the USA, the aim is for 40% of all new vehicles to be hybrids or other energy-efficient cars (over 25 miles per gallon) while, in Europe, the switch will be towards 100% diesel-powered vehicles with mandatory particulate filters. The estimated net cost over the five years until 2010 is US$23 million, mostly because the loss of lease incentives is larger than the savings on fuel cost.

Implementation challenges

Two types of implementation challenges have arisen, the first relating to obtaining accurate data and the second relating to gaining internal support at a site level for energy projects.

The information difficulties lay in determining the appropriate baseline figure and then capturing accurate emissions data from over 200 sites worldwide. The baseline was particularly difficult as, in 1990, greenhouse gas emissions were not typically part of reporting systems and the Novartis of today did not exist. In 1990 Ciba-Geigy and Sandoz were still two separate companies and, since their merger in 1996, many other acquisitions and disposals have occurred. The baseline was, therefore, calculated on a site-by-site basis by obtaining estimates of each site's emissions in 1990 from site HSE officers. Emissions were estimated mostly by analysing the available data on fuel types and energy consumption in 1990.

The rollout of a new HSE data management system in 2006 was used as an opportunity to improve the reporting data definitions and to provide training to all site HSE officers in calculating and entering the data items. A dedicated Corporate HSE Data Manager coordinates and reviews the reporting process, and is responsible for the continuous improvement of this process.

Although the new system and the training have improved data accuracy, there are continual changes in site staff, leading to data entry mistakes (e.g. reporting energy use in kilowatt-hours rather than gigajoules). Even mistakes of this magnitude can sometimes evade detection at a division level, where the consolidated results from many sites (sometimes over 50) are reported.

Despite these difficulties, Novartis is of the view that these uncertainties are not significant enough to impair the overall data accuracy. This confidence is supported by the fact that the Scope 1 and Scope 2 emission figures presented in the company's Annual Report are audited and signed off by accountants PricewaterhouseCoopers.

The second key implementation challenge has involved 'winning the hearts and minds' of the Novartis staff. As noted above, Novartis has a decentralised business structure, which means that operating divisions and business units have significant decision-making autonomy. The strategic advantages of this structure are large, but it has the drawback that head office cannot force immediate action on new corporate priorities. The approach taken has been to include energy efficiency targets in the annual goal-setting and assessment process of the divisions, thus pushing these topics into their realm of responsibility.

Energy efficiency is, however, just one goal among many for the divisions. Thus, progress on energy efficiency (as indeed on any other corporate goal) depends crucially on commitment from staff at the site level, as opportunities to improve energy efficiency or to implement renewable energy projects cannot easily be identified from Head Office.

Staff commitment can be low if the goal is not seen as important to the key business of that site, in particular if the site manager is not passionate about the cause. Ensuring that energy efficiency becomes a site priority has been a significant implementation challenge because energy costs make up only 1% of the company's operating expenses and greenhouse gas emissions do not pose an immediate health hazard for staff or the local environment. Similarly, personal convictions also play a role in the progress on the vehicle emission target. For example, some of the sales force prefer large cars and do not wish to switch to small energy-efficient vehicles.

Notwithstanding these challenges, the delivery of the greenhouse gas emissions reduction goal has proceeded reasonably well. The implementation of the revised HSE data management system has provided corporate headquarters with more reliable site-specific data. This has enabled HSE staff to identify which sites are lagging on energy efficiency measures and to address this in annual meetings with the division heads.

Despite the senior management commitment and the progress that has been made, it is clear that there is still inertia within the organisation on the subjects of energy efficiency and emissions reduction. For example, emission reduction projects require significant encouragement from head office despite the projects implemented to date quickly providing net financial savings. Although this is likely to be the outcome of picking the 'low-hanging fruit' and cannot be expected to hold true for all emission-saving projects, the fact remains that the projects have been profitable, suggesting that there are likely to be other profitable projects that have not yet been implemented.

From a budgeting perspective, no external measures should be taken until the easy internal options are exhausted. However, this approach would not lead to the Kyoto target being met by 2012, as organisational inertia is preventing the internal energy savings from occurring at the necessary speed. This is despite the wide array of measures taken by the company to support and promote energy efficiency.

Emission reductions achieved

While the 10% vehicle emission reduction target is expected to be met through changes in the vehicle fleet, the company's target for Scope 1 non-vehicle emissions will not be achieved purely through internal measures. The impact of internal measures is expected to be a reduction of the company's 152,000 tonnes per year target overhang for the five years 2008–2012 by 25–30% down to a remaining annual gap of 106,400–114,000 tonnes. The remaining gap will be met by external emission reduction projects.

In the long run, the goal is for all reductions to be achieved internally so as to ensure the continued viability of the emission reduction programme. If Novartis had to buy an ever increasing amount of external emission credits each year, this would lead to high expenses without any improvement in its own operations.

A trend analysis was conducted to assess the success to date of the internal emission reduction measures. This analysis involved removing the effects of business acquisitions and disposals in 2001–2006 so as to create a 'same business' comparison for the six years. This showed that the company's emissions had stabilised since 2004 despite a nearly 30% increase in sales in the period 2004–2006.

This stabilisation was the result of the interplay of a number of factors, which provide a good insight into the complexity of managing emissions. Energy consumption, the source of this energy (internal or external) and the fuels used to generate the energy all played a part. Energy-saving projects caused total energy consumption to almost plateau at Novartis, with only a 3% increase over the three years. In addition, the small increase in energy consumption occurred in the 'purchased energy' category, which is not part of Scope 1 (i.e. outside the scope of the company's target). Lastly, the energy generated on-site has had approximately the same carbon intensity since 2001, as its composition has remained the same (90% gas, 10% oil and other).

In conclusion, the internal measures taken at Novartis to raise energy awareness and promote emission reduction projects have succeeded in slowing down the growth in energy consumed. While this is a positive trend that needs to continue, the company has recognised that even greater efforts will be required to meet its likely targets beyond 2012. It is likely that the next series of major actions will involve changing the composition of fuel sources towards a higher percentage of carbon-free renewable sources.

External measures

Clean Development Mechanism projects

The Clean Development Mechanism (CDM) is a tool under the Kyoto Protocol whereby emission reduction projects in developing countries generate emission credits which can be bought by countries/companies to compensate for exceeding their emission limit. Novartis intends to offset its target gap predominantly through two CDM projects, both of which it initiated.

The first is the afforestation of 30 km² of former farmland in Argentina, with the long-term aim of establishing a sustainable forestry business with Forestry Stewardship Council certification and 75% native trees.

The second project also involves a large-scale planting programme, this time through planting the jatropha plant on 120 km² of deforested land in Mali. As well as providing temporary certified emission reductions from carbon sequestration, the seeds of the jatropha plant can be used to replace fossil fuels for vehicles and electricity generation, thus generating permanent emission reductions. The seeds can be pressed and esterified to make biodiesel, while the shell and seedcake can be used to produce electricity through combustion or anaerobic digestion.

Such afforestation and reforestation projects are rare in the CDM world, with expectations that only 0.3% of certified emission rights by 2012 will come from these types of projects (UNEP Risoe Centre 2007). However, the projects appealed to the management of Novartis as they are in countries in which Novartis has experience (Novartis has over 400 employees in Argentina and has Novartis Foundation projects in Mali) and both projects offer long-term business prospects for local people.

The total carbon sequestration and emission reductions achieved by these two projects in the period 2008–2012 are difficult to quantify in advance, as the success of the projects depends on variables such as how many square kilometres can be planted with trees and jatropha, and how quickly they grow. Extreme weather conditions in either country (e.g. a severe drought) could significantly slow plant growth. In the best case scenario, these projects will provide significantly more emission credits in the period 2008–2012 than Novartis requires to fill its remaining annual gap (106,400–114,000 tonnes) after internal measures.

Not wishing to rely on good luck, the Corporate Environment and Energy Manager has requested advice from a CDM consulting company regarding the possibilities for a third CDM project for Novartis. Figure 15.1 shows a conservative estimate of the sequestration contribution of the CDM projects to achieving the Novartis target.

To ensure the credibility and reliability of the emission credits, Novartis aims to obtain project registration from the UNFCCC CDM Executive Board. If this involves too much administrative work, an external audit by a UNFCCC-accredited emissions auditing company will be carried out to provide confidence in the emissions reductions being achieved.

To provide scientific expertise and oversight of the projects, a Carbon Project Advisory Council is being created from external carbon market experts, forest management experts and non-governmental organisation (NGO) representatives. This Advisory Council will also provide general feedback on the energy strategy and implementation at Novartis.

FIGURE 15.1 Scope 1 emissions from operations and offsets from CDM projects at Novartis

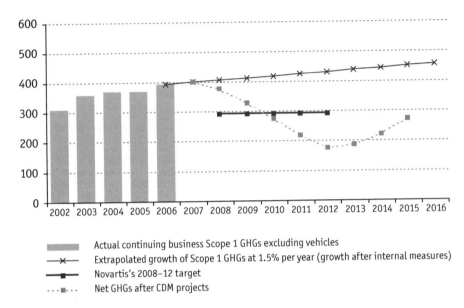

GHG emissions (thousand tonnes)

Actual continuing business Scope 1 GHGs excluding vehicles
Extrapolated growth of Scope 1 GHGs at 1.5% per year (growth after internal measures)
Novartis's 2008–12 target
Net GHGs after CDM projects

EU ETS

In the second phase of the EU ETS (2008–2012), Novartis expects that it will have an overall permit deficit of 160,000 tonnes CO_2 (32,000 tonnes per year). This deficit could be filled by purchasing permits on the market for a cost of up to US$4.4 million. However, the intention is to fill the deficit with some of the Certified Emission Reductions (CERs) gained from the company's CDM projects.

Overall Novartis prefers to engage in CDM projects, rather than participating in emissions trading, to make good the projected shortfall in meeting its voluntary target. This is due to the additional business benefits associated with the CDM projects, such as providing good-news stories for the company by demonstrating its long-term commitment to greenhouse gas reductions and to supporting sustainable development.

Business implications of taking action

When it set its internal Kyoto target in 2005, Novartis embarked on a significant long-term project. Apart from the environmental benefit of reducing emissions by nearly

800,000 tonnes CO_2(eq) by 2012, has this project resulted in other benefits for Novartis? One way of answering this question is to examine the effects of the emission reduction programme on the four measures of shareholder value below; see Schaltegger (2006) for further information on this approach to assessing the implications of environmental management for shareholder value.

Investments in fixed assets and working capital

Future free cash flows will be higher if investments in capital-intensive pollution-reduction solutions (i.e. non-productive investments) are minimised. To reduce its emissions, Novartis has made many investments in fixed assets. While these investments tie up capital in the short term, they often reduce energy costs in the medium to long term, thus freeing up working capital.

As yet, Novartis has not carried out a complete assessment of the financial costs and benefits of the implemented projects. However, the submissions for the Energy Excellence Awards provide an initial indication of the financial outcomes achieved. The 32 projects submitted to the 2006 Awards process had very positive cash flow outcomes; overall, they will result in net savings of US$50 million in five years. Interestingly, the average payback time for these projects was two years, despite an average project lifetime of 12 years. However, these figures represent the best projects implemented so it cannot be assumed that all emission-saving projects undertaken achieved such good financial results or that such results will be replicated in the long term.

It is also important to note the other side of the equation: while the money spent on internal reduction projects may be paid back through lower energy costs, it appears unlikely that this gain will be larger than the costs of the CDM projects (US$8 million start-up costs, US$2–3 million annual running costs) and the net US$23 million cost of changing the vehicle fleet.

Sales revenue growth, profit margins and the cost of emissions

Reducing greenhouse gas emissions, and indeed implementing a complete environmental improvement programme, is expected to have a minimal impact on sales revenues for Novartis. This is because an individual's choice of medication is not influenced by a company's environmental reputation; the choice of medication is dominated by factors such as the clinical effectiveness of the drug, its price and its side effects.

However, emissions reductions as part of a comprehensive corporate social responsibility (CSR) programme could enhance the attractiveness of Novartis as an employer, thereby lowering employee turnover. It is estimated that the full cost of replacing an employee who leaves is 1.5–2.5 times their annual salary (Cascio 2006), leading to a conservative estimate of the cost of the company's staff turnover of 12.6% in 2006 of US$1,147 million. Although the evidence for the link between CSR and employee recruitment and retention has been classed as 'modest' (Porter and Kramer 2006), the high cost of employee turnover means that even small changes can make large financial differences. For example, a reduction in employee turnover from 12.6% to 12.1% could save the company US$46 million per year.

CSR performance could also provide benefits through improving the company's reputation[5] and thereby lowering the cost of doing business. A good reputation has a multitude of flow-on effects including the goodwill of company stakeholders such as shareholders, local governments, local communities and employees. This goodwill can make business easier, such as by speeding up regulatory approval for new sites and providing a willing pool of new employees.

Unfortunately, the 'reputation capital' awarded by non-profit organisations (NGOs) or other interest groups to companies when they undertake positive activities is not as large or long-lasting as it could be. Research has shown that negative events (e.g. criminal activities) pull corporate reputation down more than positive events pull it up (Williams and Barrett 2000; Fischer and Khoury 2007). This weakens the business case for CSR activities such as greenhouse gas emission reductions, as these actions are often not sufficiently rewarded with an improved reputation. This is part of a broader debate about the relationship between 'big business' and its critics (Porter and Kramer 2006; Leisinger 2007). However, the growing attention directed at sustainability indexes such as the Dow Jones Sustainability Index (DJSI) and FTSE4Good is helping to strengthen the link between company actions and public recognition.[6]

But, although the emissions reduction programme seems to have no effect on sales and a marginal effect on profit margin, could it potentially have an impact on the cost of emissions? To date regulatory programmes (e.g. the Swiss CO_2 tax, the EU ETS) cover only a quarter of the company's emissions and the implied cost of carbon has been minimal. However, the general trend is towards full internalisation of the external cost of GHG emissions which means that, by reducing its emissions profile, Novartis is lowering its exposure to potential future costs.[7]

Cost of capital

The cost of capital is the price a company pays for its external funding. While it is generally considered unlikely that reducing emissions would have a sufficient impact externally to reduce the cost of capital (with the obvious exception of sectors such as

---

5 The importance of CSR for company reputation can be seen in the variety of reputation indices that include CSR factors in their ratings. The annual 'Most Admired Companies' list from *Fortune* magazine is compiled by interviewing financial analysts and executives from large companies for their impressions of corporations across a number of factors, one of which is community/environment. Novartis also engages an external company to conduct a reputation survey each year. In 2007 the 2,500 people interviewed in seven countries from a variety of stakeholder groups rated CSR as one of the six most important reputation-building factors.

6 The corporate citizenship actions of Novartis have been recognised through its inclusion in the DJSI and the FTSE4Good indices, and through its designation in 2006 as the 'Healthcare Supersector Leader' by the DJSI (Dow Jones Sustainability Index World – Supersector Leaders [2006/2007]; www.sustainability-indexes.com/07_htmle/reviews/DJSI_World_supersectorleaders_2006.html, 11 September 2007 [accessed 20 August 2008]). Nearly 10% of the evaluation form for entry into the DJSI comprises questions on climate change strategy.

7 Clearly the size of this effect depends on the regulatory structure implemented. If the future regulation places a price on gross emissions by Novartis (thus not considering the emission offsets through CDM), the savings through its strategy to date would be small. Such a regulatory structure would, however, be quite a departure from existing systems such as the EU ETS and is therefore considered unlikely.

electricity generation), it is clear that investors are increasingly concerned about climate change. This is evidenced by the level of investor support for initiatives such as the Carbon Disclosure Project (CDP). Although investor support for CDP cannot be interpreted as a sign that investors will reward companies that reduce their greenhouse gas emissions, investors in Novartis have greeted its emissions strategy positively. Indeed, some boutique investment funds have invested actively in Novartis upon receiving satisfactory answers to their questions about its emissions strategy.

Duration of value growth

The longer that lower operating costs and a lower cost of capital persist, the higher the net present value of the future free cash flows will be. The duration of the value-creating elements of the emissions reduction programme depends on dynamic factors. These include how long and well the programme is implemented in Novartis and on what competitor companies in the markets for customers, employees and finance undertake. The current programmes are estimated to have positive effects on the other three drivers of shareholder value listed above until at least 2012.

Conclusion

This discussion of the implications of taking action may be frustrating for those who would prefer a list of tangible business benefits from the greenhouse gas emission reduction programme implemented by Novartis. The analysis presented above provides only a broad suggestion of the likely magnitude of effects, mainly due to insufficient data on the relevant factors.

For Novartis, the business case for emissions reduction is not clear cut; the critical driver for its efforts to reduce emissions is its conviction that reducing greenhouse gas emissions is the right thing to do. For companies where the business case is not clear, the views and commitment of senior management will be critical determinants of the actions taken and the emission reductions achieved.

Reflecting more broadly on this case study, the most important point is that the voluntary greenhouse gas emission reduction programme at Novartis is resulting in significant emission reductions. This shows that it is not always necessary to have strict regulations to ensure action. Corporate voluntarism therefore has an important role to play in delivering emissions reduction beyond compliance, but it is only one tool in the necessary set of international agreements, national laws and emission reduction incentive systems.

References

Berry, M., and D. Rondinelli (2000) 'Environmental Management in the Pharmaceutical Industry: Integrating Corporate Responsibility and Business Strategy', *Environmental Quality Management*, Spring 2000: 21-35.

Böheim, M., M. Pfaffermayr and K. Gugler (2000) 'Do Growth Rates Differ in European Manufacturing Industries?', *Austrian Economic Quarterly* 2: 93-104.

Cascio, W. (2006) 'The High Cost of Low Wages', *Harvard Business Review*, 1 December 2006 (online version; harvardbusinessonline.hbsp.harvard.edu/hbsp/hbr/articles/article.jsp?ml_action=get-article&articleID=F0612D&ml_page=1&ml_subscriber=true).

Christensen, J., B. Hewitson, A. Busuioc, A. Chen, X. Gao, I. Held, R. Jones, R. Kolli, W. Kwon, R. Laprise, V. Magaña Rueda, L. Mearns, C. Menéndez, J. Räisänen, A. Rinke, A. Sarr and P. Whetton (2007) 'Regional Climate Projections', in S. Solomon, D. Qin, M. Manning, Z. Chen, M. Marquisn, K. Averyt, M. Tignor and H. Miller (eds.), *Climate Change 2007: The Physical Science Basis. Contribution of Working Group I to the Fourth Assessment Report of the Intergovernmental Panel on Climate Change* (Geneva: Intergovernmental Panel on Climate Change): 847-940.

Cragg, G., D. Newman and K. Snader (1997) 'Natural Products in Drug Discovery and Development', *Journal of Natural Products* 60: 52-60.

Fischer, K., and N. Khoury (2007) 'The Impact of Ethical Ratings on Canadian Security Performance: Portfolio Management and Corporate Governance Implications', *Quarterly Review of Economics and Finance* 47: 40-54.

Franzen, A. (2003) 'Environmental Attitudes in International Comparison: An Analysis of the ISSP Surveys 1993 and 2000', *Social Science Quarterly* 84.2: 297-308.

IPCC (Intergovernmental Panel on Climate Change) (2007) *Climate Change 2007: Impacts, Adaptation and Vulnerability — Summary for Policymakers. Working Group II Contribution to the Intergovernmental Panel on Climate Change Fourth Assessment Report* (Geneva: IPCC).

Jasper, K., P. Calanca, D. Gyalistras and J. Fuhrer (2004) 'Differential Impacts of Climate Change on the Hydrology of Two Alpine River Basins', *Climate Research* 26: 113-29.

Lehner, B., G. Czisch and S. Vassalo (2005) 'The Impact of Global Change on the Hydropower Potential of Europe: A Model-Based Analysis', *Energy Policy* 33: 839-55.

Leisinger, K. (2007) *Corporate Philanthropy: The 'Top of the Pyramid'* (Basel: Novartis Foundation for Sustainable Development).

Porter, M., and M. Kramer (2006) 'Strategy and Society: The Link between Competitive Advantage and Corporate Social Responsibility', *Harvard Business Review*, December 2006: 78-92.

Rosenzweig, C., G. Casassa, D. Karoly, A. Imeson, C. Liu, A. Menzel, S. Rawlins, T. Root, B. Seguin and P. Tryjanowski (2007) 'Assessment of Observed Changes and Responses in Natural and Managed Systems', in M. Parry, O. Canziani, J. Palutikof, P. van der Linden and C. Hanson (eds.), *Climate Change 2007: Impacts, Adaptation and Vulnerability. Contribution of Working Group II to the Fourth Assessment Report of the Intergovernmental Panel on Climate Change* (Geneva: IPCC): 79-131.

Schaltegger, S. (2006) 'How Can Environmental Management Contribute to Shareholder Value?', in S. Schaltegger and M. Wagner (eds.), *Managing the Business Case for Sustainability: The Integration of Social, Environmental and Economic Performance* (Sheffield, UK: Greenleaf Publishing): 47-61.

SustainAbility, UNEP and Standard & Poor's (2004) *Risk and Opportunity. Best Practice in Non-financial Reporting. The Global Reporters 2004 Survey of Corporate Sustainability Reporting* (London: SustainAbility).

UNEP Risoe Centre (2007) 'CDM/JI Pipeline Analysis and Database August 2007'; www.cdmpipeline.org (accessed 29 August 2007).

WBCSD (World Business Council for Sustainable Development) and WRI (World Resources Institute) (2004) *The Greenhouse Gas Protocol: A Corporate Accounting and Reporting Standard* (Geneva: WBCSD/WRI, rev. edn).

WHO (World Health Organisation) (2007) 'Climate Change and Health'; www.who.int/globalchange/climate/en (accessed 22 August 2007).

Williams, R., and J. Barrett (2000) 'Corporate Philanthropy, Criminal Activity, and Firm Reputation: Is There a Link?', *Journal of Business Ethics* 26: 341-50.

16

Climate change solutions at Vancity Credit Union

Ian Gill and Amanda Pitre-Hayes
Vancity, Canada

When the cabinet of British Columbia's provincial government in Canada wanted to be briefed on climate change issues, they called in the experts – David Suzuki (a well-known Canadian environmental scientist) and Vancity. What's interesting about this is that Vancity isn't a think-tank or environmental non-profit. It's a credit union – and a tremendous force for positive climate change action on the west coast of Canada. This chapter outlines how acting on climate change became a priority for Vancity, provides an overview of its strategy and related programmes, and offers an analysis of some of the key success factors that have led to Vancity's contribution to climate change action.

About Vancity

Formed in 1946, Vancity is now Canada's largest credit union with $14 billion,[1] 381,000 members and 60 branches in Vancouver, the nearby Fraser Valley, Squamish and in British Columbia's capital, Victoria.

As a cooperative, Vancity is guided by a commitment to corporate social responsibility (CSR) and is dedicated to improving the quality of life in the communities where its members live and work. This does not mean that Vancity's focus is exclusively local. Some local impacts (specifically those relating to climate change) have sources or causes

---

1 All dollar values in this chapter are in Canadian dollars.

that extend beyond the community. Therefore, in principle (and in practice, as we discuss below), climate change falls squarely within the scope of Vancity's objective of improving the quality of life in the communities where it works although, clearly, the specific actions and initiatives taken by the organisation reflect its local priorities and its ability to make a meaningful difference.

Vancity's objective is to create strong financial, environmental and social returns. This focus is enabled by its cooperative ownership structure, which makes it fundamentally different to its publicly owned competitors. Unlike publicly listed banks, Vancity is owned by its members, so its owners live and work in the communities where it operates and are therefore directly impacted by the organisation's actions. Because of this, its members have a vested interest in ensuring that Vancity delivers not only positive financial returns, but also positive environmental and social returns. This is not merely a theoretical assertion: Vancity's members have consistently supported and elected directors who support this 'triple bottom line' — the creation of strong financial, environmental and social returns.

While the phrase 'triple bottom line', coined by John Elkington in 1994, is relatively new to the lexicon of business, the *idea* of a triple bottom line is certainly not new to Vancity. Back in 1946, 14 Vancouver residents, who were irritated by the lack of access that ordinary people had to financial services, signed a charter and formed the Vancouver Savings and Credit Union. From the day it started, Vancity became known for its progressive 'firsts'. It was the first financial institution to offer mortgages on the east side of Vancouver (at that time known as the wrong side of town), the first to offer loans to women without a male co-signer and the first to reinvest a third of its profits back into the community where its members live.

Vancity's environmental leadership started some 40 years later in 1987, driven by the visionary leadership of Linda Crompton, then Vancity's Vice President of Human Resources, and David Levi, then Chair of Vancity's Board of Directors. Linda Crompton had focused her graduate research work on alternative economic theories, especially those that took the environment into account, while David Levi was heavily influenced by the work of environmental economists and was an advocate of initiatives that focused on both social and environmental responsibility.

In 1989 Linda Crompton and David Levi set the stage for the organisation to generate positive environmental results when Vancity joined Ceres, an international network of businesses and investors working to address sustainability challenges. The 'Ceres Principles' commit its signatories to minimise environmental impacts from their operations and to publicly disclose their progress toward this goal.[2]

Signing on to the Ceres Principles brought with it a change in the way that Vancity reported on its results. In the early 1990s, Vancity included environmental metrics in its annual report for the first time and worked with Ceres to develop and launch the Global Reporting Initiative — now the international standard for corporate reporting on environmental, social and economic performance.[3]

With that shift in reporting came a change in what was measured at the credit union — and, therefore, what was top-of-mind for employees in the organisation. In addition to measuring financial results, Vancity's executive and line employees began to measure

2 For more information see www.ceres.org.
3 www.globalreporting.org

— and be held accountable for — their ability to achieve environmental goals. These changes had the effect of embedding the environment as a core business issue through formally establishing the environment as an issue with its own bottom line that had to be met.

A wave of environmental innovation ensued. In 1990, Vancity launched the enviro-Visa — a Visa card product that donates a portion of its profits to local environmental projects. In 1998 Vancity launched Citizens Bank of Canada — a national, ethical, online bank that provides access to triple-bottom-line products to members across the country.

In 1995, this focus was crystallised in a cohesive strategy — Vancity's Corporate Social Responsibility Plan. In 2004, this evolved into Vancity's Community Leadership Strategy, the organisation's current strategy for:

- Acting on climate change

- Facing poverty

- Growing the social economy

- Being accountable

The framework for Vancity's Community Leadership Strategy is illustrated in Figure 16.1.

FIGURE 16.1 Vancity's community leadership strategy framework

As Vancity's strategy became more focused, the organisation started more initiatives to support the strategy and the number of people throughout the organisation engaged in this work grew accordingly. Today, all areas of the organisation are actively working to deliver Vancity's Community Leadership Strategy. Initiatives range from short-term projects to reduce a specific aspect of internal emissions (e.g. an initiative to install solar hot water heating at Vancity's head office) to multi-year projects such as Vancity Capital Corporation's effort to support British Columbia's clean technology sector with much-needed capital.

Has all this concern for the community and the environment come at a cost to the business? On the contrary, Vancity has achieved enviable growth. Over the past five years, Vancity's membership base has grown 20% faster and its asset base has grown 16% faster than the largest bank in Canada. In 2006, a full 30% of Vancity's earnings of $45 million were reinvested in the community. Although it is impossible to draw a direct link between Vancity's Community Leadership Strategy and its rapid growth, the fact that this strategy is the organisation's core differentiator suggests that there may be a strong relationship. Climate change action has now emerged as another core differentiator — and, potentially, a new source of growth.

Vancity's climate change solutions strategy

Vancity's commitment to environmental action began as a broad attempt to protect the environment though a multifaceted approach including support for environmental education, conservation and restoration. Throughout the 1990s, however, it became clear that the issue of climate change had eclipsed all other environmental priorities. It has in fact become the most pressing environmental issue of our time, with potentially devastating implications not only for the environment, but for the economy and the human race.

Though a majority of Canadians today rank the environment as the number one national issue, both Liberal and Conservative federal governments over the past 20 years have acted slowly, inconsistently or not at all. While Canada has signed the Kyoto Protocol, the Canadian government has shown embarrassingly scant interest in meeting its Kyoto Protocol targets. Some Canadian provinces have decided to act unilaterally and, indeed, British Columbia has provided unexpected leadership in setting aggressive targets for reducing greenhouse gas (GHG) emissions. Some businesses have also stepped into the leadership vacuum around climate change, and Vancity is among them.

Vancity sharpened the focus of its environmental efforts and resolved to take action on climate change through a focused Climate Change Solutions Strategy (Fig. 16.2). The strategy supports action on four fronts:

- Vancity taking action itself to reduce the environmental impact of its operations

- Helping its members to act by providing education, products and services that support the reduction of environmental impact

- Enabling the community to act by providing access to capital for projects that reduce impact on the environment

- Encouraging governments to act by advocating for public policy changes that lead to a reduction of environmental impact

FIGURE 16.2 Framework for Vancity's Climate Change Solutions Strategy

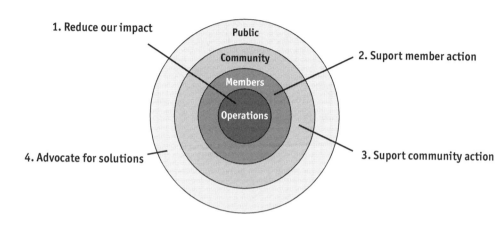

Vancity's objective is to address the root causes of climate change and reduce GHG emissions by taking action in four ways

At the beginning of 2008, over 60 related initiatives were under way. Each initiative addresses one of the four strategic goals — 20 initiatives focus on reducing Vancity's own impact, 15 initiatives on helping members to act, 15 initiatives on providing access to capital in the community and ten initiatives on advocacy. Examples of these initiatives are given in Table 16.1.

Ownership of these initiatives is spread throughout the Vancity Group, with each line of business or subsidiary using its unique resources to take action. For example, Vancity Capital contributes to the climate change strategy by providing growth capital to small and medium-sized green businesses in the province, whereas Vancity's Purchasing Department delivers on the strategy by combining with other organisations to make bulk purchases of environmentally friendly products, so as to increase demand for these products, reduce their cost and make these products more accessible to other firms.

Each of the four strategic objectives, related initiatives, and results achieved, are discussed in more detail in the following sections.

Reducing the environmental impact of our operations

Vancity's work to reduce its own impact is focused through its Carbon Neutral Program. At the end of 2007, Vancity became the first North American-based financial institution to be carbon neutral. This means that the greenhouse gases that the organisation emits are equal to the greenhouse gases that are not being emitted due to investments in projects that are reducing emissions elsewhere.

Vancity has baseline carbon dioxide (CO_2) emissions of 6,000 tonnes. Three environmental impacts are included in this baseline: energy use; employee travel (including employee commuting); and paper use.

TABLE 16.1 Examples of Vancity's current climate change initiatives

Strategic objective	Initiative examples
Reduce our impact	• **Carbon Neutral Program:** work to make the Vancity Group carbon neutral and to maintain this status by measuring Vancity's emissions, reducing organisational emissions and offsetting remaining emissions • **Cut the Carbon Campaign:** an internal campaign focused on changing employee behaviour to reduce GHG emissions • **Solar Hot Water Heating:** installation of photovoltaic panels on the roof of Vancity's head office to heat water on-site
Support member action	• **Energy Manager Audit Services:** free audits of business members' facilities to provide recommendations on how their physical plant can be retrofitted to save energy • **Eco-efficiency Term Loan:** a loan at 1% below market for business members to finance the cost of energy efficiency renovations to their facilities • **Green Business Growth Capital:** growth capital for profitable small to medium-sized green businesses based in British Columbia
Support community action	• **Climate Smart Program:** Vancity is a founding sponsor of Ecotrust Canada's work to support small businesses to shrink their carbon footprint. • **Green Building Grant Program:** grants of up to $50,000 to support green building initiatives and education • **Vancity Carbon Offset Program:** grants of up to $100,000 to qualified projects that reduce GHG emissions
Advocate for solutions	• **Take the Lead BC:** Vancity is a founding sponsor of Take the Lead BC, a multi-sectoral advocacy group supporting climate change action in the province of British Columbia, Canada • **Provincial Government Advisory:** in 2007, Vancity's then CEO provided advice to the Provincial Cabinet on climate change mitigation strategies • **Community Outreach:** presentations advocating for action on climate issues to groups across British Columbia

Over the past ten years, Vancity has made significant progress in reducing these three environmental impacts (Fig. 16.3).

While continuing the work to reduce its emissions, Vancity is also developing sources of high-quality local offsets. In 2007, it established the Vancity Carbon Offset Program which makes investments in renewable energy or energy efficiency projects in exchange for emission reduction credits. Through the programme, Vancity not only creates a supply of local high-quality carbon offset projects that it can use for its carbon neutral pro-

FIGURE 16.3 Percentage reductions in key environmental impacts 1997–2007

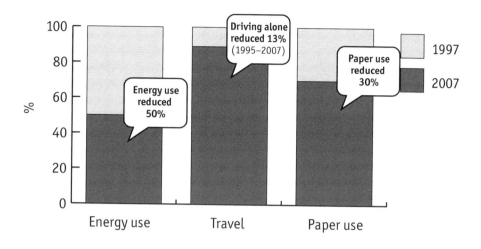

gramme, but it also helps to stimulate the development of a local carbon offset market (more detail on the development of this programme is provided below).

Reducing the climate change impact of our members

Through evolving and expanding the range of solutions available to its members, Vancity is determined over time to transform the way people live locally and to contribute to the reduction of climate change risk globally. The range of support for climate change action offered by Vancity to its members can be seen in the scenario presented in the Box overleaf. The scenario is based on a hypothetical couple but on real product offerings at the credit union.

This scenario illustrates the range of solutions that Vancity offers through its Climate Change Solutions Program. Even though it is unlikely that any one Vancity member benefits from as many of its Climate Change Solutions, the very range of products on offer means that all Vancity members have an increased awareness about climate change issues and are offered concrete opportunities for action.

Vancity's work in this area has led to significant results. Total Community Leadership funds under administration are over $1 billion dollars and membership growth topped 5% in 2006. Broader business benefits have also been realised. In a recent survey, for example, 80% of Vancity members trust Vancity to make decisions that are socially and environmentally responsible, and 87% of members and 40% of non-members indicated that they are 'likely' or 'very likely' to consider Vancity for their next financial product or service.

A member's experience of Vancity's climate change solutions

May and Roy Yuang live in Burnaby, a Vancouver suburb. May runs a cooperative that sells organic produce. Roy is an elementary school teacher.

Because their combined income is quite modest, May and Roy qualify for and buy an affordable housing unit in **The Verdant**, a LEED* Gold development financed by Vancity Enterprises. They finance their purchase using **Vancity's Climate Change Mortgage**, a mortgage product that invests in solutions to climate change.

As a resident of Burnaby Mountain, May is eligible for a **Vancity Community Transit Pass** (a discounted transit pass). Roy cycles to work on **The Greenway**, a cycling path running from New Westminster to Vancouver, funded in part by a $1million investment by Vancity. May and Roy also have a Toyota Corolla so that they can visit May's mother in Aldergrove on the weekends and stop at the local plant nursery on the way home. They finance the car with a **Vancity Clean Air Auto Loan** at prime, which they qualify for because the vehicle has low CO_2 emissions.

May's cooperative owns a small warehouse near trendy Commercial Drive. It is taking part in Ecotrust Canada's **Climate Smart** — a service that provides support for businesses to work toward reducing and neutralising their carbon emissions. Vancity is a founding financial partner of the programme.

Through the programme, May learns about Vancity's **Energy Manager**, and invites the Energy Manager to visit her cooperatives's warehouse to perform a free energy audit on the facility. The Energy Manager recommends changes to the lighting and ventilation systems. In just three years, the energy savings will cover the cost of the retrofit. May visits **Light House Sustainable Building Centre** on Granville Island to see samples of the materials the Energy Manager recommended. Vancity is a founding financial partner of Light House.

Next, May works with her dedicated account manager at Vancity to get an **Eco-efficiency Business Loan** at 1% less than the rate she would normally be required to pay. Talking with her account manager, she learns that her parents may qualify for a **Bright Ideas Home Renovation Loan** to make energy efficiency upgrades to their draughty bungalow in Aldergrove. Something to speak with them about next time they visit . . .

---

* LEED stands for Leadership in Energy and Environmental Design (www.usgbc.org/LEED).

Note: Vancity's climate change-related products and services are indicated in bold.

Supporting community leaders

Vancity has, over the past five years, donated more than $5 million to local environmental non-profit organisations (ENGOs) and, in the process, has established relationships with hundreds of organisations. It is through these relationships that Vancity leverages its resources to achieve the greatest positive environmental impact. Vancity's view is that such partnerships enable all the organisations involved to achieve results with a far greater impact than what each organisation could accomplish individually.

Vancity's long history of investment in community has led to the development of mutually beneficial partnerships with like-minded ENGOs. Vancity's primary 'product' is the redistribution of capital in the community. The organisation has a reciprocal relationship with the ENGOs, social enterprises and mission-based businesses in which it invests. Through them, capital is redeployed in communities to achieve environmental and social benefits. For example, as the founding financial partner of Light House Sustainable Building Centre, Vancity is helping to establish a green building industry on the west coast of Canada. Located on Granville Island, in the heart of Vancouver, Light House offers green building information and consulting services to individuals and industry professionals.

Vancity has more than 1,000 such relationships with ENGOs, social enterprises and mission-based businesses, ranging from strong multi-year partnerships (e.g. Vancity's investment in the local transit authority to provide free transit passes to the 58,000 students who study each year at the area's two local universities) to smaller-scale experiments such as the Solar Hot Water project lead by Solar BC which, in its pilot year, installed 50 solar hot water heating systems on residential rooftops in BC.

These relationships with local entities create a virtuous circle. As a result of these organisations using Vancity as their primary financial institution, Vancity thrives and prospers, and in turn invests back into community so that the community continues to thrive and prosper. The capital stock stays within the community and moves through the hands of those who share common values, beliefs and ideals. Since 1974, $25 billion dollars of equity has flowed in this fashion, providing significant underwriting for social and environmental benefits.

The ENGOs that Vancity supports range in size from grassroots to national, and the nature of the support ranges from pure philanthropy to more engaged models. These relationships are arguably at their finest when they achieve fourfold returns — for Vancity, the member, the ENGO and the community at large.

This is illustrated in just one example. In 2007, Citizens Bank engaged its members in climate change work by making a donation equivalent to 1% of all deposits made during its annual RRSP[4] campaign to the ENGO known as Markets Initiative. Markets Initiative is an environmental organisation that works to protect carbon-rich old-growth forests by shifting publishers and printers to environmentally responsible, low-carbon papers. Thirty per cent of the world's carbon dioxide is stored in old-growth forests; protecting these forests is essential to stopping climate change. Through the campaign, Citizens Bank donated $50,000 to Markets Initiative, enabling it to hire a full-time campaigner to work with leading publishing and print conglomerates on environmental paper issues. One result was that the latest edition of *Harry Potter* was printed on old-

---

4 RRSP stands for Registered Retirement Savings Plan. It is a tax-deferred way for Canadians to save for their retirement.

growth-forest-free paper in 22 countries, making it the greenest book in history. Another was that two of Canada's leading magazine print conglomerates announced new paper purchasing policies that, over time, will see titles such as *Canadian Living, Canadian Geographic* and newspapers such as the *Globe and Mail* shift away from paper that originates from carbon-rich old-growth forests.

Advocating for climate change action

As Vancity became known for its action on climate change, the organisation began to receive frequent invitations from its non-profit partners to join them in advocating for public policy changes at the provincial level. Historically, the organisation's approach to effecting policy change had been to fund the advocacy work of ENGOs. However, many of these organisations felt that, as a large, successful, locally based business, Vancity's voice would add strength to the chorus calling for change.

In February, 2007 Vancity sent a letter to the Premier requesting that the provincial government set targets for reducing GHG emissions. The letter was one of many the Premier received from non-profit, labour, faith and aboriginal organisations. In May 2007, British Colombia set a target to reduce emissions by 33% (from 1990 levels) by 2010. Next, Vancity partnered with the groups involved in this advocacy work to form 'Take the Lead BC', a multi-sectoral advocacy group supporting climate change action in British Columbia.

Key success factors

What is visible at Vancity today is the result of four decades of fine-tuning. Key to Vancity's success is not only a sound strategy, but also the optimisation of elements of its structure, workforce and culture so they serve its triple-bottom-line mission in general and its climate change strategy in particular. What follows is a brief analysis of how Vancity aligned these elements to achieve triple-bottom-line success using its Climate Change Solutions portfolio as an example.

Vancity's structure is optimised to achieve triple-bottom-line success primarily because, in keeping with its overarching philosophy, Vancity's directors take a stakeholder view as opposed to simply a shareholder view in governing the organisation. This means that, instead of seeking only to maximise shareholder profit, they seek to maximise the prosperity of all stakeholders who are directly or indirectly affected by the organisation.

To do this, the organisation must generate financial, environmental and social value. Not only do Vancity's directors hold the organisation accountable for delivering on a triple bottom line, they also use a longer time horizon for measuring outcomes. This enables Vancity to invest in researching and developing new opportunities for leadership in the emerging field of climate change solutions.

For example, in 2006, Vancity invested a total of $250,000 in cash and in-kind donations to develop Canada's first carbon offset granting programme. The project is focused on long-term returns, providing patient capital to develop the North American carbon offset market. Vancity teamed up with three postgraduate students from The Natural Step School in Sweden to explore ways that Vancity might be able to source high-quality carbon offsets from its existing investment portfolio. The team worked to:

- Analyse the GHG emissions that had been saved from projects Vancity had funded in the past

- Study the emerging field to determine what criteria should be set around the organisation's acquisition of offsets

- Make recommendations to Vancity to develop its own Carbon Offset Program[5]

Over the next year, Vancity invested $100,000 to:

- Develop the programme

- Establish relationships with experts qualified to verify and validate the quantification of carbon emissions reductions of potential projects

- Conduct a pilot project

Once launched, the fund was seeded with $100,000 in capital. All told, Vancity's investment and in-kind donations to this sector development work totalled $250,000 — an investment that would be difficult to make at a public company obligated by its shareholders to maximise profits each quarter. It is investments such as these that enable Vancity to continue to lead the way in the development of climate change solutions.

This groundwork made two things possible. First, it provided the capital required to support the launch of economically viable local emissions reductions projects. Second, it provided a prototype for a replicable funding programme. In fact, eight months after the Vancity Carbon Offset Program was launched, some of the programme's advisors worked with the Canadian corporation Aeroplan to launch a similar programme called the Carbon Reduction Fund.[6]

Not only is Vancity's structure key to its achievement of triple-bottom-line objectives, but its workforce and culture contribute to these results. Vancity's mission attracts a niche group of highly skilled leaders and employees who, because they have many options open to them in the labour market, have the luxury of choosing to work for an organisation that can not only offer them an interesting job and a competitive wage, but whose values are also aligned with their own. This results in benefits for both the organisation and its employees. The organisation benefits from attracting top talent in a tight labour market. The employees benefit from the opportunity to:

- Do work that is aligned with their values

- Work with like-minded colleagues

- Enjoy the reputational benefits of working for an organisation with a highly regarded brand

5 www.vancity.com/carbonoffsets (accessed 20 August 2008).
6 www.carbonreductionfund.org

Further, because Vancity is focused on achieving gains for the community, it is more likely to receive in-kind donations of time and expertise by subject matter experts who can help the organisation advance its environmental mission. For example, when Vancity as part of its work to become carbon neutral began to explore options for sourcing high-quality carbon offsets to neutralise its remaining emissions, global experts in the field such as Dr Karl Henrik Robèrt, founder of The Natural Step International, volunteered their time to sit on an advisory group to establish standards and a programme for Vancity to carry out this work.

Challenges

Vancity's success at differentiating itself in the marketplace through its climate change strategy and its focus on the triple bottom line is evident in its rapid growth. Paradoxically this success has exacerbated one of the major business challenges faced by the organisation: namely, the ability of its technology and processes to keep pace with the rate at which Vancity would like to deliver new innovative financial products and services.

Vancity's rapid growth has led to technology challenges: namely, that organisation has outgrown its core banking system. A replacement system that is affordable to the credit union yet robust enough to handle the complex transactions that an organisation of Vancity's size must perform is not yet fully deployed. This constraint leads to process challenges. Many banking requirements that would be best met by making a change to the banking system are met instead by creating a manual workaround to perform the necessary process.

Such workarounds consume large amounts of staff time and are more prone to error because of the human intervention involved. The result is that it is now very difficult for Vancity to effectively deploy new products (including innovative environmental products designed to help members reduce their emissions). An example of this is the Bright Ideas Cashback Program — a cash rebate for members to improve the energy efficiency of their home. When this programme was developed, technical staff were not available to support the implementation and so the process for awarding this rebate was set up manually. It was this manual process that ultimately led to the programme's demise because, even though the programme design was sound, the fulfilment of the offer through manual processes was cumbersome. When the programme was reviewed, it was decided that the cost to the organisation in terms of staff time to process the transactions outweighed the benefits gained by the community.

These workarounds make it difficult for staff to serve members and cause the organisation to operate less efficiently than its competitors. Overcoming these technology and process challenges is important both for the continued health of the organisation and for its continued ability to deliver on its climate change strategy and triple bottom line.

Conclusion

Despite current challenges with its technology and processes, Vancity has brought its climate change solutions strategy to life through optimising elements of its strategy, structure, workforce and culture. Its related programmes are now top of the list of the solutions that experts point to when advocating for action on climate change. By focusing on delivering on a triple bottom line and in generating financial, environmental and social returns, Vancity has differentiated itself from other financial institutions and created a successful niche in a crowded market.

Vancity's status as a member-owned credit union puts it in a unique position to act. When polled, 80% of Vancity members say they are 'extremely concerned' about climate change and 85% of members believe that Vancity is concerned about this issue. This presents the organisation with a unique opportunity (and obligation) to deliver on its triple bottom line. Vancity's role in addressing climate change is clear:

- Create innovative solutions that support action
- Address root causes
- Provide working prototypes for replication by larger companies

It is through this replication that broader change can be realised and climate change can be slowed.

To date, the solutions that Vancity has developed have been replicated in a piecemeal fashion by publicly traded financial institutions. An important question, as yet unanswered, is whether a publicly traded institution can replicate what Vancity has created in an environment where quarterly financial returns must be optimised and where shareholder value trumps social value. Vancity's experience suggests that a focus on the triple bottom line generally or on climate change in particular can add real value to an organisation's brand and reputation and can create substantial business opportunities. However, to date, these benefits have accrued over longer time horizons than are the typical focus for publicly listed companies and their shareholders. This raises the critical question of the role and contribution of government in changing the incentives for corporate action on climate change.

Much has been achieved through the actions of business leaders, both in terms of reducing their own emissions as well as enabling their suppliers, customers and stakeholders to reduce their emissions. However, the reality is that government needs to harness its influence and direct its moral authority towards creating the policy framework (regulations and incentives) that encourage all companies — leaders and laggards — to take action on climate change. It is here that Vancity believes its contribution is most significant: through demonstrating that taking action on climate change can provide real benefits for business and through calling for governments to take action on reducing greenhouse gas emissions, Vancity sees its role as supporting and encouraging action on this most important of issues.

Large-scale, deliberate, engaging and affordable strategies to address the crisis of climate are what people have begun to demand of governments around the globe. There is rising impatience for politicians who claim it can't be done because it might be bad for the economy. The economy, as Herman Daly once said, is a wholly owned subsidiary of the environment. To that end, we are all shareholders in a cool planet — and it is time we started demanding a return on our investments. What happens in the next quarter is nowhere near as important as what happens in the next quarter-century.

17

The Pole Position project: innovating energy-efficient pumps at Grundfos*

Joan Thiesen and Arne Remmen
Aalborg University, Denmark

The Grundfos Group is one of the world's largest manufacturers of pumps, with around 15,000 employees and an annual production of more than 16 million pump units. In this chapter we describe an innovation initiative established by the company in 1992 to challenge the state-of-the-art performance and energy efficiency of circulation pumps used in household heating systems. The case study shows how Grundfos engaged with its industry peers, its business association and policy-makers to overcome the main market barriers to the wider uptake of energy-efficient circulation pumps. The product innovation turned out to be a success story that enabled Grundfos to maintain and strengthen its market position.

This chapter is divided into three main sections. The first is a description of the process whereby Grundfos developed its Alpha Pro circulation pump and how the company contributed to the establishment of the EU Energy Classification Scheme. The second is an evaluation of the energy-saving benefits that are expected to result from the increased uptake of energy-efficient circulation pumps. The third section discusses some of the wider implications of the case study, in particular the critical relationship between innovation and public policy in developing and implementing sustainable solutions to climate change.

* We would like to thank Lars Enevoldsen (Innovation Manager, Grundfos), Carsten H. Pedersen (Chief Project Manager, Grundfos), Niels Bidstrup (Chief Engineer, Grundfos) and Nils Thorup (Chief Technical Advisor, Grundfos) for the information on the Pole Position project and the circulation pumps market which they provided in the course of interviews in September 2007.

The Pole Position project

The Grundfos Group

Grundfos was founded in 1945 and has its headquarters in Jutland in Denmark. Grundfos produces pumps for a variety of applications including circulation pumps for heating and air conditioning, and centrifugal pumps for industrial, water supply, waste-water and dosing applications. Grundfos also produces a range of associated products including pump motors and electronic systems for monitoring and controlling pumps and other electrical systems. Some key data about the Grundfos Group are presented in the Box below. In the context of this chapter and in particular in relation to the motivations for the 'Pole Position' Project, it is important to note that Grundfos is recognised as one of the most proactive Danish companies in terms of environmental performance (Grundfos 2007a).

The Grundfos Group

- Net turnover: DKK 15.4 billion (~€42.1 billion)
- Profit before tax: DKK 1.5 billion (~€200 million)
- Number of employees: 14,782
- Annual production of approximately 16 million pump units
- Covers approximately 50% of the world market for circulation pumps
- Represented in 41 countries, pumps being manufactured in 14 different countries
- All production companies are certified to ISO 14001. Companies in Europe are also EMAS-registered
- Life-cycle assessments were carried out on all Grundfos products at the end of the 1990s

---

The technical energy project: how it began

In 1992, an internal technical energy project was initiated at Grundfos with the aim of investigating how the state of the art regarding performance and energy efficiency[1] of circulation pumps for household heating systems could be improved. Three industrial researchers as well as a number of engineers at Grundfos were assigned to the project, which was supported by public funding.

The analysis showed that a more energy-efficient circulation pump would:

- Incorporate a motor with permanent magnets

---

1 From a life-cycle perspective, 70–98% of the total impact on the environment is caused by the use phase of the circulation pump (Grundfos 2007b).

● Include speed control

● Adapt itself to current heating demand in the home

None of these was a feature of circulation pumps at the time. In general, such pumps either ran at full speed regardless of demand or were turned on and off manually depending on the heating demand. Notwithstanding the potential performance improvements, these technologies were considered too costly to adopt.

This conclusion (i.e. that the technology improvements would be too expensive) was challenged by the Group President, Niels Due Jensen.[2] He argued that Grundfos should try its utmost to take advantage of the technical findings from the energy project. His view was that technological innovation was critical to the company's ability to maintain its position as the market leader in circulation pumps for household heating systems.

The process of technology development therefore continued, resulting in a number of patents. In 1998, the first spin-offs from the technical project were launched but, at that time, there was no market demand for circulation pumps incorporating these new technologies. There were a number of reasons for this. The first was the lack of consumer awareness of the pump and its energy consumption. The second was that, even though the new technologies would significantly reduce the energy consumption of a circulation pump, it would also make the product more expensive. Grundfos sought (unsuccessfully) to change customers' attitudes by conducting and publishing a life-cycle cost[3] analysis that demonstrated that customers would recoup the additional investment within a few years due to the savings on energy. However, customers continued to focus on the initial capital costs. The third was that, in the case of new buildings, the developer decides which components to integrate but is often not the one to gain the long-term economic benefits from choosing a more efficient pump. The consequence was that initial capital costs tended to dominate the investment decision.

Lobbying for energy efficiency

Even though a circulation pump can consume up to 15% of the energy consumption in an average European household, circulation pumps are rather invisible products and are generally noticed only when the heating season starts or when a pump breaks down and the inhabitants have no heating. Furthermore, the installation contractor has a large influence – both in terms of deciding which pump to install and in making the end-user more aware of energy efficiency.

To establish a market for energy-efficient pumps, Grundfos had to find a way to raise awareness among both consumers and installation contractors. A business reality was that developing energy-efficient products is not enough; companies such as Grundfos must also have markets where these products can be sold.

In 1998, Grundfos decided to embark on a political process with the objective of seeking a ban on the most inefficient circulation pumps. Grundfos chose to work through the European Association of Pump Manufacturers, Europump, to have a larger impact on politicians and officials. Grundfos had not previously been an active player in Euro-

---

2 Now Chairman of the Grundfos Group.
3 A life-cycle cost analysis is used to calculate the cost of a system over its entire lifetime covering initial purchase price, installation, maintenance, energy consumption and disposal.

pump, but a strong collaboration was built. Through Europump, Grundfos made sure that the issue was raised in relevant committees and councils. Grundfos also tried to encourage Danish politicians to raise the issue in some of their meetings in Brussels (Borring and Rønnov 2001).

In 2000, the Vice President of the European Commission and Commissioner for Transport and Energy, Loyola de Palacio, became interested in introducing legislation on energy-efficient circulation pumps and an examination of the energy-saving potential of such pumps was initiated. This investigation became part of the EU SAVE II Programme.[4]

Grundfos was commissioned, with two partners, to carry out three studies over the period 2000–2001 (Borring and Rønnov 2001):

● A market structure and statistical analysis

● A technical/economic analysis

● An impact analysis at various scenarios

The investigation highlighted a significant potential for energy savings and greenhouse gas emission reductions if more energy-efficient circulation pumps were introduced. The report (Bidstrup *et al.* 2001) recommended two policy tools:

● Labelling

● Minimum efficiency performance standards for circulation pumps

The report further recommended that these two tools were integrated in an Energy Classification Scheme, which would set standards for energy efficiency and send a clear signal to customers that some circulation pumps have lower energy consumption than others. This approach would build on the EU Energy Label, which places products in categories from A (most energy-efficient) to G (least energy-efficient).[5] Energy labels had previously been applied to light bulbs (since 1984) and home appliances (since 1995).

Creating the energy classification scheme

Later in 2001 a working group was established consisting of representatives from a number of pump manufacturers. The working group developed a proposal for a classification scheme, which classified the pumps according to average power consumption.

The EU Energy Classification Scheme operates with seven categories, based on an Energy Efficiency Index (EEI). The EEI is calculated by dividing the weighted average power consumption for a specific pump with a reference power consumption based on the majority of circulation pumps in the market. This calculation also takes, for instance, the size and hydraulic power of the circulation pump into account; for the precise calculation of EEI, see Bidstrup *et al.* 2003. The idea behind the EEI is that the lower a value

4 SAVE stands for Specific Actions for Vigorous Energy Efficiency. It is a EU-wide programme that aims to improve energy efficiency and thereby reduce the environmental impact of energy use in transport, industry, commerce and the domestic sector. SAVE plays a key role in the EU's response to the Kyoto Protocol. SAVE II (1998–2002) succeeded the original SAVE programme (1991–1995).

5 www.energy.eu

the circulation pump scores on the EEI, the higher its energy efficiency. The group's work concluded in February 2003.

From 2003 to 2005, negotiations took place within Europump to establish a voluntary agreement between pump manufacturers, which would oblige them to mark their circulation pumps according to the classification scheme. For Grundfos, working through Europump was critical to achieving support from the rest of the pump manufacturers for a voluntary classification scheme. Grundfos recognised that the influence of such a scheme would be limited if it was the only company using the classification scheme.

In early 2005, the Energy Classification Scheme was launched by Europump in consultation with the European Commission. Europump is responsible for the monitoring and publicity of the commitment made by a number of the European manufacturers of circulation pumps.[6] The classification scheme is presented in Figure 17.1.

FIGURE 17.1 **Classification scheme for the energy labelling of circulation pumps including the Energy Efficiency Index (EEI)**

Source: www.energypluspumps.eu

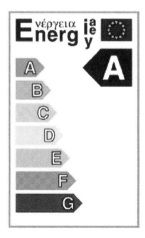

A	< 0.40	high-efficiency drive
B	0.4 < EEI < 0.6	variable-speed drive
C	0.6 < EEI < 0.8	unregulated
D	0.8 < EEI < 1.0	unregulated
E	1.0 < EEI < 1.2	
F	1.2 < EEI < 1.4	
G	1.4 < EEI	

The index for the label was calibrated so that most circulation pumps on the market at the time (2003) would fall into the categories 'D' or 'E'. When the calibration was made, there were no circulation pumps for household use on the market that would qualify for an 'A'. In order to achieve an A-label, a pump needed to incorporate state-of-the-art energy-efficient technologies.

As a result of its public policy lobbying, there was now a tool that would allow Grundfos to demonstrate the energy efficiency of its pumps to customers. It is important to note that the technologies developed by Grundfos were not the only way that an A-label could be achieved so the scheme did not preclude other manufacturers from achieving such a rating for their pumps.

When the Energy Classification Scheme was launched in spring 2005, Grundfos also had a product ready for market launch which qualified for the A-label. The market launch

6 www.europump.org

was made possible by the Pole Position project, which is described in the following section.

The Pole Position project

In April 2004, Grundfos initiated its product development project. It was clear that energy labelling would become a reality in the near future and Grundfos therefore needed a product that could qualify for an A-label. The project was named Pole Position, indicating the importance that Grundfos assigned to its efforts to demonstrate its leading position in circulation pumps – the company's most important product in terms of market share and turnover. The project was a top priority for the company and a number of highly skilled employees participated in it.

When the project was initiated, the various technologies were more or less fully developed; the major challenge was to integrate them into a new product in time for the launch of the Energy Classification Scheme.

Within 15 months Grundfos was ready to go into full production with Alpha Pro, a circulation pump with an energy efficiency that would qualify it for an A-label. The product was launched in autumn 2005 (i.e. at the beginning of the heating season). The launch time was crucial as most circulation pumps are sold at the beginning of the heating season and none of the company's competitors was ready with A-labelled circulation pumps at that time.

The launch of Alpha Pro was supported by an intensive marketing campaign that highlighted both the new classification scheme and the benefits of choosing the Alpha Pro circulation pump. Grundfos also launched an Energy Project website[7] providing information to homeowners and installation contractors on the energy-saving potential of more efficient pumps; the website also has a section on global warming. Compared with the more technical energy project launched in 1992, the Energy Project website is mainly for marketing and information purposes.

Alpha Pro is an 'intelligent pump' in the sense that electronic controls, which can assess current heating requirements, are built into the pump and the performance of the pump can be adapted according to the actual heat demand. This means that the pump is not, unlike unregulated pumps, running at full speed all the time (Grundfos 2007b). New and advanced technologies that play an important role in reducing the energy consumption of the pump were also incorporated.

What is the contribution to reducing energy consumption?

In the context of this book, one important question to ask is whether the Alpha Pro pump (or other similarly energy-efficient pumps) actually contributes significantly to reducing greenhouse gas emissions. There are two distinct ways of analysing the contribu-

7 www.energyproject.com

tion. First, one could assume that all less efficient pumps are changed into A-labelled pumps and then estimate the energy savings, or what greenhouse gas emissions this could abate. Second, one could look at the practical evidence, i.e. is there actually a move towards more energy-efficient pumps?

Estimating the energy saving potentials

To estimate the full potential of all circulation pumps being A-labelled, the first step would be to determine the energy consumption of circulation pumps. Based on figures from 1995, the EU SAVE II project estimated that:

- 87.3 million pump units were installed in the EU15[8] in 1995

- Between eight and nine million new units are installed each year

The annual energy consumption of these pumps was estimated to be 41 TWh (Bidstrup *et al.* 2001). Taking the enlargement of the European Union in 2004 into consideration, Bidstrup (2007) estimated that the total annual energy consumption of all circulation pumps in the EU25 amounted to 60 TWh in 2005.

On the Energy Project website, Grundfos estimates that replacing all D-labelled units in the EU25 with A-labelled alternatives would result in energy savings of around 40 TWh. The energy-saving potential for EU25[9] is therefore estimated to be larger than the annual energy consumption of Denmark, i.e. 33.5 TWh in 2005 according to Statistics Denmark (2007).

Practical evidence for a move towards energy-efficient pumps

The figures presented above show that the potential energy savings from installing more efficient circulation pumps are significant. But to what extent have these potential energy savings been realised?

Europump has evaluated the influence of the classification scheme based on data provided by the companies that have committed to the classification scheme (Europump 2007). Together, these companies account for more than 80% of the European market for circulation pumps. The evaluation investigated the evolution of the market following the introduction of the classification scheme. The results are reproduced in Figure 17.2.

Figure 17.2 shows that A- and B-labelled pumps have increased their share of production since the introduction of the Energy Classification Scheme, while the share of C- and D-labelled pumps has fallen. A few circulation pumps already qualified for an A- or a B-label in 2004, but their share of production did not increase significantly until the scheme was introduced. The evaluation showed that the classification scheme played an important role in moving the market towards choosing A- and B-labelled pumps — though the move towards B-labelled pumps was the most significant.

---

8 The 15 countries making up the EU between 1 January 1995 and 31 April 2004. A further ten countries joined the EU on 1 May 2004 to give a total of 25 members.

9 From 1 January 2007, 27 countries have been part of the EU. Taking this enlargement into account would potentially increase the improvement potential even further.

FIGURE 17.2 Percentage share of each energy label category by those companies committed to the EU Energy Classification Scheme

Source: Europump 2007

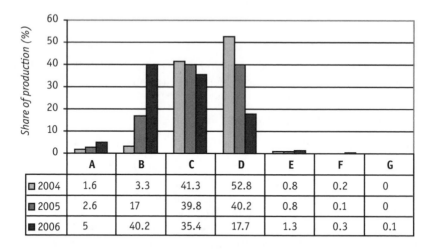

	A	B	C	D	E	F	G
2004	1.6	3.3	41.3	52.8	0.8	0.2	0
2005	2.6	17	39.8	40.2	0.8	0.1	0
2006	5	40.2	35.4	17.7	1.3	0.3	0.1

The Europump evaluation also noted that there had been a relative reduction of the EEI by 20% from 2004 to 2006, meaning that pumps on average were achieving a lower score on the EEI (i.e. pumps were becoming more efficient).

The move towards more energy-efficient pumps was assessed to have delivered total energy savings of around 580 GWh (Europump 2007). If the use of A-labelled circulation pumps becomes more widespread in the future, much higher energy savings will be achieved.

Future innovation

Another aspect to consider in this regard is technology development. Over time, new technologies might significantly improve energy efficiency (as in the case of the Alpha Pro pump). In fact, it is not inconceivable that improved technologies might allow some pumps to achieve an EEI value significantly below 0.4 (i.e. the value at which a pump presently qualifies for an A-label). If this happens, it may be necessary to subdivide the A category to reflect state-of-the-art energy efficiency technologies; otherwise it would not be clear to customers which A-labelled pumps are the most efficient.[10] For companies such as Grundfos, the potential for the 'A' category to be subdivided provides an ongoing incentive to continue its processes of innovation and technology development.

---

10 This has already happened in the case of refrigerators and freezers where the categories A+ and A++ have been added to reflect the most efficient technologies.

Discussion of wider implications of the case study

Business implications for Grundfos

Grundfos has experienced a number of direct benefits as a result of the innovation process which it initiated in 1992. First, a number of new technologies have been developed. These have enabled Grundfos to maintain and strengthen its position as a market leader. Second, the market demand for Alpha Pro has been much larger than first expected and Grundfos has sold more of these pumps than expected. Third, the innovation process has generated a number of spin-off benefits in terms of new products. One example is the SQFlex[11] which was launched in 1999. Another is ALPHA2, which is the successor to Alpha Pro and was launched in November 2007. ALPHA2 is based on the same technologies as Alpha Pro but is more compact, more user-friendly and designed for mass-production. ALPHA2 also qualifies for the A-label and its energy efficiency is slightly better than that of Alpha Pro.

The innovation process has also generated a number of intangible benefits, in particular the building of new knowledge. The innovation process has allowed Grundfos to develop internal expertise related to technology development and product development processes: for instance, how to manage product development projects that need to be completed in short time periods. The innovation process has also given the company valuable knowledge in terms of the environmental benefits and energy-saving potential of choosing more efficient pumps, which it can use to position the company and its products. Furthermore, the project has increased the willingness of Grundfos to take a strong position in the climate change debate (in particular in Denmark). Together, these intangible benefits have also helped strengthen the Grundfos brand as an innovative company with a strongly proactive approach to environmental issues.

The project has also helped Grundfos to develop a comprehensive knowledge of how to harness public policy to support the development of its business. Grundfos has continued to actively contribute to and influence future legislation and policies. As an example, Grundfos has participated in stakeholder meetings and provided data for one of the preparatory studies for the Directive 2005/32/EC on the eco-design of Energy-using Products (EuP Directive).[12] This preparatory study focused on electric motors (a category that includes circulation pumps). The company's reason for participating in this work is to ensure that the upcoming eco-design requirements will reflect state-of-the-art technology and exclude less efficient products from the internal market.

---

11 SQFlex is a water supply system based on renewable energy. It consists of a submersible pump and an energy supply based on either wind or solar power, or a combination of these. This makes the system suitable for use in areas where there is no power supply or the power supply is unreliable.

12 The preparatory studies investigated relevant provisions of the EuP Directive and recommended ways of improving the environmental performance of specific product groups. The studies included an investigation of market characteristics, relevant environmental aspects, technological/economic potential for improvement, relevant existing legislation, self-regulation by industry and standards, and the need for standards to be developed. By the end of 2007, 19 preparatory studies had been initiated covering different groups of energy-using products. These preparatory studies will form the basis for the implementing measures of the EuP Directive, which lay down a number of eco-design requirements for defined energy-using products. If a product is covered by an implementing measure, it must comply with the requirements in order to be placed on the internal market of the European Union (EC 2005).

Initiating th ... .e leadership position for Grundfos in the intern ... , also positioned Grundfos as a brand that can de ... : high-end energy efficiency product range. But how in ... :en as a business driver for Grundfos?

In the technical energ ... .ige was not the primary driver; the primary focus was on reducing en ... .on as a means of reducing operating costs. Furthermore, the wish to challenge ... -the-art technologies and to strengthen its market position was also strong. However, as climate change received more political and media focus through the 1990s, climate change became an important additional argument for Grundfos and the 'keyword' for entering the relevant political discussions in the European Commission.

Climate change is now considered an important parameter, when doing business. Grundfos sees the development and improvement of energy-efficient technologies to be a key priority in order to meet the challenges presented by climate change. Grundfos recently established Grundfos New Business to develop new business areas to supplement the core business related to pumps and pump systems. These new business areas deal with processes and technologies related to water, water purification and energy efficiency, where climate change may be one of the drivers for product development (Grundfos 2007b). However, Grundfos does not see environmental and energy performance as the only important parameters in new product development. Other parameters such as quality, function, noise, comfort and costs are equally important characteristics of new products.

Drivers for innovation of energy-efficient technologies

A precondition for the commercial success of Alpha Pro was, without doubt, the development and introduction of the energy labelling scheme. This demonstrated the importance of the interplay between drivers such as policy and regulation, innovation and commitment inside a company, and voluntary initiatives from business stakeholders. In the case of Alpha Pro, Grundfos had the technologies ready at an early point, but the market was not ready for these technologies. Grundfos therefore needed to influence the 'rules of the game'; the introduction of the Energy Classification Scheme created the necessary market for introduction of energy-efficient technologies and also helped increase the environmental awareness of customers. The fact that the scheme was developed in collaboration with the industry association, Europump, and in close consultation with the European Commission has given high credibility to the energy label.

So far, normative regulation such as environmental performance standards for pumps has not played any role in the innovation dynamics. That is, innovation has led regulation rather than vice versa, though this will probably change as the EuP Directive is implemented. Energy labels such as the Energy Classification Scheme are a voluntary initiative that seeks to provide advantages to the best-performing products. In contrast, the eco-design requirements under the EuP Directive are mandatory and are expected to result in bans for the worst energy performing products. The details of the requirements for circulation pumps given by the EuP Directive will be specified in 2008. These will build on the Energy Classification Scheme, although they may include requirements to consider environmental performance across the whole product life-cycle. It appears likely that the A-label will probably be subdivided into A, A+ and A++. Furthermore, the demands on energy-efficient pumps will be tightened since it is likely that all circulation pumps will be required to have a minimum allowable energy performance of A+ within five years (Falkner 2007). This suggests that the EuP envisages significant technological forcing of the energy efficiency of pumps. It is probable that the Energy Classification Scheme will need to review its criteria on a regular basis to keep pace with technological development.

The Alpha Pro case shows that no single driver has been decisive for the success of the innovation process. Different drivers have played a key role at different stages of the innovation process, depending on the progress and challenges in the specific phase. The mix of drivers will vary from project to project; knowledge of the product, technological opportunities, the potential for market creation, regulatory requirements and the interests of different stakeholders are therefore important. However, the commitment and creativity of the managers and the engagement of staff are also preconditions for success in such innovation processes.

Challenges and interplays between innovation and public policy

Even though public policy can support business innovations, numerous challenges can be encountered. From this case study, three issues were of particular importance:

- Policy uncertainty
- Policy weaknesses or gaps in policy
- How to use policy to support innovation

The long-term direction of policy might not always be clear and themes that at one point seem significant might turn out to be less important in the long run. This can be characterised as policy uncertainty; a consequence of this uncertainty is that companies may have doubts about whether the issues raised on the political agenda are persistent or not. As regards climate change, the level of political support for action seems quite strong — at least for the moment — with a steady stream of policy initiatives and measures from the EU in particular.

When the innovation process was initiated, there was no policy to support more energy-efficient circulation pumps. However, Grundfos figured out where the gap was and how it could be overcome. Even though policy was used to support energy efficiency, there are also some weaknesses in this. In the case of the classification scheme, and despite the significant outcomes that have been achieved, the risk is that, in the long term, the scheme could actually constrain technology development as there are, at present, no incentives for improving the energy efficiency of a pump significantly below the EEI value that qualifies for an A-label (though the EuP proposals may address this particular issue). Another unintended consequence is that, because of the structure of the classification scheme which divides performance in a relatively crude manner, two pumps could be given the same label even though their energy efficiency could be quite different.

Grundfos does not make the same intensive efforts to influence policy in all innovation projects. In the case of Alpha Pro, however, public policy was particularly important because the product challenged existing technologies and provided a more energy-efficient alternative than previously known. Policy can be used to promote technologies which in the long run can provide significant energy and environmental benefits, even if there is limited customer interest in these technologies. Furthermore, businesses often have comprehensive product knowledge and, if this technological knowledge is made available to policy-makers, it may form the basis for policy instruments that promote efficient state-of-the-art technologies. Traditionally, business lobbying has tried to avoid policy demands and new regulations. However, as seen from this case, front-runner companies on innovation and product development have a strong interest in lobbying for tighter regulations and other policy instruments that benefit products with improved environmental performance.

Finally, both innovation and policy formulation are long-term processes. This raises the question of how policy best can support innovation. As seen from this case study, the technologies were actually ready quite early in the process. However, technology development is often fenced in and kept secret to gain competitive advantage when introducing new products on the market. Therefore, companies are reluctant to inform policy-makers about best available technologies before they are launched on the market. This makes it difficult for policy-makers to ensure that policy reflects these best available technologies. A company can overcome this by establishing a close relationship with policy-makers; they can then influence policies regarding product performance and provide policy-makers with information on state-of-the-art performance standards — though without revealing the technologies used.

Conclusions

Grundfos initiated the technical energy project in 1992 with the aim of challenging state-of-the-art circulation pumps in order to strengthen and maintain its position as market leader. Even though new technologies with a major potential for improving energy efficiency were developed, there was an absence of market demand. The innovation process behind Alpha Pro is therefore a story of how Grundfos engaged with industry peers and policy-makers to overcome this market barrier by creating regulation to support energy-efficient circulation pumps.

Two general points can be highlighted from this case study.

First, technological innovation is a necessary, but not a sufficient, condition for the development and diffusion of new sustainable solutions. A public policy framework is also needed to create market demands and raise awareness of new technological possibilities when it comes to the diffusion of energy-efficient products. Seeking support from public policy might be particularly relevant in the case of climate change, where a long-term perspective is needed. Regulation is therefore important to ensure that innovations in areas such as energy efficiency actually get implemented and adopted.

Secondly, the case study shows the importance of top management commitment to provide the internal impetus for innovation, understand the policy barriers for the project and push for the change in the regulatory framework. Without this drive from senior management, Grundfos could never have sustained its focus on technological innovation and changing the policy framework over the 15 years since the internal energy project was initiated.

References

Bidstrup, N. (2007) 'Energy Labelling of Circulator Pumps in the EU', paper presented at the *CO₂action Conference*, Aalborg, Denmark, 6 March 2007.

—, M. Elburg and K. Lane (2001) *EU II Project: Promotion of Energy Efficiency in Circulation Pumps, especially in Domestic Heating Systems — Summary — Final Report* (Bjerringbro, Denmark: Grundfos).

—, G. Hunnekuhl, H. Heinrich and T. Andersen (2003) *Classification of Circulators* (Brussels: Europump).

Borring, G., and F. Rønnov (2001) 'Danske Virksomheder Går Direkte til EU efter Indflydelse' ['Danish Company Goes Directly to the EU for Influence'], *Monday Morning*, June 2001 21: 1-6.

EC (European Commission) (2005) 'Directive 2005/32/EC of the European Parliament and of the Council of 6 July 2005 Establishing a Framework for the Setting of Ecodesign Requirements for Energy-using Products', *Official Journal of the European Union* L 191: 29-58.

Europump (2007) *First Briefing Document of Europump's 'Industry Commitment to Improve the Energy Performance of Stand-alone Circulators through the Setting-up of a Classification Scheme in Relation to Energy Labelling'* (Brussels: Europump).

Falkner, H. (2007) *Appendix 6. Lot 11 — Circulators in Buildings. Report to the European Commission* (Didcot, UK: AEA Energy and Environment).

Grundfos (2007a) *Annual Report 2006: The Grundfos Group and the Poul Due Jensen Foundation* (Bjerringbro, Denmark: Grundfos).

— (2007b) *Sustainability Report 2006* (Bjerringbro, Denmark: Grundfos).

Statistics Denmark (2007) *Statistical Yearbook 2007* (Copenhagen: Statistics Denmark).

18
Responding to climate change: the role of organisational learning processes

Marlen Arnold
Technische Universität München, Germany

One of the critical questions for policy-makers and other stakeholders is the importance of organisational learning in influencing corporate responses to climate change. Although there is an extensive literature on the actions that have been taken by companies (e.g. developing new modes of production, developing new products, initiating new modes of participation or stakeholder engagement), less attention has been paid to the role that organisational learning plays in informing the specific actions taken by companies. The wide variations in corporate responses to climate change (see, for example, Chapter 2) open up a series of questions around:

- When and why companies pursue processes of learning and change to integrate sustainability and climate protection into their business practices and strategies

- What effect these innovations have (in terms of reducing companies' greenhouse gas emissions)

- What factors promote or inhibit learning

Even though the management studies literature provide answers to these questions in the broader context of corporate strategy, little has been written specifically on climate change (or sustainable development more generally).

Climate change presents fundamental challenges to prevailing business models as companies trying to implement sustainable and climate-oriented requirements may find their conventional way of operating fundamentally challenged (Walsh 2006):

- Processes and products need to be changed fundamentally

- Completely new sets of data may need to be integrated into management decision-making processes

- There may be completely new groups of stakeholders (with consequent implications for external and internal communications)

- There may be a need to change the organisation's own basic values and knowledge systems

These challenges will, for the majority of companies, require new or recombined knowledge or ideas and will require fundamental changes in corporate strategy and objectives (Teece *et al.* 1994; Arnold 2007). Therefore, organisational learning becomes a key element of any effort to implement climate protection effectively within organisations.

Given the weaknesses in the theoretical frameworks, this chapter seeks to advance the literature by examining empirical evidence from six European companies (mostly German-based) in the fields of housing, construction, transport, and information and communication technology (ICT) that have developed and implemented sustainability strategies. Although these sectors are all exposed to climate change-related risk (in particular through regulation), they have yet to be as affected by climate change-related regulation as, for example, the electricity sector has been. The case studies allow us to examine the drivers for corporate action (in a situation where there is a regulatory threat but relatively weak regulatory action), the strategies adopted and the manner in which factors such as organisational learning, culture and values have influenced these strategies.

This chapter has four major sections. First, some key concepts from the organisational learning literature are presented with the aim of providing a frame of reference for the analysis. The research methodology is briefly described in the second section. This is followed by a description of the outcomes (in terms of climate change performance) achieved by the different companies. Finally, the fourth section draws together the information on the performance outcomes achieved and the literature on organisational learning to draw some broad conclusions about the influence of organisational learning on corporate decision-making and corporate actions.

Theoretical background and conceptual framework

Is there an organisational learning theory for climate change?

While much of the writing on sustainability management acknowledges the role of learning, new knowledge and its diffusion in the company, the literature for the most part neglects the role of learning and change processes in the implementation of new concepts. Most approaches concentrate on instrumental aspects and the development of new management concepts or tools rather than highlighting the dynamics of how such instruments and concepts can be successfully realised in companies. Unfortunately, as noted by Müller and Siebenhüner (2007) and Arnold (2007), the empirical research on organisational learning has not kept pace with conceptual advances such as:

- The triple bottom line (Elkington 1997)

- Corporate sustainability (e.g. see Sroufe and Sarkis 2007)

- The sustainable corporation (Shrivastava and Hart 1995)

According to the organisational learning and the learning organisation literature,[1] the focus on sustainability-related learning processes is hard to find and very few articles address this phenomenon directly (Brentel et al. 2003; Cramer 2005). While a variety of studies explore corporate learning processes concerning technological and market-oriented innovations, environmental and/or sustainability-oriented or even climate-oriented learning processes have rarely been the focus of empirical analyses; notable exceptions are:

- Finger et al. (1996) who examined ecological learning processes and explanatory factors in Swiss companies

- Cramer (2005) who studied the diffusion of a particular corporate social responsibility (CSR) programme in different companies by using a learning perspective

However, there is no coherent empirically based understanding of the process dynamics and the causal factors that trigger and govern sustainability-related organisational learning, or motivate individual or collective corporate initiatives.

Some definitions

According to Argyris and Schön (1996), organisational learning is a change in the behaviour of the organisation or its members that is triggered by a change in the underlying 'theory in use', i.e. the often tacitly used set of values and causal beliefs that the members of an organisation share. In their systems theory view, Probst and Büchel (1997: 15) define organisational learning as:

> the process by which the organisation's knowledge and value base changes, leading to improved problem-solving ability and capacity for action.

This definition integrates the outcome perspective by asserting that organisational learning has to serve a specific purpose.

In the context of this chapter, sustainability-oriented learning will be defined as a process 'in which organisational and behavioural changes appear based on a change in knowledge and values which are supported by reflexive processes, and the concept of sustainability serving as a fundamental framework'. This definition also draws on the findings from socio-psychological literature that emphasises the necessity of actual behavioural changes as a result of changes in knowledge (Schunk 1996).

---

1 See, for example, Fiol and Lyles 1985; Huber 1991; Dodgson 1993; Argyris and Schön 1996; Klimecki and Lassleben 1998; Dierkes et al. 2001; Nonaka et al. 2001; Hertin et al. 2002.

The reference to the concept of sustainability provides a guideline for the direction of the learning and change processes within organisations. This chapter uses the framework proposed by Argyris and Schön (1978, 1996) for the analysis of organisational learning processes. Broadly, this model proposes three forms of learning:

● **Single-loop learning.** This type of learning is characterised by the incorporation of sustainable or 'climate-friendly' changes in operations and outputs (products or services), but where the prevailing strategy and organisational culture remain relatively unchanged. That is, the focus of change is confined to specific operational or product related issues. Furthermore, the driver or stimulus of change is generally perceived by the organisation as coming from outside the organisation (e.g. from new regulations or from specific stakeholders' requirements)

● **Double-loop learning.** The double-loop mode involves not only changing operations or outputs but a more fundamental reflection regarding corporate processes and outcomes, potentially resulting in concrete new actions and real changes. In such situations, the essential values, corporate strategies and culture are transformed. In addition, an active formation of learning processes as a process for the generation of new knowledge occurs, which extends beyond the diffusion of known or accepted knowledge. The triggers for this change may be either external or internal factors (e.g. conflicts, crises, general dissatisfaction, or insight and visionary thinking). That is, the triggers may be same as those for single-loop learning (e.g. regulation, stakeholder expectations) but may also come from within the organisation itself

● **Deutero learning.** While not the explicit subject of this article (although elements are canvassed in the discussion), Argyris and Schön (1978, 1996) also identified deutero learning as a third form of learning. Deutero learning aims to improve the organisational learning process (learning abilities and mechanisms) itself (Bateson 1972; Yuthas *et al.* 2004)

The Argyris and Schön framework is essentially a process model. However, in the context of this book, the critical question is how these processes translate into performance outcomes. According to the sustainability-oriented learning definition presented above, these changes are of interest in so far as they result in direct or indirect contributions to reducing greenhouse gas emissions as well as broader contributions such as the introduction of resource-efficient technologies, sustainability reporting schemes or climate-compatible products or processes.

Factors influencing climate change-related learning processes

The goal of the empirical research presented in this chapter was to examine through a series of case studies how sustainability- and climate-oriented knowledge can be absorbed, generated and disseminated by an organisation. In addition, the research sought to:

● Identify the influencing factors that seem, from the literature, to be causal for organisational learning processes

● Understand whether and how they influenced corporate actions in practice

A conceptual framework was developed for use in identifying the conditions for the emergence of such learning processes (Fig. 18.1).

FIGURE 18.1 **Analytical framework for climate protection learning processes**

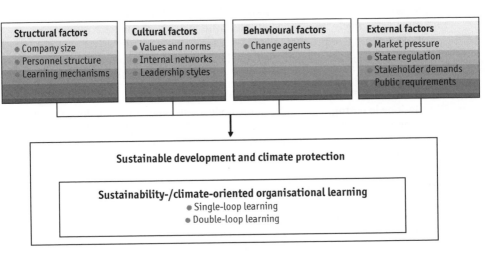

The research project identified (and subsequently examined) eight potential influencing factors:

● **Company size** (as proposed by Schumpeter 1954 and Scherer 1992). The research used the number of employees as a proxy for size

● **Personnel structure** (Czarniawska 1997). The organisation's personnel background (e.g. profession, gender) may influence sustainability-related organisational learning processes

● **Learning mechanisms** (see further Popper and Lipshitz 1995, 2000). The research examined the manner in which organisational processes for reflecting on experiences and for initiating and implementing learning processes (e.g. learning laboratories, evaluation units, workshops) influenced the actions taken

● **Values and norms** (Schein 1984; Rothman and Friedman 2001). These studies examined the manner in which changes in organisational values and norms may influence the direction and type of organisational learning processes, and the outcomes from these learning processes

● **Internal networks** (Crampton *et al.* 1998; Barabási 2002; Mutch 2002; Blatter 2003). The research examined the role of internal networks (formal and informal) in disseminating sustainability-related knowledge through the organisation

- **Leadership styles** (Robertson and Langlois 1994; Arce 2006). These studies examined whether different leadership styles (e.g. directive, consultative, participatory) initiated more sustainable learning and results due to different levels of employee participation in the decision-making process

- **Change agents** (Hartley *et al.* 1997). Change agents are individuals in organisations who initiate innovations and/or keep innovation processes ongoing. They may have a critical role to play in triggering organisational learning and the diffusion of new knowledge through the organisation.

- **External factors** (Huber 1991; Dutton and Duncan 1987; Rajagopalan and Spreizer 1996; Mintzberg *et al.* 2005). Sustainability-oriented learning processes and climate-related outcomes frequently depend on factors external to the company. The research examined the influence of factors such as market pulls and pushes, new regulation, stakeholder expectations and public opinion on organisational learning processes and on the performance outcomes achieved

The empirical research

The research involved empirical analyses of six European companies as follows:

- A large German-based and internationally operating transportation company

- A large German-based public transport company

- A large Netherlands-based international computer manufacturer

- A large Netherlands-based international electronics company

- A medium-sized German-based construction company

- A medium-sized German-based apartment management company

The companies were selected because of their track record of proactively taking action to respond to climate change and sustainable development issues. In addition, all the companies are in sectors with a significant exposure to regulation directed at reducing greenhouse gas emissions, though at the time of conducting the research (2004/2005) regulation was emerging rather than being applied (i.e. the companies were facing the threat of regulation or policy action).

The field studies of the six companies were carried out using semi-structured and thematically focused interviews, supported by desktop studies of related documents. This case study design or methodology followed that set out by Yin (1994) and Mayring (2002, 2003).

From May 2004 to April 2005, a total of 21 individuals from research and development (R&D) departments, sustainability or environmental units and general management were interviewed. The main topics covered in the interviews were:

- The process of integrating sustainable and climate requirements into business planning and operations

- Central influencing or triggering factors for organisational sustainable and climate change-related learning

- The formal and informal organisational processes for sharing knowledge or exchanging information

- The manner in which projects and other sustainability-related activities were carried out

- Management structures and the contribution of the organisation's senior management and of other individuals or departments within the organisation

- External factors that influenced sustainability and climate protection activities

In most cases, an additional written survey was conducted to address follow-up questions.

The research also sought to identify changes in the organisations' processes, products and/or services directed at reducing greenhouse gas emissions such as:

- Actions to save energy

- The development of concepts and technologies for the use of renewable energies

- The establishment of management systems to reduce the climate change impact of processes and products

A particular area of focus was whether the changes made were incremental or radical. While incremental changes can be seen as limited adaptations in processes (e.g. the introduction of checklists, or ecological improvements of existing product concepts), radical improvements generally require far-reaching changes in processes and product concepts which may ultimately entail changing to completely new forms of satisfying customer needs in a more sustainable way (Schneidewind 1997; Mogalle 2000; Beschorner *et al.* 2005).

Findings

Overall findings

Table 18.1 summarises the key findings from the research. There are a number of points to be made about the results. The first is that, interestingly, all bar one of the companies could be described as having achieved double-loop learning. That is, the research findings suggest that organisational size or activity need not be an obstacle to companies moving beyond the single-loop learning model. Second, the form of organisational learning does not necessarily provide a good indication of the actions taken or the performance outcomes achieved. Rather, the actions taken seem to be strongly contingent on

TABLE 18.1 Summary of climate change-related actions and organisational learning processes

Type of change	Single-loop learning	Double-loop learning
Incremental	*German-based and internationally operating transportation company* ◉ Adoption of technical measures for reducing fuel consumption ◉ Project 'AirRail' (diversion of short-range trips to railway) ◉ Publication of sustainability reports	*Large international computer manufacturer* ◉ Adoption of a range of sustainability-related product innovations relating to waste management and disposal, energy efficiency and resource conservation ◉ Development of a sustainability mission statement *Large Netherlands-based electronics company* ◉ Implementation of various sustainability programmes, e.g. 'Green Flagships' ◉ Increased use of life-cycle analysis and eco-accounting ◉ Achievement of ISO 14001 certification ◉ Publication of sustainability reports
Radical		*Medium-sized German-based construction company* ◉ Renovation of houses to significantly reduce energy use ◉ Construction of eco-residential areas using solar energy ◉ Development of a range of eco-friendly product innovations (e.g. eco-houses fitted with solar panels, improved insulation and laminated or triple pane-windows) ◉ Adoption of ecological–social construction principles ◉ Development of operational management tools including checklists, whitelists, quality manuals and user guides ◉ Reporting and communication on sustainability *Large German-based public transport company* ◉ Adoption of ecological–social service business model, including cooperation with car-sharing programmes and allowing the free transportation of bicycles ◉ Establishment of stakeholder dialogues ◉ Increased use of natural-gas buses *Medium-sized German-based apartment management company* ◉ Energy-saving focus for the modernisation of pre-war buildings by developing a 'three-litre building' ◉ Development of sustainable concepts for living ◉ Increased stakeholder involvement in decision-making processes

organisational size, with the medium-sized companies appearing more likely to implement radical changes whereas the larger companies appear more likely to make incremental changes. Third, some organisations focused their efforts on actions related to environmental performance whereas others adopted a broader approach where social issues were seen as equally important as environmental or climate change-related outcomes.

Performance outcomes

Each of the companies achieved a range of performance outcomes (or set performance-related targets) directed at reducing their direct and indirect greenhouse gas emissions and/or contributing to innovations to reducing greenhouse gas emissions. Some of the key outcomes are examined here.

Although the two companies in the transport sector both took significant actions, the most notable difference is the focus on relative versus absolute emissions. The large transport company is aiming to improve its fuel efficiency by 38% between 1991 and 2012 through measures such as reducing the age of its airborne vehicles and shifting short-range trips from air to rail. In contrast, the large German-based public transport company has the objective of reducing its absolute level of greenhouse gas emissions. It intends to achieve this through a range of measures including the use of natural-gas buses (instead of diesel), the shortening of traffic light waiting times for buses, the adoption of an 'ecological–social service' business model that includes cooperation with car-sharing programmes, the free transportation of bicycles and the installation of photo-voltaic (PV) modules at its premises.

For the construction company, its most important activities (in terms of greenhouse gas emissions) are the renovation of houses to a level of passive or low-energy house usage and the construction of eco-residential areas with homes using solar energy or solar parks. Despite the absence of strong market demand, the company proactively developed numerous eco-friendly product innovations such as 'eco-houses' with innovations such as solar panels, high levels of insulation, laminated or triple-pane windows, and improved balconies. Even though the company has offered free information or orientation meetings to clients on these innovations, only a few clients have adopted these ideas. More generally, the company has implemented a series of other eco-friendly processes, including:

● Continuously updating its quality manuals regarding eco-guidelines which procedures and decisions must comply with

● Publishing a sustainability report and various user guides with information relating to resource consumption and environmental impact

● Establishing whitelists (i.e. lists of accepted items) for construction

● Separation, recycling and re-use of building wastes

The medium-sized German-based apartment management company, which restores and modernises pre-war buildings, has adopted a somewhat different strategy. In the course of its renovations, the company seeks to improve the energy efficiency of these buildings to achieve a 30 kWh/m^2 heat energy demand by equipping them with fuel cell technology to operate the multi-stage ventilating system, as well as other energy-saving measures such as heat recovery, improved insulation and latent heat storage systems. Through the adoption of these measures, greenhouse gas emissions are typically reduced by 15% compared with the pre-restored state.

In the ICT sector, the use phase of a product's life-cycle is generally the most significant from a climate change perspective. Most companies therefore seek to optimise the energy efficiency of their products. Beyond legal requirements (e.g. on the recovery and recycling of electrical and electronic equipment), both the large international computer manufacturer and the large Netherlands-based electronics company seek to implement ecological criteria in product development.

The Dutch electronics company runs extensive sustainability programmes. In its 'Green Flagships' programme, for example, eco-friendly criteria are established for particular areas of environmental focus such as weight, use, end-of-life disposal of hazardous substances, energy consumption, recycling and disposal, and packaging. Products achieve Green Flagships status based on their performance in a benchmark against their commercial competitors. Moreover, the company constantly seeks to improve its processes and products through the use of tools such as life-cycle analysis, eco-accounting and sustainability reporting. In addition, the company seeks to proactively communicate the importance of energy efficiency theme through actively engaging with its stakeholders and through responding to stakeholder needs and requirements in its products.

In a similar manner to its Dutch peer, the international computer manufacturer invests significant time and resources in green product development. The company has achieved certification to ISO 14001[2] and seeks to produce environmentally friendly computers through ensuring that they are totally recyclable, contain non-halogen units and have an energy-efficient power management system.

Linking performance outcomes and organisational learning processes

Despite the small sample size, the research allows some important conclusions on the relationship between organisational learning and climate change performance to be drawn.

Company size

Climate change-related actions seem to be strongly contingent on organisational size. From the research, the medium-sized companies surveyed seem to be better at implementing radical changes whereas the larger companies (even those that showed a double-loop learning mode) tend to focus on incremental changes.

From the interviews, it seemed that, while the large companies generally allocate more resources for R&D activities, the sheer size of the organisation is a challenge for change; a significant numbers of personnel, departments and locations need to be involved in the change process, which usually takes time and meets resistance. In contrast, the shorter communication lines and leaner administration generally found in medium-

---

2 Specification for Environmental Management Systems.

sized companies appears to enable companies to diffuse new knowledge and change processes rapidly.

This conclusion should not be taken as implying that medium-sized organisations are, as a whole, more likely to adopt far-reaching changes; the research focused on organisations with a track record of taking a proactive approach to action on climate change and which had well-evolved organisational learning processes.

Personnel structure

While the companies studied vary significantly in terms of the professional backgrounds and gender mix of their employees, the research did not identify any clear links between personnel profiles and either the organisational learning processes or the outcomes achieved. One point raised by a number of the interviewees was that interdisciplinary cooperation was beneficial for sustainability and climate change-oriented learning processes — in particular for understanding the relationship between economic, ecological and social issues.

Learning mechanisms

Interviewees from all six companies identified the importance of formal learning mechanisms in their organisational learning processes, though the different organisations structured these in very different ways. The specific structures adopted were strongly dependent on the pre-existing organisational structures and cultural attributes.

The following learning mechanisms were found in the sample companies.

- Goal- or guideline-oriented learning mechanisms where managers either set ambitious goals at the beginning of learning processes or focus on guidelines to achieve relevant improvements. Examples identified in the course of the research included:
  - The managers of the medium-sized apartment management company defined the energy standards to be met and specified several indicators for modernisation. These were communicated within the company and were set as a benchmark for various company units. These standards and performance indicators were supplemented by more detailed objectives and an implementation timeline
  - The large transportation company developed clearly defined mission statements and comprehensively communicated these across the entire company, thereby triggering company-wide learning processes
  - One of the electronics companies integrated ecological objectives into its product development processes. In this company, ecological criteria such as recyclability and the use of environmentally friendly materials are now key requirements for all new products
- Project work and self-organised working groups for learning processes, where companies use project and group designs to implement sustainability-related objectives. Examples included:
  - In one of the electronics companies, sustainability projects were used to support the development of new knowledge and opportunities

- One of the transport companies used self-organised working groups for the sustainability-oriented restructuring of one of its sites. Employees were asked to create and join groups for discussing and participating in the restructuring process. While the restructuring process itself was decided by the management, concrete implementation was delegated to the working groups

- Formalised instruments of communication where managers develop formal communication tools for supporting sustainability issues. For example:
  - The medium-sized construction company anchored ecological and social standards in the company mostly through top–down methods. Managers were involved in developing and communicating a handbook explaining company-specific ecological criteria. The handbook included a series of checklists that could be used by employees in the course of their day-to-day work. These standards were disseminated among employees, who were also provided with training on how to implement the standards

Values and norms

Sustainability-oriented values and norms manifest themselves in mission statements and well-developed (sustainability) reporting schemes (Alvesson 2005), particularly in large companies.

All six of the companies covered by this research were characterised by their willingness to:

- Take stakeholder demands seriously

- Address central public issues and concerns

- Provide constructive answers to these challenges

A number of the interviewees stated that the strongest driver for this stakeholder orientation in their company was the fear of damage to their reputation; this concern was particularly strong in former state-owned companies. When coupled with the tradition of these companies of accepting responsibility for common welfare, this fear led to distinctive learning and innovation efforts. It also led to two distinctive outcomes (Beschorner *et al.* 2005):

- These companies saw the social dimension as being as important as the environmental dimension of sustainability

- They tended to be more transparent than other companies in the same business field

A common theme across all six companies studied was that, while most sustainability-related values and norms were established at the top of the organisation, these top-down guidelines did not preclude bottom-up sustainability-related initiatives from emerging. In the international transport company as well as in both large electronics companies, for instance, environmental requirements were transformed into technological challenges by engineers, product developers and service personnel. These, in turn, led to the development of new ideas and approaches for responding to the environmental requirements as well as delivering on the organisation's core products or services. Thus, new values could be merged with existing traditions of technologically focused organisational cultures, forming the basis for further learning and developments.

An important question is whether an organisational focus on sustainability (or climate change specifically) will displace other organisational values or norms (e.g. commitments to public services, technical excellence, etc.). While it is clearly impossible to draw a general conclusion on this point from such a limited sample, there is some evidence that sustainability values can complement and integrate with other existing business values and priorities. This suggests that it should, at least in principle, be possible to align sustainability commitments with pre-existing organisational commitments and values.

Internal networks

Internal networks appeared to be an important vehicle for organisational learning, particularly in the large companies. Joint projects, conferences and virtual communities were all identified as important in allowing information to be exchanged and new sustainability-related knowledge to be created (see also Nonaka 1994; Tsang 1999; Morrison and Mezentseff 1997).

The importance of internal networks was highlighted in particular by the large international transport company and the electronics company, both of which noted that many environmental management activities were driven by a network of environmental change agents and employees concerned with environmental issues. These networks were used as a platform for the dissemination of new concepts, technologies and ideas, thereby allowing higher levels of learning to emerge.

Leadership style

The quality of sustainable learning and climate change-related improvements did appear to be related to a participatory leadership style, i.e. the more team members were involved in sustainability-related work and associated decision-making processes, the richer the ideas and solutions in a sustainable content were.

Senior managers at all six companies were actively involved in setting strategy and ensuring the importance of climate change as a business issue. With the exception of the medium-sized construction company, the leaders in all of the companies involved employees actively in the decision-making and change processes focussed on sustainability and climate protection. Although the management of the medium-sized company was less proactive in involving employees, the research did reveal cooperative relations between senior managers and other employees. That is, the overall effect was that employees in this company were also involved actively in the decision-making process – albeit on a less formal basis than in the other companies.

While employee participation was an important influence on the quality of sustainable learning in the six organisations studied, this does not mean that less participatory approaches would not deliver similarly positive outcomes; the sample size was not sufficiently diverse to allow such an analysis to be conducted.

The case studies suggest that management commitment is essential for the integration of sustainability aspects in the corporate culture. However, the research did not allow us to examine the converse question, i.e. whether management hostility to sustainability is an insurmountable barrier to sustainability.

Change agents

Confirming the findings of Klimecki (1997), the research suggests that change agents play a leading role in corporate sustainable learning processes — even if they are not in leading functions. In all the companies, individual change agents were of vital importance to the sustainability learning processes and the observed results.

In the medium-sized companies, change agents were mostly found in management positions, while in the large companies they were located in the sustainability or R&D departments. In all cases these individuals initiated — within the realms of possibility and their competences and abilities — concrete measures that, in turn, stimulated further learning processes and environmental programmes.

External factors

Across all six companies, external factors were important both in catalysing internal processes and for reinforcing actions and commitment — though the specific drivers for action (and the importance of the drivers) varied between the companies.

The large companies highlighted actual or anticipated stakeholder requirements as a critical driver for action, whereas the medium-sized companies suggested that internal factors were critical to accelerating their sustainability learning processes and actions (though acknowledging the importance of external drivers). Interestingly, regulation was highlighted only by one company — the public transport company — as a key driver, although this may reflect the timing of the interviews as European and associated national climate change policies were still emerging. The one other notable point in relation to the external drivers was that the former public companies noted that they continue to feel strong pressure from the general public and non-governmental organisations (NGOs) to take action on climate change.

The research highlighted that, at least to date, market (or customer) demands have had minimal influence on processes of learning and change. The electronics companies complained that their sustainability-related actions and their ecologically improved products were neither noticed nor appreciated by most consumers; their perception was that price is what matters to their customers and other quality factors are of secondary importance.

This lack of market demand may reflect the manner in which companies market their products. While large companies frequently possess significant sustainability-related knowledge, they often confine the marketing and sale of their 'green' products to specific market niches or segments. Most mass markets (still) do not show a sizeable demand for sustainability and climate change-compatible products, though in some segments the turnover is rising. This observation is not universally true as, for example, in Scandinavian markets, the electronics companies acknowledged that consumer buying decisions did appear to consider (and reward) green product characteristics.

As noted above, new laws and policies did not consistently play a vital role in initiating corporate sustainability in the six companies studied, although there were some cases where regulations helped reinforce corporate actions in particular EU regulations on the recycling of electronics (which supported the efforts of the electronics companies to introduce ecological product recovery strategies) and regulations on new energy and heating standards in housing that supported the construction companies' efforts in these areas.

Care should be taken with drawing more general conclusions on the role of regulation. The fact that the six companies are leaders in terms of climate change action means that they are in an early mover position and changes in regulation are less likely to present a significant challenge to them. This conclusion on the role of regulation would probably have been different if the study was focused on companies that were less proactive on climate change and sustainability issues.

Conclusions

The focus of this research was on how leadership organisations have responded to the impending threat of climate change regulation and the role of organisational learning in influencing these responses. From the research, four major conclusions emerge.

First, organisational size does not appear to be an influence on whether a company has single- or double-loop learning. However, organisational size is an important influence on whether an organisation has radical or incremental innovation, with smaller companies — if they decide to adopt a leadership position — more likely to have radical innovation.

Second, sustainability-oriented learning requires formal learning processes and structures, though the research did not allow conclusions to be drawn on which is best or most effective (e.g. goal-oriented, project work).

Third, change agents are of vital importance in corporate sustainability learning processes and in initiating climate-related improvements. These individuals exercise their influence through formal and informal internal networks, and can generate lasting impacts in particular within participatory styles of leadership.

Finally, external factors (threat to reputation, stakeholder expectations) are relatively important in starting and catalysing internal processes and climate-related actions, and for reinforcing actions and commitments.

The relative weakness of consumer demand was identified by a number of companies as a fundamental barrier to making their businesses and operations more sustainable. This conclusion suggests that policy-makers should not only consider regulatory (or command-and-control) responses to climate change but also the provision of incentives (e.g. through market-based instruments), the establishment of multi-stakeholder groups and similar processes to initiate dialogues and capacity-building, the development of consumer awareness of environmental and sustainability-related issues, and the encouragement of leadership initiatives and voluntary approaches as part of the overall policy mix.

References

Alvesson, M. (2005) *Understanding Organisational Culture* (London: Sage Publications).

Arce, D. (2006) 'Taking Corporate Culture Seriously: Group Effects in the Trust Game', *Southern Economic Journal* 73.1: 27-36.

Argyris, C., and D. Schön (1978) *Organizational Learning: A Theory of Action Perspective* (Reading, MA: Addison-Wesley).

—— and D. Schön (1996) *Organizational Learning II: Theory, Method, and Practice* (Reading, MA: Addison-Wesley).

Arnold, M. (2007) *Strategiewechsel für eine Nachhaltige Entwicklung: Prozesse, Einflußfaktoren und Praxisbeispiele* (Marburg, Germany: Metropolis).

Barabási, A. (2002) *Linked: the New Science of Networks* (Cambridge, MA: Perseus).

Bateson, G. (1972) *Steps to an Ecology of Mind* (New York: Ballantine).

Beschorner, T., T. Behrens, E. Hoffmann, A. Lindenthal, M. Hage, B. Thierfelder, and B. Siebenhüner (2005) *Institutionalisierung von Nachhaltigkeit: Eine vergleichende Untersuchung der organisationalen Bedürfnisfelder Bauen & Wohnen, Mobilität und Information & Kommunikation* (Marburg, Germany Metropolis).

Blatter, J. (2003) 'Beyond Hierarchies and Networks: Institutional Logics and Change in Transboundary Spaces', *Governance: An International Journal of Policy and Administration* 16.4: 503-26.

Brentel, H., H. Klemisch, H. and H. Rohn (eds.) (2003) *Lernendes Unternehmen: Konzepte und Instrumente für eine zukunftsfähige Unternehmens- und Organisationsentwicklung* (Wiesbaden, Germany: Westdeutscher).

Cramer, J. (2005) 'Company Learning about Corporate Social Responsibility', *Business Strategy and the Environment* 14.4: 255-66.

Crampton, S., J. Hodge and J. Mishra (1998) 'The Informal Communication Network: Factors Influencing Grapevine Activity', *Public Personnel Management* 27.4: 569-84.

Czarniawska, B. (1997) 'Learning Organising in a Changing Institutional Order: Examples from City Management in Warsaw', *Management Learning* 28.4: 475-95.

Dierkes, M., A. Berthoin Antal, J. Child and I. Nonaka (eds.) (2001) *Handbook of Organizational Learning and Knowledge* (London: Oxford University Press).

Dodgson, M. (1993) 'Organizational Learning: A Review of some Literatures', *Organization Studies* 14.3: 375-94.

Dutton, J., and R. Duncan (1987) 'The Creation of Momentum for Change through the Process of Strategic Issue Diagnosis', *Strategic Management Journal* 8: 279-95.

Elkington, J. (1997) *Cannibals with Forks: The Triple Bottom Line of 21st Century Business* (Oxford, UK: Capstone Publishing).

Finger, M., S. Bürgin and U. Haldimann (1996) 'Ansätze zur Förderung Organisationaler Lernprozesse im Umweltbereich', in M. Roux and S. Bürgin (eds.), *Förderung umweltbezogener Lernprozesse in Schulen, Unternehmen und Branchen* (Basel: Birkhäuser): 43-70.

Fiol, C., and M. Lyles (1985) 'Organizational Learning', *Academy of Management Review* 10.4: 803-13.

Hartley, J., J. Benington and P. Binns (1997) 'Researching the Roles of Internal-change Agents in the Management of Organisation Change', *British Journal of Management* 8.1: 61-73.

Hertin, J., F. Berkhout and R. Haum (2002) *Business and Climate Change: Measuring and Enhancing Adaptive Capacity. Progress Report: Preliminary Results from the House Building and Water Sector* (Brighton, UK: SPRU/Tyndall Centre for Climate Change Research).

Huber, G. (1991) 'Organizational Learning: The Contributing Processes and the Literatures', *Organization Science* 2: 88-115.

Klimecki, R. (1997) 'Führung in der Lernenden Organisation', in H. Geißler (ed.), *Unternehmensethik, Managementverantwortung und Weiterbildung* (Neuwied, Germany: Luchterhand): 82-105.

—— and H. Lassleben (1998) 'Modes of Organisational Learning: Indications from an Empirical Study', *Management Learning* 29.4: 405-30.

Mayring, P. (2002) *Einführung in die Qualitative Sozialforschung* (Weinheim, Germany: Beltz).

—— (2003) *Qualitative Inhaltsanalyse, Grundlagen, Techniken* (Weinheim, Germany: Beltz, 8th edn).

Mintzberg, H., B. Ahlstrand and J. Lampel (2005) *Strategy Safari: A Guided Tour through the Wilds of Strategic Management* (New York: Prentice Hall).

Mogalle, M. (2000) *Der Bedürfnisfeld-Ansatz: Ein handlungsorientierter Forschungsansatz für eine transdisziplinäre Nachhaltigkeitsforschung* (St Gallen, Switzerland: Institut für Wirtschaft und Ökologie der Universität St Gallen).

Morrison, M., and L. Mezentseff (1997) 'Learning Alliances: A New Dimension of Strategic Alliances', *Management Decision* 35.5: 351-7.

Müller, M., and B. Siebenhüner (2007) 'Policy Instruments for Sustainability-Oriented Organisational Learning', *Business Strategy and the Environment* 16: 232-45.

Mutch, A. (2002) 'Actors and Networks or Agents and Structures: Towards a Realist View of Information Systems', *Organization* 9.3: 477-96.

Nonaka, I. (1994) 'A Dynamic Theory of Organizational Knowledge Creation', *Organization Science* 5: 14-37.

——, R. Toyama and P. Byosière (2001) 'A Theory of Organisational Knowledge Creation: Understanding the Dynamic Process of Creating Knowledge', in M. Dierkes, A. Berthoin Antal, I. Nonaka and J. Child (eds.), *Handbook of Organizational Learning and Knowledge* (Oxford, UK: Oxford University Press): 491-517.

Popper, M., and R. Lipshitz (1995) *Organizational Learning Mechanisms: A Structural/Cultural Approach to Organizational Learning* (Haifa, Israel: University of Haifa).

—— and R. Lipshitz (2000) 'Organizational Learning: Mechanisms, Culture and Feasibility', *Management Learning* 31.2: 181-96.

Probst, G., and B. Büchel (1997) *Organizational Learning: The Competitive Advantage of the Future* (London: Prentice Hall).

Rajagopalan, N., and G. Spreitzer (1996) 'Toward a Theory of Strategic Change: A Multilens Perspective and Integrative Framework', *Academy of Management Review* 22: 48-79.

Robertson, P., and N. Langlois (1994) 'Institutions, Inertia and Changing Industrial Leadership', *Industrial Corporate Change* 3.2: 359-78.

Rothman, J., and V. Friedman (2001) 'Identity, Conflict, and Organizational Learning', in M. Dierkes, A. Berthoin Antal, I. Nonaka and J. Child (eds.), *Handbook of Organizational Learning and Knowledge* (Oxford, UK: Oxford University Press): 582-97.

Schein, E. (1984) 'Coming to a New Awareness of Organizational Culture', *Sloan Management Review* 25.2: 3-16.

Scherer, F. (1992) 'Schumpeter and Plausible Capitalism', *Journal of Economic Literature* 30: 1,416-33.

Schneidewind, U. (1997) *Wandel und Dynamik in Bedürfnisfeldern- Wesen und Gestaltungsperspektiven. Eine strukturalistische Rekonstruktion am Beispiel des Bedürfnisfeldes Ernährung. IP Gesellschaft I – Diskussionsbeitrag Nr. 2* (St Gallen, Switzerland: Institut für Wirtschaft und Ökologie an der Universität St Gallen).

Schumpeter, J. (1954) *History of Economic Analysis* (New York: Oxford University Press).

Schunk, D. (1996) *Learning Theories: An Educational Perspective* (New York: Merrill, 2nd edn).

Shrivastava, P., and S. Hart (1995) 'Creating Sustainable Corporations', *Business Strategy and the Environment* 4: 154-65.

Sroufe, R., and J. Sarkis (eds.) (2007) *Strategic Sustainability: The State of the Art in Corporate Environmental Management Systems* (Sheffield, UK: Greenleaf Publishing).

Teece, D., R. Rumelt, G. Dosi and S. Winter (1994) 'Understanding Corporate Coherence: Theory and Evidence', *Journal of Economic Behavior and Organization* 23: 1-30.

Tsang, E. (1999) 'A Preliminary Typology of Learning in International Strategic Alliances', *Journal of World Business* 34.3: 211-29.

Walsh, T. (2006) *Climate Change: Business Risks and Solutions* (Risk Alert 5.2; London: Marsh; global.marsh.com/risk/climate/climate/documents/climateChange200604.pdf [accessed 6 June 2008]).

Yin, R. (1994) *Case Study Research: Design and Methods* (Thousand Oaks, CA: Sage).

Yuthas, K., J. Dillard and R. Rogers (2004) 'Beyond Agency and Structure: Triple-loop Learning', *Journal of Business Ethics* 51: 229-43.

19

Fasten your seatbelts: European airline responses to climate change turbulence

Christian Engau, David C. Sprengel and Volker H. Hoffmann
ETH Zurich, Switzerland

Even though climate change has been an environmental concern for many years, its impact on economic activities has only recently been fully recognised. In this context, the launch of the European Union Emission Trading Scheme (EU ETS) marked an important turning point (Hoffmann 2007). For the first time, the entire business community was systematically exposed to the costs of emitting greenhouse gas emissions — whether directly through their own greenhouse gas emissions or indirectly through increased costs of energy and other inputs. This, in turn, has created pressure for companies to manage their emissions as a financial asset (Egenhofer 2007). In parallel, public awareness of climate change issues has been increasing, with increasing stakeholder demands for companies to take action to reduce their greenhouse gas emissions. Companies have responded to these climate change-related pressures in markedly different ways (Kolk and Levy 2001), though the reasons for these different responses are not clear. Yet, for the design of future climate policies such as the further development of the EU ETS, it is crucial to understand both how companies have responded and the reasons why greenhouse gas emission reduction performance has varied.

In this chapter, we use the specific case of the European airline industry to examine the actions companies take in response to climate change-related pressures and the reasons why these actions vary between companies. The European airline industry represents a particularly interesting case study for three reasons.

First, the climate change debate in early 2007 arguably focused on the airline industry more than on any other industry in Europe, with airlines finding themselves caught in the crossfire between various stakeholders such as governments, non-governmental organisations (NGOs) and the media for significantly contributing to climate change.

Second, as emissions from aviation were not included in the EU ETS in mid-2007, the time of our analysis, and there still was uncertainty regarding a future inclusion of the airline industry in the scheme, the role of stakeholder pressures was of high importance for climate change activities of airlines.

Third, the companies in the sector have adopted a large diversity of positions and activities ranging from vehement opposition to enthusiastic support. For example, while Ireland's Ryanair opposed the extension of the scope of the EU ETS to include aviation, UK-based Virgin Atlantic heavily promoted it and even announced its intention to spend its entire profits over a period of ten years on climate change research and the development of green technologies. Similar differences were seen in customer communications: the Dutch airline KLM, in its in-flight magazine, highlighted the effects of carbon dioxide (CO_2) emissions released in the upper atmosphere and the effects of other aircraft emissions on global warming, whereas Germany's Air Berlin, in its in-flight magazine, challenged the benefits of reducing the aviation industry's CO_2 emissions on the global climate.

This chapter is divided into three main sections:

- An overview of the debate around European airlines and climate change

- A brief description of the research we conducted

- The key findings from our research

European airlines and climate change

New challenges such as climate change, the inclusion of aviation in the EU ETS and pressures to reduce CO_2 emissions have conquered the top agenda of European airline executives almost overnight. Owing to the success of the low-cost business model, resulting in declining air fares and, in turn, an increasing demand for air travel, aviation is one of the fastest-growing sources of CO_2 emissions in Europe (EC 2006b). Aviation emissions have increased by almost 90% since 1990 and today account for about 3% of total CO_2 emissions in the EU (EC 2007). However, due to indirect effects, scientists estimate aviation's total contribution to climate change to be at least two times as high (Sausen *et al.* 2005). These indirect effects are caused by other aircraft exhaust gases such as nitrogen oxides (NO_x) or water vapour which, if released at high altitudes, can form contrails[1]

---

1 An aircraft's NO_x emissions increase the concentration of ozone in the atmosphere, while at the same time reducing that of methane. Both ozone and methane are greenhouse gases and changes in their atmospheric concentration affect global warming. Despite the contrary influence of NO_x emissions on ozone and methane, there is agreement among scientists that the net effect is negative and amplifies global warming. Similarly, aircraft contrails have two opposing effects on the climate. While they reduce the Earth's temperature by reflecting the light of the Sun, contrails also increase the greenhouse effect by reducing radiative forcing to space. However, the net effect of contrails on climate change (which depends on many parameters including their particle composition, structure, cover and location) and the influence of aviation on the formation of cirrus clouds are highly uncertain and still subject to research.

(IPCC 1999). Furthermore, European air traffic is expected to grow by 4–5% annually over the next years, with airlines' fuel efficiency improvements only partly compensating for this growth (IATA 2007c). As a result, the contribution of aviation to total CO_2 emissions in the EU will multiply, cancelling out a large part of the emission reductions achieved by other sectors (EC 2006a).

An intense debate has therefore developed on the measures that could be adopted to reduce the climate change impact of aviation. These range from abolishing the tax exemption for jet fuel to overriding the exclusion of aviation from the Kyoto Protocol and extending the scope of the EU ETS to aircraft emissions. Even though the European Commission signalled its intention to act by announcing plans to include the airline industry in the EU ETS in September 2005 (EC 2005), the controversy continued with the Commission, the European Parliament's Environment and Transport Committees, airline associations, individual airlines and NGOs taking different positions towards the design of such a regulation. The debate provisionally culminated in the adoption of a first reading report by the European Parliament in November 2007, corroborating the intention to include CO_2 emissions from the airline industry in the EU ETS but leaving many uncertainties regarding the final details (EC 2007).[2]

Research objectives and methodology

Policy-makers face the complex task of designing climate policies that effectively regulate the contribution of companies to climate change but do not significantly interfere with their economic competitiveness. In turn, companies have to respond appropriately to pressures from climate change without endangering their long-term financial performance.

Hence, developing a detailed understanding of corporate responses to climate change is crucial for both policy-makers and companies seeking sustainable performance. Our research sought to contribute to this understanding by analysing the responses of different airlines to climate change and to identify underlying rationales for the diversity of these responses.

In order to obtain comprehensive insights into the determinants of airline responses to climate change, we focused our attention on passenger airlines covered by the latest Commission proposal, which provides for the extension of the EU ETS to any flight departing from or arriving at an airport in the EU (EC 2007). We examined the public pressures and the political attention toward all airlines with respect to their direct CO_2 emissions as well as the actual perception of pressures by an airline's managers of being in the midst of public attention.

We conducted our research in two steps. First, we prepared background reports for each airline based on both publicly available information from various data sources and discussions with industry experts. In addition, we collected advertisements, leaflets and other communications distributed by each airline to generate a comprehensive overview

2 As our analysis was conducted in mid-2007, any subsequent developments regarding aviation and the EU ETS are excluded.

of its activities related to climate change and a possible CO_2-related emission regulation. This background research strongly confirmed the initial impression of very diverse activities and positions of airlines with regard to the climate change debate. Second, in order to investigate underlying motivations, we conducted in-depth interviews with executives from each of the airlines.

Our final sample comprised nine airlines from five European countries. In order to cover a broad range of different stakeholder groups and study a diverse sample of airlines, we chose cases encompassing different business models. These included national flag carriers as well as leading and niche players from the low-cost, regional and charter sectors.

We conducted interviews with 2–4 representatives from each airline during the period mid-July to early October 2007. To cover different aspects relevant to an airline's decision on how to respond to climate change-related pressures, we interviewed managers from three functions — usually executives responsible for strategy (either a member of the management board or a respective direct report), heads of communications or of the departments responsible for external and regulatory affairs, and heads of environmental departments or, if such a unit did not exist, technical divisions. The results from these interviews and our preparatory research are described in the next section.

Results

Our analysis of the results is divided into two parts. In the first we illustrate the actual responses of airlines to pressures to reduce their CO_2 emissions. In the second part we analyse possible reasons for differences in their responses.

Responses to pressures to reduce CO_2 emissions

Even though airlines had only been exposed to pressures to reduce their CO_2 emissions for a short time when we conducted our research, most of the emission reduction measures were not new to them. In fact, all airlines were pursuing a variety of emission reduction measures, although these were typically directed at reducing fuel costs by improving their fleet's fuel efficiency.

But with every tonne of burnt fuel resulting in the emission of 3.16 tonnes of CO_2,[3] measures to reduce an aircraft's fuel consumption also reduce the amount of greenhouse gases emitted. In general, fuel costs constitute an airline's second largest cost after labour costs, usually representing 10–15% of total operating costs (Doganis 2006) though the upsurge in oil prices has driven this share up to 20–25% in 2007 (IATA 2007a). Since the fuel price is set externally and virtually impossible to control by an airline in the long term, the only way to respond to resulting fuel cost pressures is to reduce fuel consumption.

3 The calculation is based on data from the US Government's Energy Information Agency and the International Carbon Bank & Exchange's carbon database for the emission coefficient for Jet A-1, which is the most commonly used jet fuel in civilian aviation in Europe.

But, despite the associated fuel cost savings, measures to reduce CO_2 emissions are not always beneficial to airlines. Some emission reduction measures actually lead to higher total costs, mainly as a result of higher equipment wear and increased maintenance requirements (Airbus 2004). Likewise, reducing flight speed typically reduces fuel burn and emissions, but also extends flight time, consequently increasing labour costs through extending the working hours of the flight crew and reducing fleet productivity by restricting the number of an aircraft's flight movements. Finally, reducing an aircraft's CO_2 emissions often leads to higher noise levels and, more importantly in the context of climate change, may also increase NO_x emissions.

Hence, while actions to save fuel costs can contribute to an airline's efforts to reduce its CO_2 emissions, there also are aspects with potentially negative effects for the business. Consequently, one would expect airlines with the reduction of CO_2 emissions as a strategic objective (or those with a more proactive approach in the climate change debate) to be pursuing more emission reduction measures than those competitors that would reduce emissions only if the measures were also cost-efficient.

In our research, we found that some of the measures apparently directed at responding to pressures to reduce CO_2 emissions (e.g. the installation of winglets to increase lift or the replacement of seats with lighter models to reduce weight) were rather in response to economic pressures from high fuel prices. This was confirmed by the environmental representative of an aircraft maintenance provider that carries out technical services for many large European carriers:

> Airlines do not consider environmental aspects when requesting maintenance services. [. . .] They are mainly cost-driven.

But, even though such measures were generally directed at reducing costs, some companies presented them as examples of their commitment to reducing CO_2 emissions, i.e. while some measures were genuinely directed at reducing emissions, others were only communicated accordingly. That is, airlines also used communications to respond to CO_2 emission reduction pressures by directly addressing their stakeholder groups. One executive stated:

> Advertisements and customer communication are part of the overall picture. [. . .] Thereby, we demonstrate quality in the environmental dimension.

Thus, the analysis of corporate responses to pressures to reduce direct CO_2 emissions needed to recognise that these responses included two distinct components:

- Measures that reduce CO_2 emissions
- Activities directed at informing stakeholders

Reduction responses

Driven by economic pressures to reduce fuel consumption and by ecological pressures to reduce CO_2 emissions, airlines typically engaged in three types of measures: technical measures (e.g. the adoption of new technologies); procedural measures (e.g. changes of in-flight processes); and lobbying for structural changes (e.g. encouraging infrastructure improvements). Table 19.1 shows a consolidated list of the major measures for each type.

TABLE 19.1 Examples of measures to reduce CO_2 emissions in the airline industry

(continued opposite)

Source: Airbus 2004; Boeing 2004; IATA 2004

Category	Activity type	Examples
Technical	Efficiency improvement	● Fleet renewal with more fuel-efficient aircrafts
	Weight reduction	● Reduction of aircraft operating empty weight, e.g. by using carbon fibre materials for cabin fittings or replacing seats ● Reduction of excess weight by maintaining surface finish, e.g. through periodic aircraft washing
	Technical maintenance	● Reduction of excrescence drag, e.g. by eliminating surface misrigging, aerodynamic seal deterioration or skin–surface mismatches ● On-wing engine washing and bleed rigging to prevent engine deterioration ● Calibration of instruments, e.g. speed measuring equipment
	Technological development	● Installation of winglets to increase lift ● Frequent updates of flight planning systems, e.g. flight management or bad weather detection ● Development of less carbon-intensive aircraft fuel
Procedural	Flight planning	● Optimisation of route through accurate flight planning based on reliable data[a] ● Application of accurate fuel demand calculation software to reduce fuel reserves carried ● Optimisation of aircraft loading to reduce lift required ● Reduction of weight of operational items, e.g. by limiting service items or removing unneeded emergency equipment ● Avoidance of fuel tankering to reduce weight
	Ramp operations	● Reduction of engine running time at the gate, e.g. by coordinating engine start-up with air traffic control (ATC) departure schedule or minimising the use of auxiliary power unit if ground power is available[a]
	Taxiing	● Reduction of taxiing, e.g. by towing aircraft closer to runway before take-off, optimising taxi routes or using minimum thrust and braking[a]
	Take-off and climb	● Optimisation of take-off flap setting to reduce drag and enhance climb performance ● Application of aircraft-specific optimal climb technique[a]

[a] The execution of these measures might require approval from air traffic control.

TABLE 19.1 (from previous page)

Category	Activity type	Examples
Procedural (continued)	Cruise	● Flight conducted in managed mode rather than in selected mode
		● Adjustment of flight management system to most fuel-efficient cost index
		● Application of fuel-saving techniques, e.g. use of well-timed step climbs to fly at optimal altitude or optimisation of speed[a]
		● Application of wind–altitude trade tables to maximise ground fuel mileage by exploiting wind[a]
		● Reduction of drag by using proper lateral–directional trim procedures and maintaining optimal centre of gravity location
	Descent and landing	● Introduction of continuous descent approach with idle thrust[a]
		● Deferment of transition to landing configuration
		● Flying at green dot speed during holding patterns[a]
	Non-flight behaviour	● Maximisation of load factors to reduce specific emissions
		● Offsetting of own emissions
		● Offer of voluntary offsetting options to passengers
		● Training of staff on emission reduction techniques
Structural[a]	Infrastructure	● Expansion of airport capacities to reduce holding patterns[b]
	Regulation	● Reduction of airspace barriers, e.g. completion of Single European Sky or Transatlantic Common Aviation Area[b]

[b] Not pursued by airlines directly, but pushed for through lobbying efforts.

All airlines in our study engaged in ongoing technical measures to reduce CO_2 emissions by increasing their fuel efficiency. Measures such as fleet renewal programmes or the installation of winglets were frequently mentioned in our interviews, and there were no notable differences between the nine airlines in our sample.

Moreover, airlines increasingly initiated procedural measures. However, most did so only if the procedures did not compromise other objectives such as profitability or punctuality. Procedural measures pursued beyond those generally implemented by all of the airlines were mainly directed at achieving mind-set changes among their employees. For example, the Environmental Director of a flag carrier explained:

> The aircraft itself is the most important component regarding CO_2 emissions. Since the pilots are flying it, they also play an important role and have to be sensitised for the topic [of reducing CO_2 emissions]. Also, all other employees have to be included as well [. . .].

But, while measures such as specific pilot training for fuel-efficient flights or employee workshops to initiate energy-saving programmes were seen as contributing to the airlines' efforts to reduce CO_2 emissions, the actual contribution tended to be small and the main objective was to increase awareness of the importance of reducing CO_2 emissions in general. A Vice President of another flag carrier stated:

> We are currently pushing for a cultural turnaround at [our airline].

And her colleague responsible for communication added:

> Our crews need to have the right mind-set to also consider environmental aspects.

In sum, however, the engagement of airlines in procedural improvements did not differ considerably, particularly when considering the overall effect of these measures on their CO_2 emissions.

Finally, all airlines also engaged in lobbying for structural measures. By doing so, airlines did not directly reduce their own CO_2 emissions, but they pushed for general preconditions facilitating or even enabling an actual reduction. For example, control boundaries in the European airspace are organised along national borders, thus forcing aircrafts to take detours which cause higher emissions. Airlines claim that establishing a single European airspace could considerably reduce CO_2 emissions from the airline industry. However, many countries are reluctant to give up control over their national airspace to an EU-wide organisation (Paylor 2007). A charter airline's Vice President responsible for external affairs told us:

> Our [airline] association has been pushing Single European Sky in Brussels for years now. But nothing has happened so far, although this would make it possible to reduce CO_2 emissions by 10–15% at once.

Even with their different business models, countries of origin and sizes, all airlines were lobbying for broadly similar structural measures.

Despite the comprehensive reduction responses pursued by European airlines, it proved difficult to determine which of the CO_2 emission reduction measures were actually initiated due to recent climate change-related pressures (as opposed to measures that were directed at improving fuel efficiency or reducing costs).

More importantly, it was not possible to measure the actual CO_2 reductions achieved due to these measures given the number and the complexity of factors influencing an aircraft's emissions and the small reductions achievable compared with the emissions still produced by any single flight. In particular, external factors such as unfavourable weather conditions, delayed arrivals or departures of other aircrafts, or unexpected hold-ups due to infrastructure bottlenecks can considerably increase fuel burn and, thus, nullify any emission reductions achievable by most of the procedural reduction measures listed above.

However, despite the lack of quantitative data, our qualitative assessment was that the reduction responses in the European airline industry did not significantly differ between the individual airlines in our study. Rather, all airlines engaged in CO_2 reduction measures to a similar extent. A board member of a flag carrier offered a similar conclusion:

Because of the high competition and the large number of regulatory requirements, the possibilities for an individual airline to deviate from the industry
standard in terms of fuel-saving activities are very limited.

Communication responses

In contrast, the communication responses we observed varied widely in terms of both
intensity and objective. First, airlines engaged in communication to a different extent.
While some airlines used a large variety of communication means including in-flight
magazine articles, newspaper advertisements, website information and sustainability
reports, others did not participate in the climate change debate at all. One Director of
External Affairs, when we asked about the relative lack of communication by his airline
on climate change, commented:

> What other airlines communicate is true for the entire industry. If we did that
> too, it would be a waste of resources.

Second, airlines pursued different objectives with some of the more active airlines
seeking to influence stakeholder opinions (and, hence, the pressure exerted by these
stakeholders). For example, a Vice President of a charter airline said:

> The reason [for the planned communication] is the pressure from the depic-
> tion in the media [. . .].

Similarly, the head of a flag carrier's environmental department stated:

> We also do [advertisements and customer communication] to be less defen-
> sive and adjust our image among customers and politics.

However, some airlines did not design their communication response to influence
pressures to reduce CO_2 emissions but instead used it to substantiate their reduction
activities by providing fact-based information. A Vice President stated:

> The reason for [our airline's] reaction and positioning is our belief that envi-
> ronmental responsibility will be a long-term driver for customer decisions, and
> that [our airline] can create a competitive advantage.

Her colleague responsible for external affairs later added:

> I am not aware of any actions that we have taken in response to pressures from
> the outside. [. . .] We have been a little defensive in communicating our efforts
> to the public [. . .], because we have always dealt with sustainability issues, for
> us it was more a hygiene factor.

Linking communications and emission reductions

As illustrated in Table 19.2, airlines have responded to climate change in two different
ways:

● Through emission reductions

● Through communicating their (emissions) performance

TABLE 19.2 Airline responses to CO$_2$ emission reduction pressures (continued opposite)

Airline	Reduction response	Communication response	Main communication objectives
A	Driven by technical efficiency improvements, in addition lobbying for elimination of structural deficits: 'We already got rid of almost all old airplanes, and that costs us a lot of money. We do this mainly for economic reasons which are then beneficial to the environment, too.'	Hardly related to climate change, mostly in conjunction with economic aspects. Reserved public statements regarding climate change in general: 'We are continuously telling our customers that we have the most up-to-date aircraft.'	Limited engagement in climate change debate and emphasis on efficiency to support its low-cost image: 'What other airlines communicate is true for the entire industry. If A did that too, it would be a waste of resources.'
B	Driven mostly by technical efficiency improvements, with behavioural changes among staff gaining momentum: 'Our pilots are also independently dealing with this topic, on a very detailed technical level.'	No external climate change-related communication, but with reports and campaigns in preparation: 'Of course we cannot sell that flying is environmentally friendly. Our communication is probably already good, but there is still room for improvement.'	Elimination of public information deficit and catching up with competitors: 'There is a huge misconception regarding the actual contribution of the airline industry to climate change and the respective counter-measures.' 'For us, the main reason to think about this is the respective activity of our competitors . . . We feel that this forces us to act.'
C	Activities only motivated by cost considerations: 'At C, we have so far considered climate change from a cost perspective as far as possible.'	No climate change-related communications, but external communications being prepared: 'Climate change has not been taken up from a marketing perspective yet.' 'Because of the ongoing public debate, we are currently preparing comprehensive information for our external communication.'	Satisfaction of external pressures resulting from climate change debate: 'The reason [for the planned communication] is the pressure from the depiction in the media, not from our customers.'

TABLE 19.2   (from previous page; continued over)

Airline	Reduction response	Communication response	Main communication objectives
D	Extensive efforts, ranging from efficiency improvements to procedural changes in both flight and non-flight activities: 'Climate change and emission reduction have strategic relevance for D.' 'Our crews need to have the right mind-set to also consider environmental aspects. This includes things like getting the aircraft off the gate on time, avoiding speeding even when running late, or proactively asking for the optimal altitude.'	Comprehensive reporting on environmental issues, but very limited external climate change marketing or advertising. Broad internal communication via multiple channels, openly supported by top management: 'We don't enter the fight in the media, we simply refer to our sustainability report . . . Therein, we communicate objective information.' 'We inform [our employees] a lot on how to save emissions, also with simple examples such as changing light bulbs.'	Reporting activities seek to satisfy pressures, mainly from corporate customers and financial markets. Internal communications seek to support mind-set change: 'The reason for D's reaction and positioning is our belief that environmental responsibility will be a long-term driver for customer decisions, and that D can create a competitive advantage.'
E	Driven mostly by technical efficiency improvements: 'The [climate change] problem has to be solved technically. We have continuously improved our energy efficiency.' 'Costs are always our most important driver.'	Thorough general environmental reporting, but only limited concrete information: 'We would not publish multiple-page advertisements in our board magazine, mainly because of the costs, but also because of the content.'	Meeting information requests from financial markets: 'For E, the financial community is the *raison d'être*. Our product is the share, and our customer is first of all the shareholder. The most important thing is to send the right signals in our annual report.'
F	Mostly driven by technical efficiency improvements. In addition, intensive lobbying for elimination of structural deficits: 'Of course we order a new fleet to realise growth — this is what we owe our shareholders. But we also purchase the most efficient aircraft.'	Comprehensive reporting and communication activities, directed both internally and externally: 'With our advertisements and brochures, we consciously took a less emotional, but more data- and fact-based approach.' 'We use a variety of channels to inform and sensitise our employees.'	Moving climate change debate to a scientific basis, based on constructive recommendations and build-up of legitimacy: 'Advertisements and customer communication are part of the overall picture, also when it comes to quality. We also do this to be less defensive and adjust our image among customers and politics. Thereby, we demonstrate quality in the environmental dimension.'

TABLE 19.2   (from previous page)

Airline	Reduction response	Communication response	Main communication objectives
G	Activities only motivated by cost considerations: 'Besides cost pressures that lead to continuous optimisation, we don't have pressure to reduce CO_2 emissions.'	No climate change-related communication: 'Until now, nothing has changed [because of the climate change debate] . . . If we had pressures, we could also take out advertisements like our competitors.'	No need for climate change-related communication: 'Climate change is not a priority, even though somebody has to deal with it [internally] now. Otherwise, nothing has changed.'
H	Driven mainly by technical efficiency improvements, but also including flight-related measures and lobbying: 'The driver for environmental activities of airlines is not the environment. The driver is clearly an economical one.'	So far only minor sustainability reporting, but external communications in preparation: 'We are preparing to launch a broad environmental campaign to communicate the most important matters . . . Our competitors are not sleeping; admittedly we are running a little bit late here.'	Improvement of environmental image to catch up with competitors and to establish political legitimacy: 'In discussions [with politicians or customers], we often come across a lack of information, which our communication certainly also is directed at.' 'If our campaign leads to a more fact-based discussion and better political decisions, this eventually also results in ecological effects.'
I	Wide-ranging efforts, including efficiency improvements, internal procedural changes and own offsetting activities: 'For example, we have a large fleet renewal programme, we have energy-saving measures for our planes, and we do engine washing to reduce our emissions.' 'We also had 60 brainstorming sessions with employees to generate ideas on how to reduce emissions.'	Detailed sustainability reporting, comprehensive internal and sporadic external communication: 'We use training, inspiration sessions and an employee magazine for our internal communication.' 'Maybe we have done too little in terms of [externally] communicating our measures in the past . . . We have to communicate the content of what we do more proactively.'	Demonstration of trustworthy approach to establish green credibility towards customers: 'If we want to become Europe's leading airline, we also have to be the leading green airline.' 'If we survey customers, they do not necessarily know what our airline is actually doing.'

Although the emission reduction activities of the airlines were very similar (due to the underlying common objective of increasing fuel efficiency), their communication responses differed considerably. Indeed, the CEO of a regional airline explained the differences in airlines' responses as follows:

> The response [to emission reduction pressures] depends on the airline's possibilities to influence politics and the pressures.

Furthermore, to emphasise the discrepancy between airlines' reduction and communication responses, the Vice President of Corporate Communications of a competitor concluded:

> I know what other airlines are communicating with respect to their activities. But I think that there is a lot of greenwashing going on.

Reasons for diversity in communication responses

Our research started with an examination of the different factors (e.g. stakeholder pressures, environmental uncertainty, company capabilities, perception of opportunities and threats) identified in the academic literature as affecting a company's environmental strategy. But, as our research evolved, we found that two factors dominated the airlines' thinking on how to respond to climate change: stakeholder pressures and the threat of environmental regulation. Before we discuss how these factors actually influenced corporate responses, we first sketch out some of the key concepts from the literature.

There is a significant literature on how different stakeholder groups influence corporate environmental responses and strong evidence showing that pressures arising from these groups may affect a company's level of environmental proactiveness and environmental strategy (Henriques and Sadorsky 1999; Buysse and Verbeke 2003; Sharma and Henriques 2005). While different stakeholder definitions exist in stakeholder theory, one of the broadest and most prominent definitions was developed by Freeman (1984: 46):

> A stakeholder in an organisation is (by definition) any group or individual who can affect or is affected by the achievement of the organisation's objectives.

Building on this definition, the stakeholder groups most commonly used in empirical studies of natural environmental issues are customers, regulatory agencies, financial markets, non-governmental organisations (NGOs) and the media (Henriques and Sadorsky 1996, 1999; Sharma and Henriques 2005). Kolk and Pinkse (2007) suggested that corporate responses to climate change issues not only depend on the perceived pressures to take respective action, but also on the perceived salience of the stakeholder groups exerting these pressures. According to Mitchell *et al.* (1997: 871):

> It is the firm's managers who determine which stakeholders are salient and therefore will receive management attention.

Hence, we expected the interviewees to view stakeholder groups differently in terms of salience. Consequently, one objective of our analysis was to identify salient stakeholder groups and to understand the managers' perceived levels of pressure to reduce CO_2 emissions from each of these groups.

In relation to environmental regulation, Aragón-Correa and Sharma (2003) noted that companies may see environmental regulations as posing either an opportunity or a threat to their business. Reflecting the argument above that it is managers' perceptions that determine corporate responses, we sought to understand the managers' interpretations of the pressures to reduce direct CO_2 emissions.

To analyse the information gathered, we drew on the strategic issue interpretation literature. According to the model presented by Daft and Weick (1984) in which scanning (data collection) is followed by interpretation (data given meaning) and learning (action taken), it is the interpretation of the data that shapes managers' actions. According to Dutton and Jackson (1987), individuals apply categorisation in order to reduce complexity, thus organising objects into meaningful groups. One categorisation dimension is that of 'opportunity and threat'. Three attributes differentiate these two categories:

> The opportunity category implies a *positive* situation in which *gain* is likely and over which one has a fair amount of *control*; in contrast, the threat category implies a *negative* situation in which *loss* is likely and over which one has relatively *little* control (Dutton and Jackson 1987: 80).

More specifically, scholars have found that managers' interpretations of environmental issues as opportunities or threats affected their company's environmental strategy (Sharma *et al.* 1999; Sharma 2000). Thus, we were interested in executives' interpretations of the pressures arising from climate change as opportunity or threat as a possible determinant of corporate activities in relation to climate change.

Given the prominent role of regulation for a company's environmental actions, we first analysed the influence of regulatory stakeholders on the diverse communication responses of the airlines. Our case studies showed that pressures exerted by regulatory stakeholders were consistently perceived to be high across airlines. This observation seems plausible as the proposal by the European Commission to include European airlines into the EU ETS was equally valid for all airlines analysed. Although airlines were not covered by an emission reduction regulation at the time of our study (July–October 2007), the threat of an imminent regulation led airlines to take action to demonstrate their environmental awareness and responsiveness in order to legitimise their lobbying activities. Hence, the airlines already experienced some regulatory pressure prior to any emission reduction regulation. However, since all European airlines were to be covered by a regulation, we did not find particular differences in managers' perceptions regarding the height or importance of the respective pressures. Consequently, we concluded that regulatory pressures did not provide a plausible explanation for the diversity in airlines' communication responses.

We also analysed the influence of other stakeholder groups. When asked about climate change and the airline industry, a Vice President of a flag carrier responded:

> The climate change topic has been a topic for the airline industry already for years. [. . .] At first, there was some pressure from politicians and NGOs, later also some from the media. [. . .] Now we even have corporate customers asking us about our CO_2 emissions.

In contrast, an executive of another airline from the same country stated:

> Neither our customers nor our investors are exerting any kind of pressure. At most there is little pressure from the public debate.

Thus, there was a large discrepancy in terms of the pressure levels perceived by the airlines' executives and the sources of pressure, i.e. the stakeholder groups exerting pressures on the airline.

Table 19.3 provides an overview of these perceptions per airline which are based on the interviewees' statements. From our research, we were able to draw a number of general conclusions:

- Overall, pressures from corporate customers emerged prior to pressures from private customers

- Pressures from private customers were perceived predominantly by airlines in traditionally more environmentally conscious countries

- Pressure from shareholders and the financial community focused mainly on large airlines with significant investment volumes and financial market activity

- Pressures from NGOs were less airline-specific but more targeted at the airline industry in general

- Generally, regulatory agencies, corporate customers and shareholders were perceived to be the main stakeholder groups exerting strong pressure on the industry

TABLE 19.3 Perceived stakeholder pressures

Stakeholder	Airline								
	A	B	C	D	E	F	G	H	I
Regulatory agencies	Very high and similar for all airlines								
Corporate customers	Very low	High	Very low	Very high	Very low	Very high	Very low	Very high	High
Private customers	Very low	Low	Very low	Very high	Very low	High	Very low	Medium	Medium
NGOs	Medium	Very low	Very low	High	Very low	High	Very low	Medium	Medium
Financial community[a]	Low	High	Very low	Very high	Medium	Very high	Very low	Low	High
Media	Medium	Low	Medium	High	Very low	High	Very low	Medium	High
Overall pressure[b]	Low	Medium	Very low	Very high	Very low	High	Very low	Medium	High

[a] Includes financial markets and shareholders.

[b] Based on interviewee assessment and individual stakeholder pressures (regulatory agencies not included).

The diversity in perceptions of stakeholder pressures (both in terms of the key stakeholders and the influence each stakeholder group could wield) helped to explain some differences in communication responses. For example, airlines D and F experienced overall high pressures from all stakeholder groups, especially from corporate customers and the financial community. In contrast, airline G perceived no pressures from its stakeholders. At the same time, airlines D and F strongly engaged in comprehensive reporting and communication activities, both internally and externally (see Table 19.2), whereas airline G did not engage in any communication response activity. Generally, higher perceived stakeholder pressure led to more communication response activities. However, not all differences could be explained by the diversity in perceived stakeholder pressures. For example, while Table 19.3 shows that airlines D and F both perceived high pressures, the type and objective of their respective communication responses were somewhat different. While both companies engaged in external advertising and communication, D strongly emphasised internal communication related to climate change with the objective to create a competitive advantage by changing mind-sets. In contrast, airline F engaged more in the external communication of facts with the objective to demonstrate its commitment to action on climate change to its stakeholders.

From the case study interviews, it became apparent that the way in which executives perceived, approached and talked about the climate change debate and the pressures to reduce CO_2 emissions differed vastly. On the one hand, a Vice President responsible for the business development of a flag carrier declared:

> Our customers would not choose us if we wouldn't care about the environment; hence we believe that taking responsibility will be a competitive advantage in the future.

This statement suggests not only that the airline expects to gain from the situation in future, but also assumes that the situation is controllable by 'taking responsibility'. On the other hand, a Vice President responsible for external affairs of a charter airline stated:

> The entire climate change and CO_2 emission reduction debate is part of a regulative overkill which we feel to be helplessly exposed to. [. . .] The topic is purely political and opportunistic [. . .] and the regulative body is incalculable.

This quote, in contrast to the previous quote, suggests that the company sees climate change as having negative and uncontrollable impacts on the business.

Table 19.4 summarises the manner in which the nine companies we surveyed viewed the risks and opportunities presented by climate change. While the factors influencing each company's perception were quite different, three broad themes emerged from our analysis:

- The companies that saw a high likelihood of a national fuel tax being introduced tended to see a European-wide CO_2 emission regulation more favourably, as this would avoid any competitive disadvantage *vis-à-vis* the other European airlines. This position assumed that the European regulation would prevent the imposition of such a national tax

- The role of environmental responsibility (especially the environmental responsibilities of companies) in a country's culture appears to have influenced managers' interpretations of 'the right thing to do', e.g. with respect to an airline's possibilities to differentiate itself from competitors

● The personal commitment to and involvement in environmental topics by top management or the CEO had a strong influence on overall corporate attitudes

TABLE 19.4 Perception of pressures to reduce CO_2 emissions

Airline	Threat/opportunity	Qualitative assessment
A	Opportunity	'It could be an opportunity to position ourselves as the efficient airline.'
B	Slight threat	'Short-term we especially see a financial risk. We doubt that any airline will be regarded as especially green.'
C	Threat	'Political discussion is unpredictable and thus a threat. We fear that air transport will lose competitiveness.'
D	Strong opportunity	'We believe that taking environmental responsibility will be a competitive advantage in the future.'
E	Threat	'The climate change topic and the resulting debate are clearly a threat to our airline.'
F	Threat	'In the short term, we view the topic as a threat as it will incur costs. In the long term, it is clear that this topic bears risks that will affect society, economy and our airline.'
G	Strong threat	'If the regulation is implemented as planned, it has a huge negative impact on our business — threatening our company's survival.'
H	Slight opportunity	'The topic will not change the business dramatically, in either direction. Maybe it's a small opportunity to promote a green image.'
I	Strong opportunity	'We want to become Europe's leading "green" airline.'

Classification of responses

The two drivers we identified as relevant — 'perceived stakeholder pressure' and 'managers' interpretation' — are represented in form of a matrix in Figure 19.1.

Each airline's position in this matrix is determined by its perception of the overall pressure level (as shown in Table 19.3) and its interpretation of the pressures displayed in Table 19.4. Comparing the position of an airline within the matrix with the corporate response described in Table 19.2 yields some interesting conclusions. Overall, our research shows that stakeholder pressures seem to initiate and drive communication response activity, while the interpretation of these pressures influences the timing, style and objective of the communication response.

Airlines within the bottom-left quadrant of the matrix (i.e. those viewing the climate change debate as a threat, yet experiencing low pressures) tended not to engage in any noteworthy communication response measures. They adopted a 'wait and see' attitude as they did not experience any direct need for action and viewed the climate change debate as a negative, uncontrollable situation in which loss was expected.

FIGURE 19.1 Airline positioning matrix

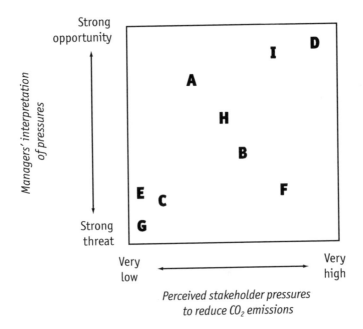

Although companies in the bottom-right quadrant of the matrix also saw climate change as a threat, they did experience stronger stakeholder pressure. As a consequence, these companies strongly engaged in communication response activities, with these communications directed predominantly at reacting to stakeholder pressures by demonstrating the company's commitment to reducing CO_2 emissions and good environmental performance more generally. The communication response of these companies could be best described as precisely targeted at specific stakeholder groups with the objective being to change (i.e. weaken) the pressures put on the company.

Companies at the top right of the matrix also engaged strongly in communication response activities. However, the difference between these companies and those in the bottom-right quadrant lay in the style and pursued objective. Whereas airlines at the bottom right sought to merely respond to and change stakeholder pressures in the short term in an informative style, airlines at the top right of the matrix aimed at crafting and establishing a longer-term image (e.g. as 'Europe's leading environmental airline') by also employing branding techniques and emotional communication.

Finally, companies in the top-left quadrant of the matrix viewed the climate change issue as an opportunity, but experienced only low stakeholder pressure. They had modest communication response activities and adopted a 'no-regret moves only' standpoint.

The two companies positioned in the centre of the matrix perceived medium levels of pressures and did not interpret the pressures strongly in either way. At the time of our study, these companies were not very active in terms of their communication responses but both companies were in the process of preparing several communication measures.

This can be explained by the fact that these companies were starting to see stronger pressures for them to act whereas, earlier, they may have been unclear on whether to interpret the situation as a threat or an opportunity.

Conclusion

In our research, we sought to analyse the responses of European airlines to pressures to reduce their greenhouse gas emissions and to identify the main determinants of their strategies to cope with these pressures. We found that the efforts of companies to reduce their CO_2 emissions did not differ significantly given that airlines have traditionally undertaken large efforts to reduce fuel consumption. As a result, additional pressures exerted by stakeholders only sporadically led to intensified emission reduction efforts. While, in theory, the differences in airlines' engagement in CO_2 emission reduction measures could be assessed by quantifying their actual emission savings, our experience was that quantifying actual emission savings or total savings potential was impracticable due to the issue's complexity and the lack of reliable and publicly available data.

As we have shown, the recent debate on climate change and CO_2 emissions has clearly affected the airline industry. Airlines have realised that strengthened efforts are required to meet their self-set mid-term target of reducing CO_2 emissions per revenue tonne kilometre by at least 25% by 2020 compared to 2005 levels (IATA 2007b). For example, a short-term reduction potential of 18% is expected to be achieved by removing infrastructural and operational inefficiencies (IATA 2007d).

Yet it has to be remembered that, even in a best-case scenario, the absolute CO_2 emission reductions that are achievable in the short term remain limited. Airlines themselves can influence only some portion of an aircraft's CO_2 emissions, as a large share depends on factors such as aircraft weight, engine technology and flight routes which are often also determined externally.

Hence, without further technological developments, total CO_2 emissions from aviation will continue to increase as any reductions will be outweighed by the ongoing growth in total air traffic. With annual industry growth rates of 4–5%, current fuel efficiency improvements of about 2.5% per year would need to double just to cancel out the additional CO_2 emissions from growth in the sector.

In order to realise the industry's ambitious vision of flying zero CO_2 emission aircrafts within the next 50 years, a long-term solution has to involve collective efforts from aircraft manufacturers, engine producers and airlines to develop innovative technologies for sustainable growth. The aviation industry is already taking action in this regard (IATA 2007d), including:

● The testing of new aircraft designs

● The development of solar-powered aircrafts

● The development of aircrafts using fuel cell technology

● The production of jet fuel from biomass

With respect to corporate communications, we found significant differences in communication content, style and objective across all company sizes, countries and business models. These differences were influenced by the perceptions of the level of pressure an airline was facing as well as the airline management's interpretation of the respective pressures. In general, communication activities were more intense the higher the perceived pressures were.

Surprisingly, our interviews showed that airlines' communication responses were not necessarily linked to their reduction responses. Rather, airlines considering the climate change debate as a threat tried mainly to counteract the pressures they were exposed to and to influence the stakeholder groups that were exerting them, thus choosing a defensive and fact-based communication approach. In contrast, those airlines that saw opportunities in the climate change debate usually did not direct their responses exclusively at the stakeholder groups imposing the highest pressures, but rather adopted a more proactive approach, using their communication for marketing purposes and to seek out long-term competitive advantages.

In the light of the present debate around the costs of climate change for business, it seems noteworthy that four of the nine airlines studied perceived the ongoing climate change debate and resulting pressures on the airline industry as an opportunity, with two even identifying strong opportunities despite, or maybe because of, the high pressures they were facing.

But as illustrated above, communications are not necessarily associated with actual CO_2 emission reductions. Rather, despite the efforts the airline industry already takes to mitigate climate change, we also found that airlines do not consistently apply all possible measures to reduce their CO_2 emissions. For example, airlines do not seem to be fully exploiting the potential of procedural measures to influence or change their employees' mind-sets, e.g. by considering CO_2 emission reduction targets in their corporate remuneration schemes.

Moreover, increasing economic pressures might be necessary to commit the industry to obligatory targets. Only when emitting CO_2 is associated with high enough costs will the measures currently not pursued for economic reasons become cost-effective — and hence be implemented. For example, taking off at full thrust compared to the typically applied derated thrust would lead to higher engine stress and consequently result in higher maintenance and equipment costs; yet it would also reduce an aircraft's exposure to the higher drag at lower altitudes and thus lower its CO_2 emissions (Airbus 2004). Likewise, airlines often apply fuel tankering, i.e. loading more fuel than required for a particular flight to avoid paying higher prices for the subsequent flight's fuel at the destination airport. Hence, while airlines can realise cost savings of up to 6% through tankering (Stroup and Wollmer 1992), the extra weight has the effect of increasing CO_2 emissions. If the costs associated with these additional emissions were to outweigh the fuel cost savings, this would probably lead to airlines abandoning this practice and thus reducing their CO_2 emissions.

Even so, all commitments made remain voluntary and the airline industry has yet to prove that it is indeed willing to act accordingly. Airlines could improve their credibility in the climate change debate by increasing their transparency, thereby allowing stakeholders to assess the effectiveness of their emission reduction efforts. This could, for example, include publishing CO_2 emissions indicators that are calculated consistently across the industry.

If the ultimate policy target is to further reduce CO_2 emissions from the airline industry, policy-makers should simultaneously pursue two objectives. First, they should implement regulatory mechanisms that impose costs on CO_2 emissions. On the one hand, these costs could motivate airlines to broaden and intensify their CO_2 emission reduction efforts; on the other hand, they could also lead to increased air fares and thus to a decline in demand for air transport. However, particularly in the light of fast-growing air traffic in Asia and Africa (often with less efficient and thus more polluting aircrafts than in Europe), a long-term regulatory solution should be global in scope and apply to airlines worldwide, in order to prevent any adverse effects on market competition. Similarly, policy has to decide whether it is desirable to promote other modes of transport in order to shift traffic.

Second, policy-makers should also promote investments that lead to significant progress in terms of fuel efficiency and the development of new low-emission aviation technologies. For example, allocating a limited number of emission allowances to the airline industry (e.g. through a regulation such as the EU ETS) and restricting the inter-sectoral trade of these allowances would not only effectively cap CO_2 emissions from aviation, but possibly also accelerate the trend to more efficient aircrafts. Moreover, to underline the seriousness of the topic and the importance being assigned to the issue of climate change by policy-makers, structural improvements such as the reduction of airspace barriers and the expansion of airport capacities to decrease holding patterns should be facilitated.

Finally, responding to climate change is not only a matter for governments. It also requires that customers take action and are willing to use their influence. By paying greater attention to the environmental credentials of airlines and, in particular, the greenhouse gas emissions performance of airlines, customers could reinforce these pressures and encourage more airlines to consider environmental issues such as the reduction of CO_2 emissions an opportunity and important rationale for sustainable performance.

References

Airbus (2004) *Getting to Grips with Fuel Economy* (Blagnac Cedex, France: Airbus Flight Operations Support & Line Assistance).

Aragón-Correa, J., and S. Sharma (2003) 'A Contingent Resource-Based View of Proactive Corporate Environmental Strategy', *Academy of Management Review* 28.1: 71-88.

Boeing (2004) *Fuel Conservation and Maintenance Practices for Fuel Conservation* (Renton, WA: Boeing Commercial Airplanes, Flight Operations Engineering).

Buysse, K., and A. Verbeke (2003) 'Proactive Environmental Strategies: A Stakeholder Management Perspective', *Strategic Management Journal* 24.5: 453-70.

Daft, R., and K. Weick (1984) 'Toward a Model of Organisations as Interpretation Systems', *Academy of Management Review* 9.2: 284-95.

Doganis, R. (2006) *The Airline Business* (New York: Routledge, 2nd edn).

Dutton, J., and S. Jackson (1987) 'Categorizing Strategic Issues: Links to Organizational Action', *Academy of Management Review* 12.1: 76-90.

EC (European Commission) (2005) *Reducing the Climate Change Impact of Aviation* (Brussels: EC).

—— (2006a) *Questions and Answers on Aviation and Climate Change* (Brussels: EC).

—— (2006b) *Aircraft Emissions Capped to Tackle Climate Change* (Brussels: EC).

—— (2007) *Air Pollution: Including Aviation Activities in the Scheme for Greenhouse Gas Emission Allowance Trading within the Community* (Brussels: EC).

Egenhofer, C. (2007) 'The Making of the EU Emissions Trading Scheme: Status, Prospects and Implications for Business', *European Management Journal* 25.6: 453-63.

Freeman, R. (1984) *Strategic Management: A Stakeholder Approach* (Boston, MA: Pitman).

Henriques, I., and P. Sadorsky (1996) 'The Determinants of an Environmentally Responsive Firm: An Empirical Approach', *Journal of Environmental Economics and Management* 30.3: 381-95.

—— and P. Sadorsky (1999) 'The Relationship between Environmental Commitment and Managerial Perceptions of Stakeholder Importance', *Academy of Management Journal* 42.1: 87-99.

Hoffmann, V. (2007) 'EU ETS and Investment Decisions: The Case of the German Electricity Industry', *European Management Journal* 25.6: 464-74.

IATA (International Air Transport Association) (2004) *Fuel and Emissions Efficiency Checklist* (Montreal: IATA).

—— (2007a) *Annual Report 2007* (Vancouver: IATA).

—— (2007b) *Building a Greener Future* (Geneva: IATA).

—— (2007c) *Passenger Numbers to Reach 2.75 Billion by 2011* (Vancouver: IATA).

—— (2007d) *Climate Change and Aviation Technology* (Geneva: IATA).

IPCC (Intergovernmental Panel on Climate Change) (1999) *Special Report on Aviation and the Global Atmosphere* (Geneva: IPCC).

Kolk, A., and D. Levy (2001) 'Winds of Change: Corporate Strategy, Climate Change and Oil Multinationals', *European Management Journal* 19.5: 501-09.

—— and J. Pinkse (2007) 'Towards Strategic Stakeholder Management? Integrating Perspectives on Sustainability Challenges such as Corporate Responses to Climate Change', *Corporate Governance: The International Journal of Effective Board Performance* 7.4: 370-78.

Mitchell, R., B. Agle and D. Wood (1997) 'Toward a Theory of Stakeholder Identification and Salience: Defining the Principle of Who and What Really Counts', *Academy of Management Review* 22.4: 853-86.

Paylor, A. (2007) 'A Fair Trade?', *Air Transport World* 44.2: 40-43.

Sausen, R., I. Isaksen, V. Grewe, D. Hauglustaine, D. Lee, G. Myhre, M. Köhler, G. Pitari, U. Schumann, F. Stordal and C. Zerefos (2005) 'Aviation Radiative Forcing in 2000: An Update on IPCC (1999)', *Meteorologische Zeitschrift* 14.4: 555-61.

Sharma, S., and I. Henriques (2005) 'Stakeholder Influences on Sustainability Practices in the Canadian Forest Products Industry', *Strategic Management Journal* 26.2: 159-80.

——, A. Pablo and H. Vredenburg (1999) 'Corporate Environmental Responsiveness Strategies: The Importance of Issue Interpretation and Organisational Context', *Journal of Applied Behavioral Science* 35.1: 87-108.

—— (2000) 'Managerial Interpretations and Organisational Context as Predictors of Corporate Choice of Environmental Strategy', *Academy of Management Journal* 43.4: 681-97.

Stroup, J., and R. Wollmer (1992) 'A Fuel Management Model for the Airline Industry', *Operations Research* 40.2: 229-37.

Part V
Closing sections

20
From good to best practice on emissions management

Ryan Schuchard, Raj Sapru and Emma Stewart
Business for Social Responsibility, USA

Rory Sullivan
Insight Investment, UK

The data presented in Chapter 2 indicate that many large companies have now established the basic systems and processes necessary for them to manage their greenhouse gas (GHG) emissions. These data also suggest that the current conception of good practice tends to be characterised by an emphasis on the management of direct emissions and energy efficiency, rather than the wider role that could be played by business in responding to climate change or the broader implications of climate change for corporate strategy.

In this chapter, we canvass what going from current good practice to best practice on emissions management might entail for companies. Our intention is not to prescribe the specific actions that companies should take, but rather to offer some reflections on how thinking on good practice needs to evolve and where companies that want to be recognised as leaders need to focus. Our expectation is that, within the next three years, the elements of what we describe below as best practice will virtually all be seen as standard management practice. Indeed, as Chapter 2 suggests, certain of these elements — in particular those relating to corporate governance and public policy engagement — may even have already achieved this status.

Before we set out our views on good and best practice, we would first like to explain how one might conceptually differentiate between them. Current definitions of good practice seem to concentrate on managing and, where possible, reducing 'internal emissions', i.e. those emissions associated with energy and electricity consumption and which, therefore, fall under the organisation's direct control. This emphasis fits quite comfortably within the conventional, efficiency-focused approach to environmental management where the goal is continuous improvement in performance relating to energy or GHG emissions. Although reducing these internal emissions is an important and significant contribution, it omits consideration of the wider strategic role that companies could play in:

● Reducing their overall footprints (i.e. including emissions from their products and supply chains)

● Creating an enabling environment for governments, other companies and individuals to take real and substantial action on reducing GHG emissions across the board

The other point we wish to make here is that our conception of leadership or best practice may not fit comfortably within conventional business cost–benefit analysis, for a range of reasons:

● The benefits may accrue to parties other than the company itself

● The time horizons over which benefits accrue may be longer than those that would normally be considered acceptable[1]

● The regulatory threat or other drivers for action may be extremely uncertain in terms of the likelihood of occurrence, the business impacts and the timing of the impacts

● Many of the benefits are intangible relating to aspects such as innovation, competitive positioning and reputation

We do not intend in this chapter to make a 'hard business case' for best practice (although some aspects of the business case are canvassed by Rory Sullivan in Chapter 21 on voluntary approaches) other than to acknowledge that the business case for best practice — as, indeed, with best practice in virtually any area — is inevitably difficult to define, in particular in advance of taking action.

Good practice

Overview

Even though this chapter primarily examines the question of best practice, it is crucial to emphasise the importance of companies managing their own GHG emissions properly — in part because it is the obvious point at which they have influence, and in part because the skills and competences developed through these activities can then be har-

---

1 See, for example, the discussion in Chapter 16 of how Vancity Credit Union's triple-bottom-line orientation has allowed it to take a longer-term view on the costs and benefits of taking action.

nessed and extended as the company extends the scope of its management focus. It is also important to recognise that many companies have yet to take substantive action on managing their emissions; the research presented in Chapter 2 suggests that, even at the end of 2007, some 20% of the 100 largest companies in the UK did not have adequate climate change or environmental management systems in place.

In broad terms, and acknowledging that the specific approaches adopted will depend on the company, its exposure to climate change-related risk and its existing approaches to managing environmental and other issues, one would expect all companies to:

- Establish clear governance and management structures for climate change including appropriately defined responsibilities at the board and senior management levels

- Publish a clear policy on climate change that acknowledges that the company's activities contribute to GHG emissions and which commits the company to emissions reductions over the short and long term. This policy should be signed by the CEO as evidence of senior management's commitment to action on the issue

- Conduct a thorough assessment of the risks and opportunities presented by climate change including the risks and opportunities presented by regulation and by the physical effects of climate change. The scope of this assessment should not be confined to direct operations but should also consider the organisation's suppliers, contractors, joint venture partners and other partners, as well as emissions from the use of the company's products or services

- Report on their historic performance and anticipated changes in emissions over time (see further the comments under corporate disclosures below)

- Set clear targets for emission reductions and energy efficiency, and develop a management plan — including the allocation of appropriate resources — for the delivery of these targets

- Contribute to the wider climate change debate through participating in leadership initiatives and supporting calls for strong and effective government policy on climate change

- Consider working with their suppliers and/or customers to encourage them to take action to reduce their emissions

Policy, objectives and targets

Written policies are widely used, particularly in large companies, as internal and external statements of the company's purpose, vision and values. However, the quality and content of these policies is variable and many have limited substantive content.[2] In broad terms, we believe that corporate climate change policies should:

---

2 A similar criticism can, as noted in Chapter 21, be directed at many voluntary approaches.

- Identify climate change as a key business concern
- Recognise the company's own contribution to climate change
- Apply to all the company's GHG emissions (including emissions from products and services)
- Have an explicit commitment to emissions reductions over the short and long term
- Explicitly support government's efforts to reduce emissions

These polices should be supported by clear measurable targets for emission reductions. These targets should reflect the broad direction of international climate change policy (i.e. reductions in absolute emissions of 20–30% by 2020 and 80% or more by 2050 against a 1990 baseline).

Reporting

Despite the efforts of initiatives such as the Global Reporting Initiative[3] and the Carbon Disclosure Project (CDP),[4] there are still systemic gaps and weaknesses in the emissions data being provided by companies which makes it difficult to meaningfully compare performance between companies even within relatively homogeneous sectors. Furthermore, reporting continues to almost exclusively concentrate on past performance rather than future trends, which makes it difficult for policy-makers and other stakeholders to assess where attention needs to be focused. We therefore suggest that all companies (see further Sullivan 2008):

- Report on all their greenhouse gas emissions — not just direct emissions from processes and operations and indirectly from electricity consumption, but also emissions from (or associated with) the products they use, transport and logistics, and the use and disposal of their products and services
- Provide a clear statement on the materiality (or financial significance) of climate change-related risks and opportunities for their business including a description of the process followed to assess materiality
- Report on at least their last five years of emissions data and, if feasible, back to 1990. This should include:
  - A description of the manner in which figures have been calculated
  - The scope of reporting, data checking and verification processes
  - The key assumptions made
  - The key data sources used (e.g. the emission factors used to calculate emissions from electricity consumed)
  - An estimate (ideally quantitative) of the uncertainty associated with the reported data

3 www.globalreporting.org
4 www.cdproject.net

- Explain how they expect their emissions to change over time and, if possible, provide at least five years' projections of expected changes in emissions

- Provide an explanation for changes in their historic emissions (e.g. due to acquisitions or divestments, changes in data collection and analysis processes) and the factors that are likely to impact on emissions going forwards

- Publish their emissions reduction targets including:
  - A clear statement on the proportion of the company's emissions that are covered by the target
  - The baseline for the target
  - The expected emissions reductions
  - The deadline for achieving the target
  - Whether the target is to be met wholly or partially through offsetting
  - An explanation of how they expect to meet these targets (actions taken, expected emission reductions associated with each action, resources allocated, key responsibilities and accountabilities, etc.) and the actions they will take in the event that the targets are not met

- Publish a clear statement of their views on the role of public policy in delivering on climate change goals (e.g. what national or global emission targets does the company consider appropriate and what policy instruments does the company think should be used to meet these targets?)

- Publish a list of the climate change-related initiatives that they participate in or support

- Consolidate their climate change-related information in a single corporate publication and/or in a single location on their website

There are two important factors that limit the quality and accuracy of the data reported by companies.

The first is that calculating emissions from many sources is not a trivial task and obtaining robust data with reasonably high reliability can be technically difficult and expensive (Sullivan and Gouldson 2007). This problem is exacerbated when calculating emissions from sources that are not owned or controlled by a company (e.g. emissions from supply chains or from a product over its entire life-cycle).

The second is that there are a series of technical issues that still need to be resolved around reporting such as the manner in which emissions are to be 'accounted for' (i.e. whether they should be consolidated on the basis of 'financial control' or of '[equity] ownership').[5] It will be some time before there is a consensus on this issue. In the interim, we suggest that companies should:

- Specify clearly the scope of their reporting and particularly any exclusions from the scope of reporting

---

5 The *Greenhouse Gas Protocol* discusses the different approaches that may be adopted but does not provide definitive (i.e. prescriptive) guidance on the answer to this question (WBCSD/WRI 2004: 26-33).

- Endeavour to calculate and report emissions from those activities or operations where they have a significant stake and/or which are a significant source of emissions

- Encourage their partners, suppliers and customers to gather and publish information on their emissions

Best practice/leadership on climate change[6]

The fact that most companies are concentrating their attention on their internal GHG emissions suggests that they may not be maximising their potential to contribute to substantive action on climate change. We, therefore, suggest that best practice or leadership on climate change requires that companies take action in three areas:

- **Initiating paradigm shifts** in the way that climate issues are managed by focusing attention on how they can deliver (or contribute to the delivery of) emissions reductions at scale

- **Developing 'climate-friendly' value chains** to internalise the most significant negative impacts including 'embedded' and 'product use' emissions. The idea is that companies look beyond simply managing emissions from sources that they control to considering emissions within their broader spheres of influence. Leaders are recognising that this is important because, although direct emissions are relatively straightforward to track, simply targeting end-of-pipe solutions just moves 'pollution around in a circle' (Esty and Winston 2006: 108). To illustrate, while the production of aluminium is relatively emissions-intensive (creating 12 tonnes of greenhouse gas emissions for every 1 tonne of finished product), that same tonne of aluminium added to a car offers the potential to save 20 tonnes of GHG emissions over the car's life-cycle (Esty and Winston 2006: 210)

- **Shaping external systems** by engaging policy-makers, individuals and other organisations to multiply positive externalities. The aims of such engagement are twofold:
  - To encourage policy-makers to encourage companies to internalise the negative externalities associated with GHG emissions
  - To create a business environment where companies can innovate and find solutions to the problems presented by climate change

In Table 20.1, we present the key approaches and a series of illustrative practices that companies may consider when taking action in each of these areas. In the following sections, we provide examples of the actions that companies — across a whole range of sectors and activities — have taken in each of these areas.

6 This section draws heavily on the Business for Social Responsibility (BSR) publication, *Beyond Neutrality: Moving Your Company towards Climate Leadership* (Schuchard *et al.* 2007). Unless indicated otherwise, the examples cited are taken from this source.

TABLE 20.1 Climate change best practice action areas

Source: Adapted from Schuchard et al. 2007

Area	Approach	Illustrative practices
Initiating paradigm shifts: building enabling environments that spur paradigm shifts in the way that emissions are managed and which target resources at the most scalable emissions reductions	Rethinking basic assumptions about business processes	• Harnessing new processes • Using clean energy and fuel • Generating clean power
	Building strong governance structures	• Communicating transparently • Ensuring high-level oversight of climate change-related issues • Aligning the organisation's climate change commitments with its overall corporate objectives and strategy
	Engaging employees as contributors	• Institutionalising a culture of conservation and innovation • Involving employees in problem solving and decision-making • Providing consistent and appropriate incentives to employees
Developing 'climate-friendly' value chains: internalising the most significant negative impacts including 'embedded' and 'product use' emissions	Reducing embedded emissions	• Undertaking due diligence on materials specifications • Giving preference to suppliers with low emissions • Reducing product miles
	Designing low-emissions products and services	• Designing products and services with low greenhouse gas emissions • Linking existing products to restoration • Deploying new products
	Educating customers	• Labelling products better • Offering product metering to provide real-time energy use feedback • Providing buyer and user guidance
Shaping external systems: shaping external systems by engaging with policy makers, individuals and other organisations to multiply positive externalities	Advancing dialogue on policy	• Stating publicly the company's support for collaborative solutions • Furthering science-based, equity-oriented policy frameworks • Building global pathways and policy frameworks directed at significantly reducing emissions
	Co-creating opportunities with other organisations	• Advancing market-based initiatives • Building informational networks • Developing operational partnerships
	Encouraging climate-friendly behaviour among individuals	• Facilitating dialogue • Enabling action • Inspiring lower-impact behaviours

Initiating management paradigm shifts

In broad terms, there are three basic approaches to initiating management paradigm shifts: namely, rethinking basic assumptions about business processes, building strong governance structures and engaging employees as contributors.

Rethinking basic assumptions about business processes

Ultimately, climate change requires companies to rethink their most basic assumptions about how products are made, what energy to use and what opportunities exist for creating net energy-*positive* processes. In broad terms this requires that they think about:

- Harnessing new processes including technologies for greater scale efficiencies, decarbonisation, and carbon capture and storage

- Using clean energy including switching to renewables and biofuels

- Generating clean power from sources such as solar, wind, biomass, landfill gas, geothermal and sustainable hydro power

In relation to the harnessing of new processes, high-profile examples have included:

- The removal by Nike of sulphur hexafluoride (SF_6) — one of the most harmful greenhouse gases — from the process it uses to fill air pockets in its shoes

- The switch by Unilever of its freezer technologies from hydrofluorocarbons (HFCs) — some of the most potent greenhouse gases — to natural gas

- Work by Vodafone with suppliers to design better network cooling systems

Many companies have taken steps to increase their use of clean energy:

- Xcel Energy has acquired 775 megawatts of new wind power in Colorado, making it one of the US's largest utility purchasers of wind energy

- Major UK high-street fashion retailer Marks & Spencer fuels its delivery vehicles with a minimum of 50% biodiesel

While much of the investment in clean power generation has been led by electricity utilities seeking to meet the renewables targets set by governments in the countries in which they operate, other companies have also sought to make these types of investments:

- Google is building a 1.6 MW solar photovoltaic panel array (the largest corporate campus solar energy project in the USA)

- Johnson Controls is investing $45 million to add green roofs, wind turbines, solar panels and a geothermal field to its international headquarters

- Tyson Foods, the world's largest chicken producer, has developed Tyson Renewable Energy to turn its annual one million tonnes of animal fat by-products into biofuels

Governance

As noted in Chapter 2, a majority of large European companies now have board/senior management representatives with explicit responsibility for climate change (whether as a stand-alone issue or as part of a wider environmental management remit).

Different companies have adopted different approaches depending on their climate change exposures and their approaches for managing and responding to this type of risks. For example:

- The Executive Council of Hewlett Packard (HP) directly oversees a Supply Chain Board, which all company businesses support with a Procurement Management Process defining supply chain social and environmental responsibility (SER) buying criteria

- Bayer has an executive Corporate Sustainability Board on climate change and a working group on renewable raw materials

- The Public Policy & Environmental Board Committee of International Paper reviews all the organisation's climate change and related policies

As a signal of the importance assigned by the organisation to the issue of climate change, having senior management representative(s) with explicit responsibility for climate change or, more generally, for environmental matters is an important statement of intent and provides a level of accountability right to the top of the organisation. However, effective climate change governance requires much more. It requires that directors and senior managers are engaged fully with the climate change debate and use their influence to ensure that climate change is integrated properly into the manner in which the company runs its business. While the specific details will vary, we would expect that senior managers and board directors (as relevant to the business as a whole and to their specific responsibilities):

- Understand what climate change is and how it may affect the business (through regulation, litigation, physical impacts, etc.)

- Know what systems and processes the company has in place to respond to climate change

- Understand what current best practice among the company's sectoral peers looks like, and how their company compares

- Communicate their personal commitment to action on climate change to their employees and to relevant external stakeholders

- Play an active role in ensuring that appropriate risk management systems and processes are in place to protect the company against the consequences of climate change

- React immediately and personally in the event that climate change impacts on the business

Effective climate change governance also requires that directors and senior managers ensure that corporate objectives and strategy are aligned properly with the organisation's climate change commitments. This is frequently delivered through setting specific climate change-related objectives at appropriate levels within the organisation. For example, General Electric (GE) has a company-wide campaign to create expectations for individual business lines to set reduction targets, while Johnson & Johnson uses an 'Environmental Dashboard' where GHG emissions are one metric that each employee's performance is assessed on.

Proper alignment requires that companies ensure that climate change commitments apply across all their activities and not just their direct emissions. Perhaps the most obvious example where this does not occur is in the financial services sector where many UK and continental European companies have made commitments to carbon neutrality in relation to emissions from their energy consumption and business travel. The reality is that for these companies, however, is that their primary impact on emissions is through their lending portfolios and other activities. Yet, despite a number having policies requiring the climate change-related risks associated with their investments to be assessed, no major financial institution has made commitments to reducing the GHG emissions from its lending portfolio. Clearly there are a series of questions around who exactly is responsible for the emissions (is it the financial institution or the client?), but this inconsistency inevitably raises questions about the sector's real commitment to action on climate change. This question of alignment not only applies to companies' operations and activities, but also to areas such as corporate lobbying where the objectives of the lobbying activity may run counter to the organisation's stated climate change commitments.

The other important strand of governance is communication with stakeholders and the public. We discuss reporting frameworks above and are of the view that meeting these expectations represents a minimum for any company. For leadership, we would emphasise that reporting is not just about performance reporting (i.e. current and future emissions), but also involves communicating the company's views on climate change and related issues. A number of companies have started to issue reports that set out their views on climate change and/or canvass specific dimensions of the climate change debate. Examples include:

- Swedish energy company Vattenfall's *Curbing Climate Change* and *Climate Map 2030* outline a policy framework for a low-GHG-emitting economy and present a series of actions that could be adopted to deliver significant GHG emission reductions (Vattenfall 2006, 2007)

- McKinsey & Company's Climate Change Special initiative which provides a range of reports and other information on cost curves for greenhouse gas emission reductions[7]

- The Ford Motor Company's recently commissioned report, *The Business Impact of Climate Change*, which discusses the company's key business challenges resulting from climate change (Ford 2005)

These types of reports are important for two reasons. First, one of the most important obstacles to government action on climate change is the concern that existing businesses interests will be damaged or, more precisely, the fear that companies will inflict political damage if their interests are not completely protected at all times. Therefore, having companies express support for government action on climate change is vitally impor-

7 www.mckinsey.com/clientservice/ccsi (accessed 20 August 2008).

tant in helping overcome this obstacle. Second, such reports help to address the information asymmetries that continue to militate against effective policy action on climate change. By providing information on cost curves and on the likely business responses to specific regulatory instruments, these reports allow policy-makers to design and implement policy that is more effective and more efficient than would otherwise be the case.

Engaging employees as contributors

Companies are in a unique position to empower individual employees and make them feel as though, through their everyday work or simply their choice of employer, they are making a positive contribution to addressing climate change or other environmental problems. Some organisations have tried to institutionalise a culture of conservation and innovation through involving employees with climate change outcomes. For example:

- Serco Leisure[8] in the UK taps employee creativity through a 'Big Ideas' staff suggestion scheme, competitions between sites and emails, workshops, staff meetings and noticeboard communications to encourage efficient energy use

- Wal-Mart is using 'Personal Sustainability Projects' to educate staff and drive cultural change towards sustainability

- Prospero Recruitment based in London has reduced waste by 90% through simple practices such as powering down IT hardware when not in use, switching to energy-efficient light bulbs, turning down thermostat settings slightly and recycling waste

- Google uses discussion forums and emailings to facilitate employee carpooling

Employee involvement may also be strengthened through broadening participation in the company's climate change and energy-related decision-making processes. This can be done in many different ways:

- 3M uses a company-wide system called Pollution Prevention Pays (3P) which encourages employees at all levels to rethink products and processes to eliminate waste

- Alcoa has had an internal greenhouse gas information system since 1998 which tracks emissions data for its worldwide operations

- Cadbury in the UK is committed to cutting its greenhouse gas emissions by 50% and encourages green activism among employees by giving them the authority to take action to reduce emissions or otherwise improve environmental performance in their areas of work

Finally, organisations can provide incentives to employees who take action to reduce their GHG emissions or minimise their environmental impacts. Examples include footwear retailer Timberland providing $3,000 to employees to subsidise fuel-efficient hybrid

---

8 Part of Serco Group plc, the company operates leisure facilities for local authorities.

cars, pharmaceutical and drug delivery specialist Alza Corporation[9] paying its employees $1/day when they bike, walk or carpool to work, and Hyperion,[10] as part of its 'Drive Clean to Drive Change' employee programme, offering $5,000 to 200 employees per year for the purchase of cars that average 45 miles per gallon or better.

Developing climate-friendly value chains

Reducing embedded emissions

Supply change management is one of the business activities that has changed most dramatically over the past ten years: supply chains have 'globalised' as evidenced by the rise of countries such China and India, many companies (particularly in the retail sector) now have tens of thousands suppliers and purchasing functions bear more responsibility for revenue and innovation. Companies have started to reap benefits (reduced costs, reduced risk, new business opportunities) by reducing their 'embedded' emissions (i.e. those emissions that occur upstream of their processes and products) through changing their procurement criteria, rationalising their logistics systems and encouraging their suppliers to become part of a climate-friendly value chain.

One way of delivering reductions in embedded emissions is to analyse and possibly change material specifications to give preference to options with lower emissions. For example, UK crisp manufacturer Walkers Snacks Ltd found that, by purchasing potatoes on the basis of low water content rather than by weight, suppliers reduced their energy inputs, transportation become more fuel-efficient, and there were fewer costs associated with post-purchase drying (Carbon Trust 2006). Another strategy could be to give preference to low-emissions suppliers by ranking suppliers on their climate change performance (e.g. Unilever gives preference to suppliers with lower emissions) or working to help existing suppliers improve (e.g. after BSkyB hosted supplier workshops on environmental performance, ten of its suppliers followed its example by making a commitment to becoming carbon neutral).

Companies could seek to reduce their food/product miles through optimising routes, localising sourcing and using more efficient distribution technologies. For example:

- Timberland, along with several other companies including furniture retailer Ikea and shipowner Maersk Company Ltd, is engaged in the BSR Clean Cargo Working Group[11] to promote industry-wide tools and methodologies for shippers to collaborate with carriers in tracking and reducing emissions

- Marks & Spencer is seeking to reduce its food miles by doubling regional food sourcing and working with growers to extend British growing seasons through new varieties and growing techniques

- Nike is committed to reducing its inbound logistics-related emissions by 30% from a 2003 baseline by 2020

9 Part of Johnson & Johnson family of companies.
10 Global provider of performance management software; acquired by Oracle Corporation in March 2007.
11 www.bsr.org/sustainabletransport (accessed 20 August 2008).

Designing low emissions products and services

Companies are increasingly tapping in-house R&D teams or collaborating with their suppliers to move beyond 'reducing, re-using and recycling' to 'redesigning and re-imagining' how their products might be used – a field known as Design for Environment (Esty and Winston 2006: 197).

There are a series of examples from widely different and differing sectors that show what can be done. As one illustration, GE launched its 'Ecomagination' project in 2005, an initiative that involved the company investing in climate change-friendly technologies with the objective of reaching $20 billion in annual sales by 2010. In addition, GE will double its investment in clean technologies (including wind turbines, high-efficiency gas turbines and hybrid diesel–electric locomotives) to $1.5 billion a year by 2010. Other companies following a similar path include: Procter & Gamble (P&G), which has doubled detergent concentrations in order to shrink the volume of packaging by 22% and thus reduced the energy needed to manufacture the packaging and transport products, and reduced the amount of waste sent to landfill; and Unilever, which has reformulated its soaps and apparel for lower-temperature washing. Companies can also use their access to customers and/or link their products to environmental restoration. UPS, for example, has used its capabilities in fleet efficiency to offer the Roadnet Transportation Suite consulting service, which helps its clients to streamline their transport routes; this service has helped client associated food stores reduce the distance travelled by goods by over 600,000 km/year.

Finally companies can deploy new products such as new technologies and new value propositions for existing technologies. For instance:

- Liftshare.com has developed a car share model that provides a national database for individuals and a for-profit business model for organisational partners to save approximately 14 million car-miles annually

- GE, through its Ecomagination programme, is developing new consumer and industrial products such as the Jenbacher biogas engine, which converts organic waste into energy

Customer education

Despite the rising profile of climate change, customers are confused about what they should do; they are discouraged by inadequate product information and 'green premiums', and sceptical regarding claims made by business (AccountAbility 2007). Until recently, customers have been given little objective and meaningful information on which to base their purchasing decisions. But this is changing. A growing number of companies are working to educate their customers on the climate change impacts of their consumer choices by providing metrics and information that customers can use to compare the climate change impact of different products.

There are a series of programmes to label products with information about embedded and product use emissions including:

- Timberland's 'Green Index' tags to communicate embedded emissions and other impacts to customers

- Marks & Spencer's introduction of symbols depicting airplanes on food products that it flies in

- Tesco's labelling of products with GHG emissions information

Companies can also offer product metering and other data to provide real-time energy use feedback. For example, G&E, in partnership with Wellington Energy, is providing smart meters to customers, which customers can use to manage their energy usage, and Easyway Insurance Brokers in Ontario, Canada, is providing customers with optional smart meters to measure fuel efficiency for their cars.

In addition, companies can provide their buyers and users with information that can be used for choosing products and optimising usage efficiency:

- Shell, which acknowledges that over 80% of emissions from fossil fuels occur when energy products are used, has developed a Fuel Stretch Campaign to help its customers understand how to make fuel go further

- Unilever's Washright programme encourages consumers to wash at lower temperatures and to use full washes

Shaping external systems

Shaping external systems involves three broad areas of action: namely, advancing dialogue on policy, co-creating opportunities with other organisations and encouraging climate change-friendly behaviour among individuals.

Advancing dialogue on policy

It is widely recognised — and, indeed, a core theme of this book — that we cannot rely on markets alone to deliver the substantial cuts in GHG emissions necessary to minimise the likelihood of catastrophic climate change. In order to ensure that companies reduce their emissions, governments need to establish a long-term policy framework that provides appropriate incentives and certainty. Without policy certainty, companies (particularly in energy-intensive sectors) will find it difficult to make decisions that account properly for the risks associated with climate change policy. Stronger and clearer public policy is also crucial to strengthen investor confidence in areas such as renewable energy where companies' business models are critically dependent on the existence of a supportive long-term policy environment.

One of the key obstacles to stronger government action on climate change has been the concern that such action will be damaging to business or that businesses will publicly criticise government for taking such action. We believe companies have a fundamental responsibility to actively support government action directed at delivering substantial reductions in greenhouse gas emissions, even if such action entails increased costs for individual companies. This is no pipe dream. For example, Chapter 12 describes how an increasing number of institutional investors have started to play a greater role in public policy debates around climate change. Investors have stressed the importance of governments specifying and committing to meeting long-term policy goals as an essential prerequisite for companies to make economically efficient investment decisions that properly incorporate climate change factors. The rationale is that investors' interests are threatened by the potential for climate change to materially and severely impact on economies as well as individual companies. Many institutional investors (asset managers and pension funds) now accept that public policy engagement is entirely consistent with their fiduciary responsibilities to maximise long-term returns.

Proactive engagement with public policy is not confined to investors. Companies are also supporting public policy measures directed at reducing greenhouse gas emissions. Initiatives such as The Climate Group,[12] the Corporate Leaders Group on Climate Change[13] and the EU Corporate Leaders Group on Climate Change[14] all involve the signatories or participants expressing clear support for emission reduction targets of the order of 60 to 80% by 2050.

Individual companies are also explicitly calling for policy action on climate change. The survey of large UK and continental European companies reported in Chapter 2 found that almost half of the companies surveyed (62 out of 125) have expressed some support for market-based instruments such as emissions trading as part of governments' response to climate change. Examples of the policy positions that have been adopted by companies include:

- BP's statement in 1997 when it became the first major oil company to publicly state that precautionary action on global warming is justified

- Duke Energy's strong advocacy for regulation that will assign a price to GHG emissions because it believes such control is inevitable and thus it makes sense to transition now to reduce 'stroke-of-the-pen risk' (i.e. the risk that policy-makers could sign carbon pricing into law and cancel the value of assets overnight)

- Ford's statement:

    It is in the interest of society and business to reduce the uncertainty and increase the predictability of policy frameworks and market conditions around the issue of climate change.

There is some evidence that this type of policy engagement can be effective. In Chapter 11, it is argued that advocacy partnerships have helped to resolve a significant Catch 22 in climate change policy: namely, that, while companies need regulatory certainty in order to invest, governments are hesitant to take action because they are uncertain about how companies would react to climate change policies.

Co-creating opportunities with organisations

An increasing number of companies are partnering with other organisations to take action to reduce emissions by developing and implementing voluntary market-based instruments. Perhaps the most important example (given the US government's hostility

---

12 See Chapter 10.

13 www.cpi.cam.ac.uk/programmes/energy_and_climate_change/corporate_leaders_group_on_cli.aspx (accessed 20 August 2008).

14 www.cpi.cam.ac.uk/programmes/energy_and_climate_change/clgcc/eu_clg.aspx (accessed 20 August 2008).

to action on climate change) has been the active support provided by companies to the Chicago Climate Change which, at the time of writing (April 2008), is the only greenhouse gas emissions allowance trading system in North America based on legally binding rules. In addition, companies have contributed to the design and implementation of state and regional greenhouse gas registers such as the California Climate Registry, which helps establish greenhouse gas emission baselines against which future reduction requirements can be assessed.

Advancing public policy is not the only contribution that such partnerships can make. Informational networks also have an important role to play in reducing emissions. Examples include:

- Recreational Equipment Inc. working with regional partners to share best practices as part of the Seattle Climate Partnership and San Francisco Business Council on Climate Change

- Vodafone commissioning *Earth Calling* — a report informing the public of the environmental impacts of the mobile telecommunications industry (Forum for the Future 2006)

- Intel sharing best practices with other companies at the working group 'Arriving at Strategic Policy Positions on Climate Change' led by Business for Social Responsibility (BSR)

Companies can also establish industry and cross-sector partnerships and investments. For example, banking and financial services organisation HSBC has developed the $100 million Climate Partnership — a five-year alliance between The Climate Group, Earthwatch Institute, Smithsonian Tropical Research Institute and WWF, which aims to:

- Create cleaner, greener major cities

- Enable 'climate champions'

- Conduct the largest-ever field experiment on the long-term effects of climate change

- Protect the world's major river systems

Encouraging climate-friendly behaviour among individuals

Some companies are using their strengths to encourage climate-friendly behaviour among individuals. There are various ways this can be done. Companies can use their technologies and skills to facilitate dialogue, including providing real and virtual venues and leading discussions. For example, Yahoo's Green web portal enables customers to meet and engage in dialogue on climate issues, and News Corporation's MySpace channel 'OurPlanet' informs users about climate change.

Companies may also be able to address the 'value action' gap through providing choices that are consistent with customers' ethical values. As an illustration, B&Q has extended its product lines to include home micro-generation technology, home solar panels and wind turbines.

Finally, companies may be able to inspire lower-impact behaviours by appealing to customers or individuals to adopt new actions and lifestyles. For example:

- BSkyB is communicating to mainstream customers on its 'Join the Bigger Picture' website[15]

- Yahoo! helps customers understand how they can address global warming in their daily actions with Yahoo! Green[16]

- MTV's Switch campaign[17] seeks to motivate more climate change-friendly attitudes among young people

Closing reflections

The aim of this chapter is not to prescribe or dictate company actions but rather to provide some initial thoughts on the actions that companies may consider as part of their efforts to make a wider contribution to the climate change debate. We would like to conclude by offering a number of more general reflections on current practice and on how best practice needs to evolve.

The first is that best practice represents a dramatic change in emphasis for many companies, moving away from the relatively comfortable issues of direct emissions and energy consumption to much more difficult areas such as 'embedded emissions' (those occurring upstream in the production of goods and services that the company purchases) and 'product use emissions' (those which occur downstream as a result of consumers using the company's products). This change will, potentially, require companies to develop or acquire completely new sets of skills and competences. It will also require significantly different mind-sets and attitudes: for example, the scope (or boundaries) of the company's corporate responsibility initiatives may need to be completely redrawn.

The second is that great care is required to ensure that the 'solution' is not worse than the 'problem' or that the rush to take action does not create a series of unintended or undesirable consequences. For example, consumer information (or the lack thereof) is a well-recognised barrier to individual action on climate change and product labelling has a potentially important role to play in overcoming this barrier. However, we are concerned that the current trend for each retailer or company to develop its own labels and performance measures may actually perpetuate this barrier. While the fact that a retailer has developed its own labels may enable customers to differentiate between the climate change impacts of that retailer's products, it does not necessarily allow for meaningful comparisons to be drawn between different retailers' products. We therefore emphasise that real leadership may — perhaps paradoxically — entail being 'part of a pack' where all the key actors work together to find a common, credible solution to this type of problem.

---

15 www.jointhebiggerpicture.com
16 green.yahoo.com
17 www.mtvswitch.org

The third is a reiteration of one of the core messages of this book. There is much that business can do within its own operations and, as the examples in the latter section of this chapter illustrate, through providing support to wider action on climate change. However, all these initiatives need to be judged on a single test: what is their contribution to reducing global GHG emissions? If their effect is not to enable, facilitate or deliver significant emissions reductions, companies need to rethink their strategies. We have moved beyond a point where good-news stories, 'greenwashing' or green branding are appropriate responses to the threat of climate change.

We would therefore like to reiterate three key points. First, reducing emissions intensity may not result in total emissions being reduced. Our view is that companies that are serious about responding to climate change will commit to reducing their total emissions not just to being more efficient. Second, the objective of climate change governance is not just to manage risks to the business; it is about how the company reduces its own emissions and maximises the contribution it makes to reducing emissions elsewhere. Third, companies need to be wary of sinking resources into trivial actions. For example, while we recognise the importance of employee engagement, we are concerned that most employee engagement programmes are simply a substitute for action (or an excuse for inaction) by the company.

In conclusion, we remain convinced that business has a critical role to play in the fight against climate change and we believe that the most successful companies in the future will be those that have responded most proactively to the challenges presented by climate change. Although we recognise that companies face practical barriers to action, we do not consider the best-practice actions set out here as being beyond the ability of companies to deliver. What is required from our business leaders is the moral courage to embrace this agenda and drive change throughout their organisations.

References

AccountAbility (2007) *What Assures Consumers on Climate Change?* (London: AccountAbility).

Carbon Trust (2006) *Carbon Footprints in the Supply Chain: The Next Step for Business* (London: Carbon Trust).

Esty, D., and A. Winston (2006) *Green to Gold* (New Haven, CT: Yale University Press).

Ford Motor Company (2005) *Ford Report on the Business Impact of Climate Change* (Dearborn, MI: Ford).

Forum for the Future (2006) *Earth Calling: The Environmental Impacts of the Mobile Telecommunications Industry* (London: Forum for the Future).

Schuchard, R., R. Sapru and E. Stewart (2007) *Beyond Neutrality: Moving Your Company towards Climate Leadership* (San Francisco: Business for Social Responsibility).

Sullivan, R. (2008) *Taking the Temperature: Assessing the Performance of Large UK and European Companies in Responding to Climate Change* (London: Insight Investment).

—— and A. Gouldson (2007) 'Pollutant Release and Transfer Registers: Examining the Value of Government-Led Reporting on Corporate Environmental Performance', *Corporate Social Responsibility and Environmental Management* 14: 263-73.

Vattenfall (2006) *Curbing Climate Change* (Stockholm: Vattenfall).

—— (2007) *Climate Map 2030* (Stockholm: Vattenfall).

WBCSD (World Business Council for Sustainable Development) and WRI (World Resources Institute) (2004) *The Greenhouse Gas Protocol: A Corporate Accounting and Reporting Standard* (Geneva: WBCSD, rev. edn).

21

Do voluntary approaches have a role to play in the response to climate change?

Rory Sullivan
Insight Investment, UK

Voluntary approaches are playing an increasingly important role in the business and policy response to climate change. Building on the material and case studies presented in this book and the wider literature on self-regulation, this chapter examines the potential for voluntary approaches to deliver significant reductions in greenhouse gas emissions. The chapter is divided into two parts:

- A brief review of the literature on voluntary approaches
- A wider assessment of the potential for voluntary approaches to make a significant contribution to mitigating greenhouse gas emissions mitigation

Voluntary approaches: a brief overview of the literature[1]

Some definitions and characteristics

Voluntary approaches can be defined as schemes where organisations agree to improve their environmental performance beyond legal requirements (OECD 1999), though the

---

1 This section is a summary of a more comprehensive review of the literature on voluntary approaches in Sullivan 2005.

term 'voluntary' may not be strictly accurate as voluntary approaches are often implemented in response to consumer and community pressures, industry peer pressure, competitive pressures or the threat of new regulations or taxes.

Voluntary approaches can be divided into four broad categories (OECD 1999), namely:

- **Unilateral commitments** (also widely referred to as self-regulation), which involve organisations — individually or collectively — defining their own environmental objectives and then communicating this information to stakeholders

- **Private agreements**, which are made by direct bargaining between polluters and those affected by pollution. These agreements are generally underpinned by contract law

- **Negotiated agreements**, which are formal agreements between public authorities and industry, and generally include targets and a time-frame within which the target is to be met. These are often underpinned by a threat of regulatory action if the conditions of the negotiated agreement are not met

- **Public voluntary programmes** (e.g. challenge programmes, eco-labelling, award or prize programmes, research and development or innovation programmes), which involve organisations agreeing to meet standards developed by public bodies

Within this typology, the features that influence the outcomes that can be achieved include:

- **The degree of government involvement.** The extreme situations are where both rulemaking and enforcement are carried out exclusively by the industry participants or exclusively by government. In practice, however, even the strictest forms of government regulation will include some voluntary elements, while voluntary approaches are frequently implemented against the backdrop of some form of government sanction or threat of regulation

- **Whether the voluntary approach is individual or collective.** The outcomes achieved from voluntary approaches are strongly influenced by the number of participants. If only a few companies are involved, it may be possible to set more stringent targets at the level of the individual company than if it is desired to have a greater number of companies participate (where there may be pressure to have lower targets)

- **Whether the approach is local or global in scope.** Global voluntary approaches may help create norms around specific issues and provide non-regulatory enforcement mechanisms (e.g. company purchasing power) to ensure compliance with these norms

- **Whether the approach is binding or non-binding.** While binding agreements are more likely to be effective than non-binding agreements, the fact that an agreement is mandatory may make companies reluctant to commit to the agreement, and may lead to the objectives being set at a lower level than would otherwise have been the case

- **Whether there is open or closed access to third parties.** Voluntary agreements are generally developed outside the standard regulatory framework and, therefore, the level of consultation with external stakeholders such as environmental groups varies

- **Whether the approach is process- or performance-based.** The distinction is important as an emphasis on process may mean that the desired performance outcomes are not achieved, whereas an emphasis on outcomes alone may mean that due process is not followed

Environmental effectiveness

While acknowledging that many voluntary approaches have failed to deliver on their stated objectives, the first point to be made about environmental effectiveness is that there is evidence that voluntary approaches can meet their goals in situations where the programme is administered appropriately and has the support of those involved. Even though this is a positive conclusion, it must be qualified by recognising that the objectives specified in voluntary approaches are generally suspected of being less stringent than those that would have been established in an equivalent command-and-control regime.

Although it is difficult to prove this point, the outcomes from voluntary approaches often do not deviate significantly from business as usual. It has been argued that the voluntary approaches that have achieved the most substantial outcomes are those that have established an effective system of sanctions or have offered the greatest rewards (e.g. tax rebates, collective benefits, simplicity of licensing arrangements). But, as noted above, the creation of effective sanctioning processes may increase the reluctance of companies to participate or may create pressure for the targets to be lowered before they will participate.

Industry has been widely suspected of using voluntary approaches to 'capture' environmental policy, so that regulation or policy favours industry at the expense of other interest groups. For example, the existence of a voluntary approach is often used by companies to argue that regulation is not required or, if regulation is seen as necessary, that the targets specified in the voluntary approach represent acceptable targets for industry.[2]

Although capture is relatively easy to describe in qualitative terms, it can be difficult to assess in practice because every form of regulation involves some degree of negotiation or dialogue between business and government. Ultimately, the potential for regulatory capture relates less to the choice of policy instrument than to the manner in which the regulatory approach is organised: in particular, the processes that ensure that all interests are represented, control the discretionary power of the regulatory agency, require the abatement objectives and the schedule for their achievement to be made explicit, mandate *ex post* public policy evaluation, and ensure credible sanctions in the event of non-compliance.

2 An analysis — using game theory and economic models of regulation — of the regulatory gains (pre-empting regulation, influencing the shape of regulation and deflecting the enforcement of existing regulation) that may accrue to the participants in voluntary programmes is presented in Lyon and Maxwell 2004.

Voluntary approaches are of particular concern in this regard because many of the necessary safeguards (e.g. the availability of sufficient information to differentiate between commitments that represent genuine abatement efforts and those that are simply business as usual, the ability to limit collusion between government agencies and industry interests) are frequently not available.

The credibility of many voluntary approaches has also been affected by 'free-riders'. That is, even though individual organisations may benefit from collective action, organisations that do not participate or that participate but merely feign compliance may also benefit. The greater the number of companies involved, the greater the temptations to free-ride, as there is a lower likelihood of detection and the benefits of cheating are likely to be greater. The ability to control free-riders depends on factors such as whether companies are aware of each other's behaviour and are able to detect non-compliance, the history of cooperative action, the ability to punish or sanction non-compliant behaviour, and the presence of market or other pressures to ensure that participants comply.

Finally, voluntary approaches tend to suffer from a lack of information. Common issues are: ambiguous targets and monitoring results, the unavailability of monitoring data, the lack of suitability of reported information, and the absence of interim targets. These information gaps make it difficult to assess the performance of the companies participating in the voluntary approach or the effectiveness of the voluntary approach as a whole.

Economic efficiency

The financial benefits of voluntary approaches potentially include improved compliance, better management of litigation risk, improved brand or reputation, better relationships with shareholders and society, and better morale and culture within the participating companies. While there have been few published economic evaluations of voluntary approaches (and those that have been completed have tended not to account for the level of environmental protection achieved), it appears that the primary benefits accrue to participants through their ability to forestall or influence regulations. That is, even though voluntary agreements offer the potential for privately efficient outcomes, these may not be the same as societally efficient outcomes (i.e. when all externalities are taken into account). Even in relation to private outcomes, the evidence is that voluntary approaches do not result in efficient outcomes for the participating companies because they tend to adopt a rule of equal burden-sharing, based on uniform standards rather than differentiating based on lowest abatement costs.

Despite this, voluntary approaches may provide financial benefits for individual companies by allowing the company to allocate pollution efforts among its facilities or to be flexible regarding when the final target is to be met. An example could be a company arranging for its pollution abatement efforts to fit with its investment cycle rather than forcing the early retirement of capital stock; this point is also raised in Chapter 7 where Jeffrey Apigian suggests that one of the advantages of the Climate Leaders programme (and similar voluntary initiatives) is that it provides companies with the flexibility to move slowly away from existing infrastructure in which they may have made huge capital investments.

Transaction costs

It has been argued that the transaction costs associated with voluntary approaches are lower than those for command-and-control instruments because the participating companies are better informed about their operating practices and processes than regulatory bodies, and so can design and implement better compliance management systems. In practice, however, many voluntary agreements include at least some prescriptive elements on how compliance is to be assessed (e.g. by reference to standard methods for measuring pollution), which means that the flexibility for companies to optimise these activities (and thereby reduce transaction costs) is often quite limited.

Competitiveness

A key question around voluntary approaches is whether they provide companies with the opportunity to collude and develop anti-competitive behaviour (e.g. through denying market access, price fixing, phasing out products that may be competitors to new products). This threat is likely to be greatest in situations where a voluntary approach concerns a concentrated sector where a small number of companies dominate the sector, though these are also precisely the conditions (few industry players, high exit costs, history of cooperation) that seem to offer the greatest potential for voluntary approaches to deliver substantial environmental and economic outcomes. Even though the potential for collusion exists, there is limited evidence available to enable its level to be judged. In addition, many countries have antitrust legislation that provides legal remedies in the event of anti-competitive behaviour which, while not guaranteeing that anti-competitive behaviour will be eliminated, frequently provides strong penalties to discourage collusion.

Soft effects

Soft effects refer to the behavioural, attitudinal and awareness changes that result from the implementation of policy instruments; providing these effects is often a stated objective of voluntary programmes. For example, voluntary approaches can:

- Disseminate information on pollution abatement techniques
- Provide a forum for collective learning and information sharing
- Develop management competence
- Develop new and improved relationships between the members of the industry and between the industry and other parties such as government

Innovation

The potential for voluntary approaches to stimulate innovation is unclear. Although learning processes (e.g. education, provision of information, experience sharing, technical support) are common objectives of voluntary approaches, it has been argued that the generally modest targets set in voluntary approaches offer limited incentive for companies to innovate. If a target can be met with a business-as-usual approach, there will be little incentive for companies to innovate.

This conclusion may be too harsh in situations where voluntary approaches are seen as the precursors to legislation. In such situations, voluntary approaches may help companies anticipate or prepare for regulatory developments, or to develop innovative technologies and approaches.

Acceptability

Opinions on the acceptability of voluntary approaches differ. Industry groups have tended to support voluntary approaches because of the potential financial savings and flexibility, the opportunity for industry to define its own standards, the potential to reduce or avoid regulation, and the reputation and public relations benefits of such approaches. However, this support is not universal, particularly where voluntary approaches are seen as the precursors of regulation or as raising the performance expectations of companies.

Governments have expressed interest in voluntary approaches as a means of reducing cost burdens on government and industry, accelerating the implementation of policy and creating the potential for win–win outcomes. However, they have also been concerned about the potential for voluntary approaches to undermine the ability of governments to implement environmental regulations.

Voluntary approaches have been criticised by environmental groups on the grounds of weak standards, ineffective enforcement, the exclusion of stakeholders and government, lack of credibility and transparency, and the potential for voluntary approaches to weaken the regulatory framework or to delay the implementation of regulations. These arguments may reflect a lack of trust in business rather than necessarily being inherent flaws in voluntary approaches as non-governmental organisations (NGOs) have themselves been involved in voluntary approaches where such approaches have enabled their specific issues and agendas to be advanced.

Voluntary approaches in climate change policy

The scale of the challenge

Before discussing the role that voluntary approaches could pay in climate change policy, it is important to step back and examine the scale of the challenge that climate change presents. Table 21.1 presents the major policy options available to reduce greenhouse gas emissions from industrial and other activities, and an assessment of the potential impact of voluntary approaches on these policy options. These policy options all require significant investments (capital and operating costs, political support, cooperation) over the medium to long term. Table 21.1 suggests that, while voluntary approaches may assist in implementing or delivering some of these policy measures (particularly those that require communications, cooperation or support), significant changes will inevitably require strong financial and regulatory drivers.

TABLE 21.1 The impact of voluntary approaches on greenhouse gas emissions

Source: Krarup and Ramesohl 2000

Policy option	Requirements	Time-frame	Impact of voluntary programmes
Changes in product design, composition of processed materials, resource use	Strategic commitment and long-term decisions with regard to a change of technical paradigms, process technologies and resource structures	Long-term	Minor
Change of energy supply structure	Strategic commitment and long-term decisions with regard to energy infrastructure and fuel input	Mid-/long-term	Some effects but depend on the policy mix
Increased technology innovation	Strategic commitment and long-term research and development investment	Long-term	Minor
Enhanced investment	Change in strategic and operative business goals as well as altered decision criteria and procurement processes	Short-/ medium-term	Some, depending on policy mix
Enhanced technology diffusion	Increased communication, exchange of practical experience, dissemination of best practice, new network links, energy-related cooperation	Medium-term	Some, depending on existing cooperation and competition
Improved energy management	Integrated approach and systematic search for improvement options, changes in organisational routines, staff empowerment	Medium-term	Some, depending on the design of the scheme
Awareness and motivation	Mobilisation of company actors, provision of information, know-how and expertise, continuous discussion of the issue	Short-/ medium-term	Some effects

A related issue is that, in primary production and basic industries, measures to close resource cycles and optimise processes and activities play the main role in reducing energy consumption and greenhouse gas emissions. Any significant changes in energy efficiency will depend on innovation in process technologies and ongoing research and development activities. These measures are mainly triggered by cost reduction pressures or by distinct environmental regulations. While voluntary programmes may foster single projects or research initiatives, they are unlikely to change underlying strategies and pressures for energy efficiency. This is of particular relevance given the major contribution of primary industry and power generation to global greenhouse gas emissions, and given that change in such industries is driven both by technological developments and by the rate of retirement of existing plant and equipment (which, for many such industries, can be over a period of 20–30 years or more).

Even outside heavy industry, responding effectively to climate change presents a fundamental dilemma for companies. For most companies in most sectors, business expansion (in the absence of a complete and dramatic change in the business) will, almost inevitably, be accompanied by increases in total greenhouse gas emissions. A recurring theme through many of the contributions in this book is that, while many companies are willing to set targets to improve the emissions intensity of their business (i.e. to reduce the greenhouse gas emissions per unit of production),[3] for most companies, the emissions abatement achieved through such measures is likely to be dwarfed by increases in the overall scale of the business. In practice, delivering absolute reductions in greenhouse gas emissions is an extremely difficult and complex challenge and, in the absence of strong incentives for emission reductions, is likely to run counter to the manner in which companies define their financial self-interest.

The changing policy environment

In many ways, the case studies presented in this book broadly confirm the strengths (relatively low cost, speed of implementation, increased management attention on greenhouse gas emissions) and weaknesses (modest environmental outcomes, lack of dependability, questionable efficiency benefits, frequently used as a barrier to stronger policy action) of voluntary approaches.

The weaknesses, in particular, reflect the fact that many voluntary approaches were introduced when the threat of regulation was seen as remote (or where the effect of the voluntary approach was to deflect calls for regulation). However, the situation has changed fundamentally; regulation directed at reducing greenhouse gas emissions is now seen as inevitable by many businesses and, in countries where such regulation is already in place, the expectation is that emissions will be reduced even further over time. Moreover, there is a growing recognition that governments are willing to take actions that may not be completely business-friendly. Given this changed policy environment, the case studies presented in this book suggest that there are two interesting roles that can be played by voluntary approaches in the climate change policy area:

- As a transitional policy instrument
- As a mechanism to address some of the limitations in corporate cost–benefit assessments

Voluntary approaches as a transitional policy instrument

Before discussing the specific question of voluntary approaches as a transitional policy instrument, it is important to note that voluntary approaches may have a policy value in and of themselves — even if they are not directed towards enabling stronger policy measures to be adopted at a later stage.

Both the literature on voluntary approaches and the case studies presented here confirm that voluntary approaches can provide a range of important soft effects including:

---

3 The literature on voluntary approaches suggests that they are more likely to be effective (in terms of meeting their targets and maximising participation) when the target is expressed in relative rather than absolute terms.

- The development of managerial and technical knowledge in areas such as the preparation of greenhouse gas inventories

- The formulation of emission reduction targets

- The development of standards, processes, technologies and business models (including new ways of financing) to reduce emissions

- The development of databases (e.g. on industry performance, emissions reductions techniques)

- The identification of opportunities for greenhouse gas emissions reductions

- The development of better relationships between business and government

These are all necessary and important in enabling companies to respond effectively to legislation related to climate change. Furthermore, experience with voluntary approaches assists policy-makers by generating essential data and information on issues such as possible emission reductions, the profitability of specific technologies, and the manner in which companies will respond to different incentives and policy instruments.

One of the criticisms of voluntary approaches has been that they have been seen (particularly by business participants) as an end in themselves rather than as enabling a transition to a more stringent regulatory regime or policy environment. The most clear-cut example of this is the Australian Greenhouse Challenge (see Chapter 8), which has been in place for over 12 years with relatively little substantive change in the programme over that time. The Greenhouse Challenge has consistently been used by Australian industry as an argument against the introduction of tighter climate change regulations. It is interesting that the case studies in this book suggest that at least some of the more recent climate change-related voluntary initiatives are not seen as an end point in themselves, but as intermediate steps towards regulation. For example, the Cement Sustainability Initiative (CSI) (see Chapter 14) sees its role as not to set performance outcomes for the industry but rather to deliver a framework emissions calculation methodology and emissions data, which can then be used by governments in setting targets for the sector. When considering the role of voluntary approaches as a transitional policy instrument, there are – in very simple terms – two distinct situations that need to be considered.

First, in countries with reasonably well-developed regulatory frameworks (e.g. many of the countries in the European Union where the EU Emission Trading Scheme (EU ETS) and national policy frameworks in areas such as renewable energy mean that climate change is a well understood and recognised business issue), Chapter 2 suggests that the majority of large companies now have established systems and processes (management accountabilities, environmental or climate change policies, greenhouse gas inventories) for managing their climate change-related issues and risks. In this context, voluntary approaches that simply seek to reinforce these characteristics seem likely to have less relevance going forwards; as we move towards mandatory emissions reporting and, as the regulatory framework tightens even further, having these systems and processes in place will simply be seen as standard business practice. However, leadership initiatives (e.g. focusing on the development of new markets or the development and testing of new technologies) will have an important role to play as they will enable companies to experiment and innovate, and to demonstrate the benefits of proactively responding to climate change.

Second, in countries where the regulatory framework is less well developed, voluntary approaches focused on establishing management systems, encouraging companies to consider the risks and opportunities presented by climate change and improving reporting are likely to continue to have an important role to play for some time to come – at least until more stringent regulatory requirements are introduced. In these countries, leadership initiatives may, over the short term, be of less relevance – or may have less of a role to play until some of the more basic management systems and processes have been established.

The Cement Sustainability Initiative (see Chapter 14) points to another potentially important contribution of voluntary approaches. The challenge faced by international climate change policy-makers is that it is likely to be some time before all countries have signed up to binding international targets. In the interim, different regulatory environments may result in competition distortions, with companies in countries with strict emission limits facing higher costs, which in turn may lead to a loss of competitiveness. This may result in 'carbon leakage' where industries with significant greenhouse gas emissions decide to relocate to countries where emissions limits are less stringent.

As a solution to this problem, it has been suggested that international climate change policy should focus on global industrial sectors (e.g. iron and steel, aluminium, cement) as well as on single national or regional economies. The CSI is one example of what a sectoral initiative could look like involving the industry making a commitment to a proactive approach to emissions management, setting targets for reducing CO_2 emissions, reporting annually on performance towards meeting these targets, and obtaining third-party verification of emissions data. The type of approach that underpins the CSI offers the advantages of involving companies from developed and less developed countries, helping address at least some competitiveness issues, accelerating technology diffusion, and allowing policy measures that are appropriate to the sector to be developed.

Despite the theoretical potential of initiatives such as the CSI, it is important to recognise that the most significant challenge (i.e. setting and agreeing emission reduction targets for the sector) has yet to be addressed. It is also important to recognise that the cement industry is seeing significant increases in demand and, notwithstanding the sector's efforts to improve its efficiency, total emissions are expected to continue to rise.

Cost–benefit assessment

In the literature on business and climate change, one of the commonly identified barriers to business action on climate change is that corporate cost–benefit assessments do not account properly for the risks and opportunities presented by reducing their greenhouse gas emissions. It appears that there are actually three distinct issues at play:

● **The climate change market failure.** Given that most companies are not required to pay anything near the full externality or damage caused by their greenhouse gas emissions, the incentive is for companies to continue to emit rather than reduce their greenhouse gas emissions

● **The significance of energy or greenhouse gas emissions as a cost for the business.** For most companies, energy represents a relatively small part of their overall cost base and so there is limited incentive for them to focus management attention in this area

● **The rates of return** required by companies on their investments in emissions abatement or energy efficiency, particularly the relatively short time-frames over which most investments are to be justified or generate a return

Turning first to the question of market failure, the economics literature suggests that the primary solution to market failure is for government to regulate to force companies to internalise the damage caused by their emissions. The literature on voluntary approaches has tended to support this contention, with scepticism that voluntary approaches can play a meaningful role in correcting market failures. However, the case studies in this book challenge this rather pessimistic view, as many of them do involve companies taking action that cannot be justified in strict cost–benefit terms. But, in sectors where internalising greenhouse gas emissions will entail significant costs (see the general comments on voluntary approaches in climate change above and the chapters about the cement and aviation sectors), the potential for voluntary approaches to deliver substantial change in performance appears limited.

Having noted these limitations, advocacy partnerships (i.e. those directed at public policy) do seem to have a very interesting and potentially important role to play. Among other chapters in this book, those describing the activities of The Climate Group (Chapter 10), the role played by some of the major climate change partnerships (Chapter 11) and the activities of the Institutional Investors Group on Climate Change (Chapter 12) all demonstrate the role that can be played by voluntary approaches directed at public policy engagement in resolving the significant Catch 22 situation in climate change policy; while companies need regulatory certainty for their investments, governments are hesitant to provide complete regulatory certainty because do not know exactly how positively companies (and thus their national economies) will react to climate policies. Advocacy partnerships have helped changed the terms of the debate from one of whether governments should take action to reduce emissions from business to one where the question is what level of emissions reductions are required and over what time-frame. These groups have also provided technical advice to regulators on how policy may be defined and implemented in a manner that provides flexibility to policy-makers while also providing a high degree of confidence in the business response.

There is another point to note in relation to the analysis of climate change as a market failure which is that, despite the fact that the market failure persists (i.e. many governments have yet to introduce regulations and, even where regulations have been introduced, the coverage is limited and the incentives provided tend to be a small proportion of the value of the externality caused by the greenhouse gas emissions in question), many companies are acting in manner that suggests that the market failure has been corrected – at least partially. The evidence presented in this book suggests that many companies see the threat of climate change regulation as real and are building the likelihood of regulatory action into their investment decisions, i.e. even though the climate change market failure is by no means close to being corrected, companies are starting to take action that suggests that they expect the market failure to be at least partially corrected over the short to medium term.

Turning to the second point above, i.e. corporate cost–benefit assessments, it is important to recognise that, for most companies, energy or greenhouse gas emissions are a very small part of their overall cost base and so there is limited incentive – from a business perspective – to focus on these areas. The lack of visibility of these costs and the significant transaction costs involved in identifying and implementing energy-saving opportunities are important barriers to companies taking action. In practice, for the majority of companies (particularly outside energy-intensive sectors), there are often easier and more readily available cost-saving opportunities available elsewhere in the business. As a result, even in organisations with sophisticated core business evaluation processes, energy-related choices may be relegated to simple payback criteria or even just taken on the basis of lowest upfront costs. Voluntary approaches can, at least partially, help overcome this barrier by making climate change or greenhouse gas emissions an explicit subject for management attention; the expectation is that management will follow through on its commitments even if this is simply to prepare an inventory of greenhouse gas emissions. Even signing up to such a relatively modest commitment can – as the case studies in this book demonstrate – increase the commitment of senior management to taking action by:

- Providing better information on the sources of emissions

- Creating a framework for accountability (e.g. through reporting, industry peer pressure)

- Helping develop skills and capacity (e.g. through education, training and support for participating organisations).

These skills and competences seem to be particularly important in enabling companies to identify 'easy wins' or 'low-hanging fruit', i.e. opportunities that may not otherwise have been identified or acted upon.

Another issue with corporate cost–benefit assessments is that there is a general focus on short-term returns rather than longer-term business sustainability. In practice, the cost–benefit tests applied to investments in energy efficiency/greenhouse gas emission reductions may mean that many cost-effective opportunities for emission reductions are not taken. Even in a low-inflation environment, it is not uncommon for companies to expect environmental or energy investments to repay the capital investment within two years (see, generally, Sullivan and Sullivan 2005). This is significantly greater than the typical investment criteria in industries such as energy, oil, gas and mining, which typically expect large capital investments (e.g. new power generating equipment) to provide a rate of return of 15–20%.

While such a comparison may not be strictly fair, it suggests that there are likely to be many opportunities for energy or environmental performance improvements that are economically viable (and relatively risk-free) that are not presently being implemented. Voluntary approaches can help address this through encouraging companies to change the manner in which they conduct cost–benefit assessments on energy-saving measures. For example, Novartis (as described in Chapter 15) allows energy-saving and energy efficiency investments to be paid back over the lifetime of new investments.

Voluntary approaches may also change the internal expectations of companies (e.g. changes in management attitudes) and/or empower other stakeholders to exert influence on companies (e.g. through voluntary disclosure initiatives). The effect is to increase the incentive (or pressure) for companies to follow through on their commitments. That is, at least at the margins, voluntary approaches may increase the incentives for companies to reduce their greenhouse gas emissions which may, in turn, alter the cost–benefit calculation for the investments. Some care is required with this conclusion as the targets that are to be met may simply be 'business as usual' targets (i.e. may not represent a substantive change in management's views). Furthermore, notwithstanding the literature on stakeholders and stakeholder theory, it is by no means clear that stakeholders such as environmental NGOs are sufficiently well resourced or informed to deliver on this type of monitoring and oversight or, even if they are, that they would be interested in discharging such a role.

Closing comments

Irrespective of their efficacy, voluntary approaches will continue to be seen as an important part of the climate change policy framework. The reasons include:

- The need to have business 'on side' in the climate change policy debate
- The reality that in many countries there is simply no foreseeable policy alternative to voluntary approaches
- The soft effects (particularly capacity-building) that voluntary approaches can provide
- The potential to encourage and reward business leadership

The material presented in this book, while confirming many of the more negative critiques of voluntary approaches (in terms of dependability, limited environmental outcomes, questionable efficiency benefits), suggests that there are two potentially important applications of voluntary approaches:

- As a transition tool – whether preparing companies for the introduction of new legislation or the stimulation of leadership and innovation
- As a means of improving the manner in which companies make investment and other decisions that impact on their greenhouse gas emissions

Despite these potentially important contributions, the inherent weaknesses of voluntary approaches, the limited environmental outcomes that have been achieved and the limited evidence of economic and competitiveness benefits should make governments very reluctant to rely on voluntary approaches as the primary policy response to climate change. That is, while they probably do have a role to play, voluntary approaches should not be seen as a replacement for the stronger policy measures that are likely to be essential to the delivery of significant reductions in greenhouse gas emissions.

References

Krarup, S., and S. Ramesohl (2000) *Voluntary Agreements in Energy Policy: Implementation and Efficiency* (Copenhagen: AKF Institute of Local Government Studies).

Lyon, T., and J. Maxwell (2004) *Corporate Environmentalism and Public Policy* (Cambridge, UK: Cambridge University Press).

OECD (Organisation for Economic Cooperation and Development) (1999) *Voluntary Approaches for Environmental Policy: An Assessment* (Paris: OECD).

Sullivan, R. (2005) *Rethinking Voluntary Approaches in Environmental Policy* (Cheltenham, UK: Edward Elgar).

—— and J. Sullivan (2005) 'Environmental Management Systems and their Influence on Corporate Responses to Climate Change', in K. Begg, F. van der Woerd and D. Levy (eds.), *The Business of Climate Change: Corporate Responses to Kyoto* (Sheffield, UK: Greenleaf Publishing): 117-30.

22
Setting a future direction for climate change policy

Rory Sullivan
Insight Investment, UK

Current business responses

The past five years have seen a dramatic change in the debate around business and climate change. The research presented in Chapter 2 suggests that we have reached a point where the governance argument has been won: companies accept that they have responsibility for managing or reducing their greenhouse gas emissions and most have — notwithstanding weaknesses in individual companies and in specific aspects of emissions management and reporting — established the governance and management systems they need to manage their greenhouse gas emissions, with many also having set emission reduction targets. An increasing number of companies and institutional investors are engaging with policy-makers to encourage them to establish a clear, long-term climate change policy framework — including emission reduction targets — as an essential part of ensuring the delivery of climate policy goals in an economically efficient manner. While some of this lobbying is, inevitably, self-interested (e.g. many companies have expressed support for emissions reductions so long as their sector is excluded from these reductions because of their own 'special circumstances'), it suggests that there is some willingness on the part of companies to accept mandatory emission reduction targets.

Notwithstanding the progress on governance and the support for public policy action on climate change, the arguments around corporate strategy have yet to begin in earnest. With the exception of a few leadership companies that recognise the fundamental challenge presented by climate change to their business models and their wider societal responsibility to significantly reduce their greenhouse gas emissions, the majority of companies are focusing their efforts on improving their energy efficiency or greenhouse gas emissions intensity per unit of production or turnover. Yet, while acknowledging the very significant time and resources being invested by companies in reducing their greenhouse gas emissions, the reality is that most expect their total greenhouse gas emissions to increase because of business growth.

Perhaps of greater concern is that, despite the scientific consensus that significant reductions in greenhouse gas emissions are required as a matter of urgency, the vast majority of companies perceive climate change as having minimal impact on their business strategy or business models. Very few companies seem to be searching for 'transformational initiatives' that will allow them to deliver step change reductions in their greenhouse gas emissions performance and in the emissions associated with their supply chains or the use and disposal of their products. Many companies seem happy to adopt a 'wait and see' approach and rather than seeking to pre-empt public policy (other than to the extent that it is justified in financial terms) are prepared to wait and respond to policy as and when it emerges.

Public policy

Although various factors — cost reductions, new business opportunities, innovation, improved brand and reputation, stakeholder pressures — have encouraged companies to take action to reduce their greenhouse gas emissions, the clear message from the case studies presented in this book is that, with the exception of those companies that have positioned themselves as climate change leaders, regulation has been the single most important driver for action.

The EU Emission Trading Scheme (EU ETS) has played a catalytic role in putting climate change on the corporate agenda. It has not only demonstrated that governments could act, but that they were prepared to act and to take decisions that were not necessarily in the best short-term interests of all companies. Furthermore, the EU ETS has shown that European policy-makers will at least countenance a carbon price of at least €30/tonne with some degree of equanimity. Yet, despite the EU ETS and the variety of other policy measures that have been adopted (in areas such as renewable energy and energy efficiency), it is clear that many companies do not have confidence in the longer-term climate change policy framework. The uncertainties in future climate change policy and the gaps and inconsistencies in existing policy approaches are critical barriers to corporate action. There are a number of points to note in this regard.

The first point is that climate change policy is not simply about single policy instrument solutions. An effective policy response to climate change will require the deployment of a range of policy instruments (command-and-control instruments such as product standards, taxes and other economic instruments, information-based approaches such as product labelling and voluntary approaches), targeted at specific sectors and focusing on different corporate motivations. While emissions trading has been a key catalyst, it clearly cannot address all greenhouse gas emissions or motivate all companies to take action.

There are also a number of practical issues to highlight from the experience to date with the EU ETS. First, the effectiveness of emissions trading depends on the cap that is set and the amount of emissions reductions that are to be delivered through offsetting as opposed to direct emissions reductions actions. But, as the analysis presented in Chapter 3 suggests, the existence of a price signal and a cap is not enough. Factors such as the duration of the price signal and policy certainty (discussed further below) are key. In relation to duration, many companies have decided to wait and see how climate change evolves before committing large amounts of capital to new projects. The second issue is that, contrary to the literature, companies do not respond 'rationally' (in an economic sense) to price signals. As indicated in Chapter 4, many companies see emissions trading as a compliance requirement (i.e. they are not active participants in the market other than to the extent required to cover any emissions shortfall that they may have). The reasons include the transaction costs associated with buying and selling permits, the view of emissions trading as a compliance requirement rather than a business opportunity, and the reality that emissions trading is not a significant cost item for many companies. The third issue is that, while the case for using market-based instruments is well known and increasingly accepted by governments, such instruments are typically effective over the medium to long term; in the short term, demand for greenhouse gas emitting activities such as electricity generation and transport tends to be relatively inelastic with supply linked to sunk capital costs in existing infrastructure.

The second point is that the policy and market conditions faced by companies remain highly complex and uncertain. The uncertainties in climate policy include the political context within which climate policy is developed (e.g. the level of government support for climate policy measures, concerns about energy security or wider competitiveness issues), the targets that are to be met and how these are to be allocated between companies, the policy instruments that are used, and the duration of climate change policy instruments. These uncertainties are a key influence on how companies conceive of and respond to climate change policy. The reality is that reducing emissions is likely to require significant irreversible investment by the private sector and the profitability of such investments is highly sensitive to climate change policy. Whether companies will invest depends on whether they will take governments at their word; faced with the political demands of elections, will governments renege on their promises to maintain carbon taxes or EU ETS, or will they maintain commitments to limit the number of permits allocated under emissions trading schemes?

One of the most important conclusions from this book is that the response of companies to policy uncertainty is to wait for clarity. Expressed another way, the incentives for early action (e.g. the carbon price in an emissions trading scheme) need to be significantly higher than would otherwise have been the case to stimulate the changes that are desired by policy-makers.

There are no 'silver bullets' for this problem. Chapter 3 offers some suggestions on the types of actions that public policy-makers need to take to address this issue of policy uncertainty including:

- Avoiding policy disconnects (e.g. avoiding a post-2012 hiatus in the international policy framework). Although the politics around the specific links between EU ETS and the Kyoto Protocol are sensitive, there should be a clear EU commitment to ensuring the two remain linked. There should also be a fallback plan in the event of any hiatus in the international process, which would be very damaging for momentum in private (e.g. emissions trading) markets

- Making it clear that emissions trading is an integral part of the policy framework for responding to climate change

- Communicating clearly the post-2012 ambition even if the policy mechanisms remain unclear. This should include the establishment of clear national and international greenhouse gas emission targets for 2020 (as the EU has done) since a 10–15 year time horizon is the key time-frame for companies' investment decision-making (particularly for large capital expenditure)

- Establishing the credibility of emissions trading and other measures directed at reducing greenhouse gas emissions through explicitly considering competitiveness issues as a key part of the design and implementation of these policy measures. In addition, governments need to be clear about the other trade-offs that will need to be made. For example, how will the goals of energy security (which may see coal as a key part of the solution) be reconciled with the need to significantly reduce greenhouse gas emissions? As another example, if the aviation sector is to grow, does this mean that other sectors of the economy will need to deliver even greater reductions in emissions?

- Accepting that action on climate change will probably cost money, at least over the short and medium term, and being clear about who will meet these costs. Without that acceptance, companies will not take government commitments seriously

Finally, in order to compensate for policy risk, policy-makers need to recognise that, in order to incentivise investment in low-emission technologies, the cost of emitting greenhouse gases may need to be substantially higher than expected under a normal discounted cash flow analysis. A specific issue is that the EU ETS on its own (even if there is a high degree of confidence that it will remain as part of the policy framework for responding to climate change) is unlikely to stimulate major investments in lower CO_2-emitting forms of power generation unless prices increase significantly from the level of around €20–25/tonne of CO_2 prevailing in the markets in early 2008. Raising the price will require strong political commitment to deal with the knock-on effects of higher carbon prices in terms of increased electricity prices and the consequences for competitiveness, fuel poverty and other policy objectives.

Concluding comments

In many ways, the material presented in this book is extremely encouraging: it demonstrates the significant actions that have been taken by companies to reduce their greenhouse gas emissions and it provides important insights into the manner in which public policy needs to be designed to enable and encourage companies to make a substantive contribution to responding to climate change.

Notwithstanding the progress that has been made, the positive messages from the book are overshadowed by the clear tension between business growth (and, indeed, economic growth more generally) and the urgent need to reduce greenhouse gas emissions. For most companies, responding to climate change is seen primarily in efficiency terms, with the consequence that business growth is frequently accompanied by increases in the absolute levels of greenhouse gas emissions. Clearly, in the absence of significantly stronger and clearer incentives for reductions in greenhouse gas emissions, we will continue to face this tension. The types of policy action being envisaged by the EU provide us with an example of what public policy in a world constrained by greenhouse gas emissions might look like. They should also provide us with confidence that we can make the transition to a low-carbon future with minimal economic dislocation.

In conclusion, ensuring that all sectors of the economy and all countries of the world significantly reduce their greenhouse gas emissions as a matter of urgency is the single most important issue for policy-makers, for companies and for us as individuals. That is a stark and maybe unpalatable conclusion. But the bottom line is that the future of our planet, not just our economy, depends on our success in delivering significant reductions in global greenhouse gas emissions.

Acronyms and abbreviations

ABI	Association of British Insurers
ACCA	Association of Chartered Certified Accountants
ACF	Australian Conservation Foundation
ACX	Australia Climate Exchange
AGO	Australian Greenhouse Office
AIGN	Australian Industry Greenhouse Network
APP	Asia–Pacific Partnership
BBA	British Bankers' Association
BC	British Columbia
BCA	Business Council of Australia
BERR	Department for Business, Enterprise and Regulatory Reform (UK)
BiTC	Business in the Community
BLS	Bureau of Labor Statistics (USA)
BSI	British Standards Institution
BSR	Business for Social Responsibility
C&G	Cheltenham & Gloucester
CaCX	California Climate Exchange
CAIt	Climate Analysis Indicators Tool (WRI)
CAR	Capital Appropriation Request (Novartis)
CCAP	Center for Clean Air Policy
CCGT	combined cycle gas turbine
CCX	Chicago Climate Exchange
CDM	Clean Development Mechanism (Kyoto Protocol)
CDM-EB	CDM Executive Board
CDP	Carbon Disclosure Project
CEO	chief executive officer
CER	Certified Emissions Reduction (Kyoto Protocol)
CESPEDES	Comisión de Estudios del Sector Privado para el Desarrollo Sustenible (Mexican Business Council for Sustainable Development)
CH_4	methane

CHP	combined heat and power
CICA	Canadian Institute of Certified Accountants
CO_2	carbon dioxide
$CO_2(eq)$	carbon dioxide equivalent
COA-RETC	Cédula de Operación Anual–Registro de Emisiones y Transferencia de Contaminantes (Annual Operation Report: Register of Emissions and Transference of Pollutants)
COP	Conference of the Parties (under the UNFCCC)
CSI	Cement Sustainability Initiative
CSR	corporate social responsibility
DCF	discounted cash flow
Defra	Department for Environment, Food and Rural Affairs (UK)
DJSI	Dow Jones Sustainability Index
DNA	Designated National Authority
DrKW	Dresdner Kleinwort Wasserstein
DTI	Department of Trade and Industry (UK; now BERR)
EAI	Enhanced Analytics Initiative
EC	European Commission
EDF	Electricité de France
EEI	Energy Efficiency Index
EMAS	Eco-management and Audit Scheme
ENGO	environmental non-governmental organisation
ERPA	emission reduction purchase agreement
ETS	Emission[s] Trading Scheme
EU	European Union
EUA	EU Allowance
EuP	Energy-using Products [Directive]
FDI	foreign direct investment
FT500	Financial Times Global 500
GAO	Government Accountability Office (USA)
GDP	gross domestic product
GE	General Electric
GHG	greenhouse gas
GM	General Motors
HAP	hazardous air pollutant
HFC	hydrofluorocarbon
HP	Hewlett Packard
HSE	health, safety and environment
IAI	International Aluminium Institute
IASB	International Accounting Standards Board
IATA	International Air Transport Association
ICT	information and communication technology
IEA	International Energy Agency
IFRIC	International Financial Reporting Interpretations Committee
IGCC	Investor Group on Climate Change (Australia/New Zealand)
IGES	Institute for Global Environmental Studies
IIGCC	Institutional Investors Group on Climate Change
IISI	International Iron and Steel Institute

IMP	Inventory Management Plan
INCR	Investor Network on Climate Risk (North America)
IPCC	Intergovernmental Panel on Climate Change
ISC	Institutional Shareholders Committee (UK)
ISO	International Organisation for Standardisation
IT	information technology
JI	Joint Implementation (Kyoto Protocol)
LPFA	London Pension Fund Authority
LULUCF	land-use, land-use change and forestry
MEMR	Ministry of Energy and Mineral Resources (Indonesia)
METI	Ministry of Economy, Trade and Industry (Japan)
MNC	multinational corporation
MoE	Ministry of Environment (Japan)
MT CO_2(eq)	million tonnes of carbon dioxide equivalent
MTBE	methyl tert-butyl ether
N_2O	nitrous oxide
NAO	National Audit Office (Australia)
NEMS	National Modeling System (US Department of Energy)
NGO	non-governmental organisation
NO_x	oxides of nitrogen
NPV	net present value
NSW	New South Wales
NWO	Netherlands Organisation for Scientific Research
OECD	Organisation for Economic Cooperation and Development
P&G	Procter & Gamble
PACIA	Plastics and Chemicals Industry Association (Australia)
PEMEX	Petróleos Mexicanos
PFC	perfluorocarbon
PR	public relations
PV	photovoltaic
PwC	PricewaterhouseCoopers
R&D	research and development
RGGI	Regional Greenhouse Gas Initiative (USA)
RRSP	Registered Retirement Savings Plan (Canada)
SAVE	Specific Actions for Vigorous Energy Efficiency (EU)
SEMARNAT	Secretaría de Medio Ambiente y Recursos Naturales (Ministry of Environment and Natural Resources, Mexico)
SENA	Suez North America
SER	social and environmental responsibility
SF_6	sulphur hexafluoride
SIP	Statement of Investment Principles
SME	small or medium-sized enterprise
SRI	socially responsible investment
$teCO_2$	tonne of CO_2
UNDP	United Nations Development Programme
UNEP	United Nations Environment Programme
UNEP FI	UNEP Finance Initiative

UNFCCC	UN Framework Convention on Climate Change
USCAP	United States Climate Action Partnership
USEPA	United States Environmental Protection Agency
USS	Universities Superannuation Scheme (UK)
VCS	Voluntary Carbon Standard
WBCSD	World Business Council for Sustainable Development
WEF	World Economic Forum
WHO	World Health Organisation
WRI	World Resources Institute
WWF	formerly the World Wildlife Fund/World Wide Fund for Nature

About the contributors

Jeffrey Apigian is a graduate student in the Environmental Science and Policy programme at the Department of International Development, Community and Environment (IDCE), Clark University, Worcester, MA. He received his master's degree in May 2008.

Dr **Marlen Arnold** is a post-doctoral research associate at the Chair of Brewery and Food Industry Management, TUM Business School. She is also a lecture mentor on several web-based educational programmes at the University of Oldenburg, and teaches and consults for several companies. Her experience includes project work in several research projects at the University of Oldenburg and at the Institute for Ecological Economy Research (Berlin) in the fields of sustainability, innovation and strategic management, organisational learning and product development covering sectors such as information and communication technology (ICT), transport, housing and construction, and forestry. Marlen holds a PhD in economics and a qualification as mediator. She has published in the areas of sustainability-oriented organisational learning and cultural change, strategic and innovation management, product development and evolutionary economics.

Dr **William Blyth** has over 13 years of experience working on energy and climate change policy analysis. He is currently an Associate Fellow of Chatham House as well as Director of Oxford Energy Associates, an independent research company. His recent research focuses on decision-making under uncertainty in decisions on energy sector investment and climate change policy. He has written various publications on this topic including *Climate Policy Uncertainty and Investment Risk* for the International Energy Agency (IEA). William worked previously at the IEA in Paris on energy security and climate change policy. Before joining the IEA, he worked for a short period at the European Environment Agency in Copenhagen and, for over eight years, at AEA Technology, one of Europe's largest environmental consultancies. William has a DPhil in physics from Oxford University.

Professor Dr **Han Brezet** holds a tenure position in the field of design for sustainability–ecodesign in the Faculty of Industrial Design Engineering at the Delft University of Technology, The Netherlands. As head of the DfS School, he has successfully supervised almost 20 PhD and more than 200 MEng students. In addition, he coordinates the Solar Product Design programme of the 3TU Cartesius Institute in Friesland Province. Professor Brezet is also a visiting professor at the IIIEE Institute of the Lund University, Sweden. The main focus of his current research is the integration of solar energy technologies and renewable materials into consumer and professional products. Professor Brezet is the author/co-author of several books and academic papers on ecodesign published through various United Nations distribution channels as well as through scientific journals concerned with design and environmental sciences. He is also a member of the jury for the main Dutch TV programme on product innovation, 'The Best Idea in the Netherlands'.

Dr **Timo Busch** is a senior research associate in the Department of Management, Technology, and Economics at the Swiss Federal Institute of Technology (ETH) Zurich, where his research focuses on carbon constraints and strategic management. From 1999 to 2005 he worked at the Wuppertal Institute for Climate, Environment and Energy addressing issues such as resource consumption, eco-efficiency, sustainable finance and climate change. Timo holds a master's degree in Business Administration and Economics and obtained a PhD from ETH in 2008. He has published in the *Ecological Economics* and *Journal of Industrial Ecology*, and is the co-editor of a book on material efficiency published in 2005 (*Materialeffizienz: Potenziale bewerten, Innovationen fördern, Beschäftigung sichern* [oekom verlag]).

Dr **Andrea B. Coulson** is a Senior Lecturer in Accounting at the University of Strathclyde. Her research and teaching interests lie in environmental accounting and accounting for risk. Her PhD focused on corporate environmental performance considerations within bank lending processes. Andrea has conducted extensive research on this subject for a range of organisations including the Economic and Social Research Council (UK), F&C Asset Management, Lloyds TSB Group, the Scottish Government, UNCTAD, UNEP Financial Institutions Initiatives and UniCredit Group. Andrea has worked as a consultant to the United Nations, developing and delivering workshops on environmental accounting in Africa, Asia, Central and Eastern Europe and South America. She has published a range of academic research papers, commissioned reports and book chapters on her research and holds a number of advisory positions including membership of Association of Chartered Certified Accountant's Social and Environmental Committee.

Rachel Crossley joined Insight Investment in 2002 to establish what is now recognised as one of the world's leading responsible investment teams. Rachel is responsible for the design and delivery of the team's engagement with companies on a wide range of environmental and social issue. She leads Insight's engagement on consumer health and obesity, nanotechnology, sustainability in the house-building sector and responsible supply chain management, and also undertakes analysis on the financial implications of these and other issues to help inform Insight's investment research. In addition, Rachel works on Insight's climate change engagement and investment research programmes, and she is a board member of The Climate Group. Rachel previously worked in the Socially Responsible Investment team at Friends Ivory & Sime, with the World Bank as an economist and environmental specialist developing natural resource and environmental management projects in India, Indonesia, China, Guyana, Belize and other countries, and with EA Capital in New York, specialising in providing strategic management consulting, financial and business development services to socially responsible businesses and international financial institutions. She holds an MSc in environmental management and policy from the University of London and a dual BA Honours degree in economics and geography from Sheffield University.

Professor **Claus-Heinrich Daub** is the Director of the Centre for Sustainable Management at the University of Applied Sciences Northwestern, Switzerland, where he is also Professor for Marketing. He heads the International Network for Sustainable Management and is the current President of the Society for Scientific Publishing in Switzerland. His previous experience includes heading the Marketing and Communications department at the European Centre for Economic Research and Strategy Consulting. He holds a PhD in Sociology from the University of Basel. In 2005 he completed a post-doctoral lecture qualification project on the role of multinational enterprises in sustainable development. He has published widely in the fields of corporate responsibility, ethics and sustainability reporting including publications in the *Journal of Business Ethics* and *Journal of Business Strategy and the Environment*.

Christian Engau is a PhD candidate at the Department of Management, Technology and Economics of ETH Zurich (Swiss Federal Institute of Technology). His research interests are in international business and strategic management, with particular emphasis on climate change and regulatory uncertainty. For his doctoral dissertation, he is investigating corporate climate change strategies in the context of the uncertainty associated with post-Kyoto Protocol climate change policy. He was previously a visiting scholar at the St Petersburg State University of Economics and Finance and the MIT Sloan School of Management. His non-academic experience includes working as a management consultant with McKinsey & Company with a focus on strategy projects in the chemical, basic materials and transportation industries. Christian holds a joint degree in mechanical engineering and business administration from Darmstadt University of Technology, Germany.

Olga Fadeeva is a PhD student in the Faculty of Industrial Design Engineering and the Design for Sustainability School at the Delft University of Technology, The Netherlands. She also works as a researcher at the Cartesius Institute. She previously studied environmental management and policy, economics and management, and applied mathematics and physics. Her professional interests include regional environmental management, regional policies for sustainable development, energy management, knowledge and change management, information management and communication. She has worked on these and related issues in Russia, Sweden and The Netherlands. She is presently working on the area of life-cycle costing for the commercialisation of photovoltaic products.

Taryn Fransen is a senior associate with the GHG Protocol Initiative at the World Resources Institute (WRI) where she oversees the design and implementation of national greenhouse gas (GHG) registries and conducts research and analysis on GHG accounting and climate change policy. Her expertise includes corporate- and project-level GHG accounting and reporting, GHG Program design, corporate climate change strategy, developing-country responses to climate change, and environmental markets. Prior to joining WRI, Taryn was an associate programme officer at the United Nations Foundation where she co-managed a portfolio of projects to engage institutional investors on climate change issues and attract financing for sustainable energy development. She has also served as a consultant on markets for ecosystem services for WWF and as Business Solutions Fellow at the Pew Center on Global Climate Change. Taryn holds a master's in Earth Systems from Stanford University. She has worked internationally throughout Latin America and Asia.

Ian Gill is a director of Vancity Credit Union. In addition, he is president of Ecotrust Canada (Vancouver, BC) and a director of Ecotrust (Portland, Oregon) — related non-profit organisations that promote the emergence of a conservation economy in the coastal temperate rainforests of North America. Before founding Ecotrust Canada in November 1994, Ian was a writer–broadcaster with the Canadian Broadcasting Corporation. He is a director of the Na na kila Institute and the Forestry Advisory Council at the University of British Colombia.

Dr **Volker H. Hoffmann** is Assistant Professor for Sustainability and Technology in the Department of Management, Technology and Economics at ETH Zurich (Swiss Federal Institute of Technology). In his research, Volker focuses on corporate strategy and climate change. His specific areas of interest include the impact of climate change policy on innovation, effects of regulatory uncertainty and carbon performance. Volker holds diplomas in chemical engineering and in business administration. He obtained his PhD from ETH Zurich with a thesis on multi-objective decision-making under uncertainty and was a visiting scholar at the Massachusetts Institute of Technology. For several years, he worked as a consultant and project manager at McKinsey & Company where he helped large multinational corporations to develop their corporate strategy with respect to climate change. Volker's research appears in various journals including *Climate Policy*, *Energy Policy*, *European Management Journal*, *Ecological Economics*, *Business Strategy and the Environment*, *Journal of Industrial Ecology* and *Environmental Science and Policy*.

Dr **Aileen Ionescu-Somers** is deputy director the Forum for Corporate Sustainability Management (CSM) — the research platform for the International Institute for Management Development (IMD) in Lausanne. Previously she was head of the International Projects Unit at WWF International and, before that, held programme management roles with its Africa and Latin America regional programmes. She holds a PhD in business administration from the National University of Ireland (NUI) and an MSc in environmental management from Imperial College London. Her doctoral thesis concerned the business case for corporate social responsibility in the food and beverage sector.

Rosa María Jiménez-Ambriz is the sub-director of the Business Council for Sustainable Development–Mexico (BCSD-Mexico; CESPEDES) where she is in charge of the department of energy and climate change. She is the coordinator of the Mexico greenhouse gas (GHG) programme and the Sectoral Approach Project on Climate Change. Her expertise includes GHG accounting and reporting, assessment of Clean Development Mechanism (CDM) projects and sectoral analysis. Before joining BCSD-Mexico, Rosa was a consultant with PA Consulting Group where she coordinated the analysis of a portfolio of projects for the Mexican Center for Cleaner Production. She holds a master's in Quality, Safety and Environment from Otto von Guericke University in Germany.

Dr **Howard Klee** works with the World Business Council for Sustainable Development (WBCSD) in Geneva where he is programme director for the Cement Sustainability Initiative (CSI). He also supports the WBCSD's work with the China Council in developing policy recommendations to promote sustainable development in China and manages several other WBCSD programmes including a recent analysis of sustainable health systems. Before joining WBCSD, Howard worked at BP where he served in a number of executive and business functions, including strategic planning, business development, environmental affairs and manufacturing. Howard received his undergraduate degree in chemistry from Williams College, MA, and his PhD in chemical engineering from MIT. He served on President Clinton's Council on Sustainable Development and was previously a Director for the Foundation for Research on Economics and the Environment.

Ans Kolk is Professor of Sustainable Management of the University of Amsterdam Business School, The Netherlands. Her research, teaching and publications are in the areas of corporate social responsibility and sustainability, especially in relation to the strategies and international policy of multinational corporations. One of the topics on which she has published extensively is business and climate change. She has also been involved in various international projects on strategy, organisation and disclosure related to social and environmental issues. Professor Kolk has published in a range of international journals including the *Journal of International Business Studies, Journal of World Business, Harvard Business Review, Management International Review, California Management Review, Journal of Business Ethics, Business and Society, Business Strategy and the Environment, World Development, European Management Journal, International Environmental Affairs, Environmental Politics* and *Business and Politics*. She has written many book chapters as well as two books, *Economics of Environmental Management* (Financial Times Prentice Hall, 2000) and, with Jonatan Pinkse, *International Business and Global Climate Change* (Routledge, 2008).

Jennifer Kozak joined Insight Investment's responsible investment team in October 2002. She undertakes thematic research and engages with companies on a range of social and environmental issues including climate change, renewable energy technologies, genetic engineering, water and waste. She is also responsible for managing the screening process for Insight's ethical funds. Before joining Insight, Jennifer worked for six years at KLD Research & Analytics, Inc. — a leading provider of corporate social and environmental research to the global financial services industry. As a senior analyst and director of consulting, Jennifer researched the social and environmental records of hundreds of US and international companies and managed the company's consulting business. She also served for four years on the company selection committee for the Domini Social 400 Index, a well-established SRI benchmark for the US market. Jennifer holds a BA in international relations and Spanish from Mount Holyoke College, MA, and an MA in urban and environment policy from Tufts University, MA.

Dr **Yoram Krozer** is the Director of the Cartesius Institute, Institute for Sustainable Innovations of the Dutch Technical Universities. He studied biology, economics, business administration and was awarded his PhD for research on the environment and innovation. Before becoming an academic, he worked with non-governmental organisations and private companies, and owned a consultancy company. Yoram's research focuses on technological innovations in public domains such as water, energy, ecological systems and city development. One of his current projects relates to the development of markets for innovations in the water chain in a number of European, Asian and African countries in cooperation with four leading Dutch water companies and a Dutch water board.

Helen Mathews recently completed a master's degree in Sustainable Development at the University of Basel, Switzerland, and is now working as a sustainability consultant. The master's degree included a placement in the Corporate Health, Safety and Environment Division at Novartis's head office. Her research focus is on climate change and emissions trading. Helen's past experience includes four years as a financial auditor at KPMG Sydney. She is an Australian Chartered Accountant and has a degree in commerce.

Takaaki Miyaguchi holds a doctoral degree from the Kyoto University in Japan specialising in the role of the corporate sector in disaster and environmental management. He is currently working at the Regional Centre in Bangkok of the United Nations Development Programme (UNDP) covering the climate change mitigation and Clean Development Mechanism (CDM) portfolio in the Asia and Pacific region. He holds a master's in Public Policy from the University of Chicago and a bachelor's degree from the University of Michigan.

Leticia Ozawa-Meida worked as the Mexico GHG Program Coordinator at SEMARNAT (Mexican Ministry of Environment and Natural Resources) from August 2005 to December 2007. She completed her PhD studies on the response of the Mexican cement industry to climate change initiatives at the School of Environmental Sciences at the University of East Anglia in the UK. Leticia has a master's in Energy Planning and an engineering degree from the National Autonomous University of Mexico. She has previously worked as a research assistant in the Energy and Environment Group at the Engineering Institute in Mexico and as a research visiting student in the Energy Analysis Department in the Lawrence Berkeley National Laboratory. Her expertise includes corporate- and project-level greenhouse gas (GHG) accounting and reporting, energy analysis and modelling, and corporate climate change strategy.

Stephanie Pfeifer is programme director for the Institutional Investors Group on Climate Change — a forum for collaboration on climate change between leading pension funds and asset managers in Europe. The group's focus is on promoting better understanding of the implications of climate change for financial performance and encouraging the integration of climate risks and opportunities into investment decision-making. Stephanie has eight years of experience in the financial sector, having worked as an economist with NatWest, Morgan Grenfell and Deutsche Bank. Stephanie holds an MA in environmental change and management from Oxford University, an MA in European economics from Exeter University and a BA in philosophy, politics and economics from Oxford University. She has published on climate change and investment issues in *Climatic Change* as well as in many trade and investment journals.

Dr **Jonatan Pinkse** is Assistant Professor at the University of Amsterdam Business School, The Netherlands. His areas of research, teaching and publications are strategy and sustainable management. His PhD thesis, which was awarded the 2006 ONE Academy of Management Best Dissertation Award, addressed business responses to climate change. Jonatan has published articles in the *Journal of International Business Studies*, *California Management Review*, *Business and Society*, *Business Strategy and the Environment*, and *European Management Journal*. He is the author with Ans Kolk of *International Business and Global Climate Change* (Routledge, 2008).

Amanda Pitre-Hayes is manager, community leadership at Vancity, where she leads Vancity's Climate Change Solutions Strategy and is responsible for Vancity's carbon neutral, carbon offset, and environmental granting programmes. Before joining Vancity, Amanda worked at REDF, a foundation supporting social enterprises, and The Body Shop, an organisation dedicated to environmental and social returns. She was previously a manager in Accenture's consulting practice. Amanda holds an MBA from the University of California, Berkeley.

Dr **Arne Remmen** is Professor of Technology and Society Studies at Aalborg University in Denmark. He has a master's in Social Science and a PhD in Technology Assessment. Since the late 1980s, he has carried out research into the development of cleaner technologies, the implementation of environmental management systems in industry, collaboration and communication in product chains, and changes in environmental policy.

Oliver Salzmann is a research associate at the Forum for Corporate Sustainability Management (CSM) – the research platform of the International Institute for Management Development (IMD). He holds a master's in Industrial Management from Dresden University of Technology and a PhD in Engineering from Berlin University of Technology. His doctoral thesis focused on corporate sustainability management in the energy sector. Since joining IMD in 2001, Oliver has conducted empirical research in areas such as private households and sustainable consumption, the business case for corporate sustainability, and stakeholders' perceptions of corporate social and environmental responsibility.

Raj Sapru is Manager, Advisory Services for Business for Social Responsibility (BSR). Raj consults with BSR's clients in a variety of sectors with a particular focus on energy and climate change issues. His work has included Corporate Social Responsibility (CSR) reporting, responsible supply chain management, environment and carbon strategies, and general advice on CSR strategy for clients such as General Electric, Caterpillar, Verizon, SK Telecom and Sprint Nextel. Raj also leads the Clean Cargo Working Group, a business-to-business collaboration of global shippers and transportation providers with a common goal of reducing emissions for the transportation supply chain. The 30 member companies include Maersk Line, Starbucks, Wal-Mart, Gap, Cisco, Coscon and Nike. Raj holds an MBA from the University of Michigan's Ross School of Business, with a focus on corporate strategy and corporate social responsibility. He also holds a Bachelor of Science in Engineering (BSE) degree in Civil and Environmental Engineering from the University of Michigan.

Ryan Schuchard is an Associate, Environmental R&D at Business for Social Responsibility (BSR) where he works with companies to develop corporate strategies for 'horizon' environmental issues and environmental leadership, with a focus on climate change. He analyses technology, energy, carbon markets, forestry and land use in order to turn macroeconomic data into actionable business advice. Ryan also leads research and development for 'de-carbonising the supply chain' – an ICT-sector initiative building common approaches to reducing emissions in supply chain systems for companies such as Dell, HP, Apple and Intel. Before joining BSR, Ryan held various supply chain and marketing roles for consumer products and services, with an emphasis on innovation and collaboration. He has served as a Peace Corps Volunteer in Kyrgyzstan and conducted field work in the Middle East, Eastern Europe, Latin America and Scandinavia. Ryan has an MBA from the Thunderbird School of Global Management and a degree in Finance from Oregon State University.

Dr **Rajib Shaw** is an associate professor at Kyoto University in Japan. His experience includes community-based risk management in the Asian region. He holds a doctoral degree from Osaka City University, Japan. Rajib has published more than 100 research papers and books, mainly in the field of disaster and environmental management with a particular emphasis on climate change adaptation and urban risk.

David C. Sprengel is a PhD candidate at the Department of Management, Technology and Economics of ETH Zurich (Swiss Federal Institute of Technology). His research focuses on corporate responses to direct and indirect climate change effects. At present, he is identifying and analysing corporate response strategies in different industry sectors to regulatory and stakeholder pressures to reduce direct CO_2 emissions.

David studied business administration at the Otto Beisheim School of Management (WHU) in Koblenz (Germany), the Warwick Business School (UK) and the Ecole Supérieure de Commerce in Bordeaux (France) and holds a Master of Science in Business. Prior to his PhD research, David worked as a management consultant with McKinsey & Company focusing on strategy projects in various industries.

Ulrich Steger holds the Alcan Chair of Environmental Management at the International Institute for Management Development (IMD) and is director of IMD's Forum on Corporate Sustainability Management (CSM). He is also the director of major partnership programs at IMD (e.g. DaimlerChrysler and the Allianz Excellence Program). Ulrich also holds an honorary Professorship for International Management at Technical University Berlin. He was previously Minister of Economics and Technology in the State of Hesse and a member of the managing board of Volkswagen, where he was responsible for the implementation of an environmental strategy across the worldwide VW group. He has published extensively in the areas of corporate social responsibility and corporate governance, most recently the book, *Inside the Mind of the Stakeholder* (Palgrave Macmillan, 2006).

Dr **Emma Stewart** is Director, Environmental R&D at Business for Social Responsibility (BSR). Emma launched and directs BSR's Environmental R&D team, which produces cutting-edge analysis on corporate environmental trends, and designs cross-sector initiatives that address 'horizon' issues in climate change, water, land use and biodiversity and toxicity and health. Emma joined BSR in 2005 with a background in strategic environmental management. She has been awarded the Bernard Siegel Award for Outstanding Research and Publication and the O'Bie Shultz Fellowship in International Studies. In 2006, she was selected as a Next Generation Fellow of the American Assembly, which counts Jeffrey Sachs, Warren Christopher and Senator Richard Lugar among its advisors. Emma has published numerous reports, case studies, articles and book chapters on the issues of climate, water, biodiversity and certification. She has been an invited speaker at the National Press Club, Retail Leadership Program, Sustainable Packaging Coalition, Yale School of Management, Haas School of Business at UC Berkeley, Stanford Public Policy Colloquium, and Boston College Leadership Roundtable. She is an advisor to the UNEP Finance Initiative and a board member of ODC/Dance. Emma holds a PhD from Stanford University and a BA degree from Oxford University.

Dr **Rory Sullivan** is Head of Responsible Investment at Insight Investment. He is responsible for leading Insight's research and engagement activities on social and environmental issues, with a particular focus on climate change. He is also a member of the Steering Committee of the Institutional Investors Group on Climate Change (IIGCC) and is the Chair of the Confederation of British Industry (CBI)'s Carbon Reporting Working Group. Rory has authored a series of major reports on the investment implications of climate change including *Taking the Temperature: Assessing the Performance of Leading UK and European Companies in Responding to Climate Change* and *Climate Change Policy and the Electricity Industry* (published by Chatham House). He also led the drafting of the IIGCC Investor Statement on Climate Change and was a contributor to the UNDP 2007/2008 Human Development Report (*Fighting Climate Change: Human Solidarity in a Divided World*).

His previous experience includes advising Environment Australia and the Organisation for Economic Cooperation and Development (OECD) on the development and implementation of pollutant release and transfer registers, advising companies on greenhouse gas emission inventories and greenhouse gas emissions management, and contributing to the Intergovernmental Panel on Climate Change (IPCC) report *Good Practice Guidance and Uncertainty Management in National Greenhouse Gas Inventories* (2000).

Rory has written over 300 articles, book chapters and papers on climate change, energy policy and investment issues, including articles in *Climatic Change, Journal of Corporate Citizenship, Journal of Business Ethics* and *Atmospheric Environment*. He is the author/editor of five books on these issues including *Rethinking Voluntary Approaches in Environmental Policy* (Edward Elgar, 2005) and *Responsible Investment* (editor with Craig Mackenzie; Greenleaf Publishing, 2006). He holds a first-class honours degree in Electrical Engineering (University College Cork, Ireland), master's in Environmental Science (University of Manchester) and Environmental Law (University of Sydney) and a PhD in Law (Queen Mary, University of London).

Joan Thiesen is a PhD student at the Department of Development and Planning at Aalborg University, Denmark. Her research focuses on how environmental considerations are integrated into innovation processes, and the prerequisites for the development and diffusion of eco- and energy-efficient technologies. Joan has a master's degree in Environmental Management from Aalborg University.

Jim Walker co-founded The Climate Group in late 2003 with chief executive officer (CEO) Steve Howard. As chief operating officer (COO), Jim is responsible for managing the Group's global operations, fundraising and research. Before establishing The Climate Group, Jim worked for six years as a consultant on a wide range of public and private sector responses to environmental and social challenges and co-authored a number of reports building the business case for sustainability. Prior to his environmental career Jim was a professional athlete, competing in the British rowing team at the 1992 and 1996 Olympic Games and at six World Championships. He holds a bachelor's degree in biology and a master's in Environmental Technology from Imperial College London.

Index